Chemical Process Technology and Simulation

SRIKUMAR KOYIKKAL

Process Design Consultant

PHI Learning Private Limited

Delhi-110092
2013

₹ 395.00

CHEMICAL PROCESS TECHNOLOGY AND SIMULATION
Srikumar Koyikkal

© 2013 by PHI Learning Private Limited, Delhi. All rights reserved. No part of this book may be reproduced in any form, by mimeograph or any other means, without permission in writing from the publisher.

ISBN-978-81-203-4709-0

The export rights of this book are vested solely with the publisher.

Published by Asoke K. Ghosh, PHI Learning Private Limited, Rimjhim House, 111, Patparganj Industrial Estate, Delhi-110092 and Printed by Mohan Makhijani at Rekha Printers Private Limited, New Delhi-110020.

CONTENTS

Preface ix

Chapter 1 INTRODUCTION 1–14
    1.1   Technologies and Design   1
            1.1.1   Development of New Technologies   2
    1.2   Planning and Preparation   3
            1.2.1   Approaches to Planning   3
            1.2.2   Time Schedule Charts   4
            1.2.3   Planning Software   5
            1.2.4   Document Production Sequence   5
            1.2.5   Environmental Planning and Statutes   6
    1.3   Simulation and Design Tools   7
            1.3.1   Conventional Methods   7
            1.3.2   Process Simulation Software   8
            1.3.3   Other Related Software (Multidisciplinary)   13
    Review Questions   14

Chapter 2 NATURAL AND BIOPROCESS CHEMICALS 15–75
    2.1   Cardanol from Cashew Nut Shell Liquid   15
    2.2   Fermentation Products (Alcohol, Food and Pharmaceuticals)   19
            2.2.1   Alcohol from Molasses   19
            2.2.2   Bakers' Yeast   22
            2.2.3   Citric Acid   23
            2.2.4   Lactic Acid   25
            2.2.5   Fructose (as High Fructose Corn Syrup)   26
            2.2.6   Penicillin   27
            2.2.7   Streptomycin   28
            2.2.8   Ampicillin   29
            2.2.9   L-ascorbic Acid (Vitamin C)   29

2.2.10 Vitamin B_2 (Riboflavin $C_{17}H_{20}N_4O_6$) 30
2.2.11 Industrial Alcohol (Rectified Spirit) 30
2.3 Absolute Alcohol with Case Studies 33
    2.3.1 Molecular Sieve Method 34
    2.3.2 Azeotropic Distillation Method 36
2.4 Products from Sea—Agar and Agarose 39
    2.4.1 Agar 39
    2.4.2 Agarose 45
2.5 Coal and Coal Chemicals 47
    2.5.1 Coal Gasification 48
    2.5.2 IGCC (Integrated Gasification Combined Cycle) Plant 50
2.6 Sugar and Starch 51
    2.6.1 Sugar 51
    2.6.2 Starch 53
2.7 Pulp and Paper 54
    2.7.1 Pulp 54
    2.7.2 Paper 57
2.8 Soaps and Detergents 58
    2.8.1 Soaps 58
    2.8.2 Detergents 60
2.9 Rubber and Leather 61
    2.9.1 Rubber 61
    2.9.2 Leather 67
2.10 Oils and Their Hydrogenation 69
    2.10.1 Vegetable Oils 69
    2.10.2 Hydrogenation of Vegetable Oils 71
2.11 Biodiesel from Jatropha Seed Oil 72
    2.11.1 Batch Process of Producing Biodiesel 73
    2.11.2 Continuous Process Technologies 73
Review Questions 75

Chapter 3 ORGANIC CHEMICALS 76–129

3.1 Methanol, Phenol and Acetone 76
    3.1.1 Methanol 76
    3.1.2 Methanol Plant—A Case Study 84
    3.1.3 Phenol and Acetone Coproduction 89
    3.1.4 Solvent (Acetone) Recovery 92
3.2 LPG and Propylene from Cracked LPG 97
    3.2.1 Liquified Petroleum Gas 97
    3.2.2 LPG Production from Gas Processing Units 97
    3.2.3 LPG Production from Refinery Gases 100
    3.2.4 Propylene from Cracked LPG—A Case Study 101
3.3 Man-Made Fibres 105
    3.3.1 Viscose Rayon 105
    3.3.2 Nylons 107
    3.3.3 Polyester 111
    3.3.4 Acrylonitrile 112

- 3.4 Petrochemicals 114
  - 3.4.1 Steam Cracking Products 114
  - 3.4.2 Vinyl Chloride and Vinyl Acetate 116
  - 3.4.3 Phthalic Anhydride 119
- 3.5 Polymerization Products 120
  - 3.5.1 HDPE, LDPE and LLDPE 121
  - 3.5.2 Poly Vinyl Chloride and Poly Vinyl Acetate 124
  - 3.5.3 Oxo-biodegradable Polyethylene 125
- 3.6 Pesticides 126
  - 3.6.1 DDT and Dicofol 126
  - 3.6.2 Malathion and Parathion 127

Review Questions 129

Chapter 4 INORGANIC CHEMICALS 130–201

- 4.1 Precipitated Calcium Carbonate 130
  - 4.1.1 Process Description 131
  - 4.1.2 Case Study—A Small Capacity Plant 133
- 4.2 Phosphorus 135
  - 4.2.1 Electrochemical Process of Manufacture 135
- 4.3 Phosphoric Acid 136
  - 4.3.1 Description of the Dihydrate Process 138
  - 4.3.2 Simulation of the Phosphoric Acid Process 139
- 4.4 Hydrogen 141
  - 4.4.1 Process Description 142
- 4.5 Ammonia 150
  - 4.5.1 Process Description 150
  - 4.5.2 Ammonia Plant—A Case Study 153
- 4.6 Soda Ash 154
  - 4.6.1 Solvay Process 155
- 4.7 Glass 156
  - 4.7.1 Glass-making Furnaces 157
  - 4.7.2 Glass Blowing and Automation 157
  - 4.7.3 Glass Rolling (for Plate and Float Glasses) 158
  - 4.7.4 Annealing and Finishing 158
  - 4.7.5 Other Types of Glasses 158
- 4.8 Caustic-Chlorine Products and Metals 159
  - 4.8.1 Caustic Soda and Chlorine 159
  - 4.8.2 Hydrogen Chloride and Hydrochloric Acid 162
  - 4.8.3 Metals 162
- 4.9 Sulphur 172
  - 4.9.1 Sulphur Production from Mines by Frasch Process 172
  - 4.9.2 Sulphur Production from Hydrogen Sulphide by Modified Claus Process 173
- 4.10 Sulphuric Acid 174
  - 4.10.1 DCDA Process 174
- 4.11 Nitric Acid 177
  - 4.11.1 Process Description of Single Pressure Process 177
  - 4.11.2 Dual Pressure Process 179

        4.11.3   Concentration of Nitric Acid   179
        4.11.4   Design and Simulation   179
4.12   Urea   180
        4.12.1   Process Description   181
4.13   Hydrazine   181
        4.13.1   Olin Raschig Process   182
4.14   Complex Fertilizer   182
        4.14.1   NPK by Pipe Reactor Process   183
4.15   Cement   184
        4.15.1   Designation of Components   185
        4.15.2   Compounds in Cement   186
        4.15.3   Beneficiation of Limestone   186
4.16   Lime   188
        4.16.1   Process and Plant Description   189
4.17   Paints   190
        4.17.1   Manufacture of Titanium Dioxide from Ilmenite   190
4.18   Varnishes   192
4.19   Water   193
        4.19.1   Water for Municipal and Industrial Use   193
        4.19.2   Potable Water from Sea Water   195
4.20   Uranium and Heavy Water   197
        4.20.1   Uranium   198
        4.20.2   Heavy Water   199
Review Questions   201

Chapter 5   REFINERY OPERATIONS   202–229

5.1   Crude Distillation   202
        5.1.1   Atmospheric Distillation Column   203
        5.1.2   Vacuum Distillation Column   209
5.2   Coking   211
        5.2.1   Delayed Coking   211
5.3   Other Types of Coking   221
        5.3.1   Fluid Coking   221
        5.3.2   Flexi Coking   222
        5.3.3   Contact Coking   222
        5.3.4   Comparison of Methods of Coking   222
5.4   Catalytic Cracking   223
        5.4.1   Fluid Catalytic Cracking Unit   223
        5.4.2   Hydrocracking   226
        5.4.3   Catalytic Reforming   227
Review Questions   229

Chapter 6   OIL AND GAS (UPSTREAM)   230–261

6.1   Phase Separation   230
        6.1.1   Four Types of Reservoirs   230
        6.1.2   Two-Phase Separators   231
        6.1.3   Three-Phase Separators   232

		6.1.4	Development of Separator Sizing 235

        6.1.4   Development of Separator Sizing   235
        6.1.5   Free Water Knockouts   238
        6.1.6   Three-Phase Separator—A Case Study   239
   6.2 Natural Gas Transportation Network   248
        6.2.1   Operation of Gas Pipelines   249
        6.2.2   Methodology for Hydraulic Simulation   254
        6.2.3   Pipeline Network—A Case Study   256
        6.2.4   Shipping of Natural Gas   261
   Review Questions   261

Chapter 7 NANOTECHNOLOGIES 262–277

   7.1 Polysilicon   262
   7.2 Polysilicon by Trichlorosilane Route   263
        7.2.1   Production of MG-Si   263
        7.2.2   Production of Polysilicon from MG-Si   263
   7.3 Alternative Routes   267
        7.3.1   Polysilicon from Magnesium Silicide   267
        7.3.2   Silicon Production from Hydrosilisic Acid Produced from Superphosphate Plant   268
   7.4 Pilot Plant and Mini Plants   269
        7.4.1   Pilot Plant   269
        7.4.2   Mini Plants   271
   7.5 Semiconductors and Solar Panels   271
        7.5.1   Semiconductors   271
        7.5.2   Solar Panels   276
   Review Questions   277

Chapter 8 MODELLING AND SIMULATION 278–298

   8.1 Reformer Modelling   278
        8.1.1   Equations for Model   278
        8.1.2   Development of Model   284
   8.2 Dynamic Modelling   292
   Review Questions   298

Chapter 9 USER-WRITTEN PROGRAM EXAMPLE 299–316

   9.1 Introduction   299
   9.2 Terrace-Walled Reformer Simulation Program   299
   Review Questions   316

Chapter 10 COST ESTIMATION EXAMPLES 317–329

   10.1 Introduction   317
   10.2 Cardanol Distillation Unit   318
         10.2.1   Reactor (Distillation Vessel)   318
         10.2.2   Condenser   320
         10.2.3   Cooler   321
         10.2.4   Receiving Vessels   321

10.3 Industrial Alcohol Distillation Unit   323
    10.3.1 Wash Column (Main material of construction—deoxidized copper)   323
    10.3.2 Rectifying Column   324
    10.3.3 Wash Preheater   326
    10.3.4 Final Condenser   326
    10.3.5 Vent Condensers   327
    10.3.6 Product Cooler   328
    10.3.7 Reflux Tank   328
    10.3.8 Rectified Spirit Plant (Plant and Machinery)—Cost Summary   328

Appendix   *331*

References   *333–336*

Index   *337–345*

PREFACE

This book is written to build an understanding of chemical technology processes, as used in the modern chemical industries, classified in the book as natural and bioprocess products, organic chemicals, inorganic chemicals, petrochemicals, polymers, metals, refinery operations, oil and gas operations, and nanotechnology products.

A technology is a result of a good and feasible design. In this book, I have, therefore, related technologies to design and cost aspects as well. This approach makes us go much deeper and helps us in understanding the logic behind various operating techniques and parameters that we otherwise take for granted.

The book has been developed primarily to fulfil the curriculum needs of the courses on Chemical Technology, prescribed for the undergraduate students of chemical engineering.

Chemical products have been grouped and subgrouped in this book as per the groupings followed in industry. A common structure is followed for each topic. However, some examples of computer process simulations, case studies, mechanical designs, costing procedures and safety aspects (only one or two of a type) have been added to make the book more useful and informative.

I have been environmentally conscious, too. The book includes the technologies of hydrogen and methanol, which are the clean fuels of the future. The book also includes biodiesel which is a carbon neutral fuel, solar panels as a supplier of clean energy, and biodegradable polymers.

It is to be noted that the subject 'Chemical Technology' is well integrated nowadays with the subject of 'Process Simulation Software', and hence the latter also finds a place in this book. The discussion on simulation consists of two parts, with practical examples.

1. Overall process flow simulations using the commercial computer software.
2. Complex or 'out of the ordinary' equipment simulations using equipment modelling.

Pilot plants and miniplants have also been discussed.

Chemical engineers have to translate new ideas into new plants. An ability to make a quick cost estimate of the process plant will be very beneficial for this purpose and hence the book has included two mechanical design and cost estimation examples as well. The intention in this book is only to show the method of cost estimation.

In a nutshell, with an evolutionary approach, the book deals with all the most important technologies with special emphasis on latest technologies and modern topics such as polysilicon.

The book mostly follows the present syllabi of different Indian Institutes of Technology (IITs), National Institutes of Technology (NITs) and technical universities, with a few more new topics added. The organization of chapters is based on products, except in the case of 'refining' and 'oil and gas' industry. In the 'refining' and 'oil and gas' chapters, the organization is based on operations because each plant produces multiple products and the same products from different plants are blended to make the final products.

Besides the degree students of chemical engineering, the book is also suited for postgraduate degree courses (Chemical Process Design) in chemical engineering. Project works are also required to be done for every course and this book will turn out to be useful for the same. The book is also expected to become a ready reference for practising process design professionals and consultants.

I acknowledge with great sincerity my indebtedness to all my colleagues and co-workers who encouraged me in my development of this humble knowledge base. I am also indebted to my wife, Maya, who supported me throughout the writing period of this book.

Srikumar Koyikkal

Chapter 1

INTRODUCTION

1.1 TECHNOLOGIES AND DESIGN

Some chemical plants remain in operation for centuries after they are built, undergoing only changes which are superficial in nature, without affecting the basic process. Yet others go off business in a few years. Still others are built but fail to be commissioned ever. While we often blame the technology for this failure, we should remember that the technology itself is the result of intelligent and innovative design efforts.

There are too many aspects to be considered in a chemical plant design. It may not be possible to fathom all those aspects. However, in adopting a practical and evolutionary approach to chemical process technologies, while we learn more about the technological aspects, we also try to understand the imponderables in the success or failure of chemical plants and projects. It is only when we look from a longer distance, both in time and space, that we get to see a larger picture. Looking from a distance (or evolutionary approach) also gives rise to an impartial outlook and unprejudiced attitude, which improves our judgement.

We start with some old technologies—cardanol from cashew nut shell liquid, alcohol produced by fermentation, and proceed to large petrochemical complexes. Similarly, from examples of simple inorganic chemicals we progress up to silicon materials and solar energy. All developments of process technology are expressed mainly through products so that the idea is better understood and retained by the user.

Cost is at the root of all process decisions and must be considered early on. Hence, costing is treated from the beginning itself. It is best to treat costing of equipment as part of initial designs. The methodology of costing is explained in detail with the help of examples in Chapter 10. However, we should not be blinded by cost and take decisions based on it alone. Sometimes cheap equipment may lead to instability in operations. Cost should be weighed against stability of operation, which is a major factor in success of technologies. Every nook and corner of the spectrum has to be examined before taking process decisions.

Process engineers need a fixed concrete path and have to go really far. Occasionally they need to meander out of it and come back. This provokes original thinking and gives rise to innovations.

Process engineers need to be fast enough to impress their clients, hence speed becomes a concern. Speed itself depends on practice; needless to say, continuous practice is an absolute necessity.

This book covers different domain areas such as organic chemicals, inorganic chemicals, refinery operations, and so on. Since chemical engineering deals with tens of thousands of chemicals and their manufacture in many forms, many diverse units such as MMSCMD(million standard cubic metres per day), MMSCFD(million standard cubic foot per day), and bbls (barrels), etc. in the oil and gas sector, are unavoidable in order to relate with particular industrial practices.

Most domains of chemical industry are covered in this book, but we also hope that others with better knowledge in different fields will take up the baton and develop this science further and also focus on environment-friendly ways. Focus is not only on knowledge of processes alone, but also on its evolution and practical application.

1.1.1 Development of New Technologies

Any new technology develops from a preliminary idea that gets converted into a crude design. Based on this crude design, a pilot plant or a mini plant (or some other form of simulation/experimentation) is built and tested, and if successful a new technology results. On the basis of this new technology, a new design is made resulting in a new process plant. These stages are shown in Figure 1.1.

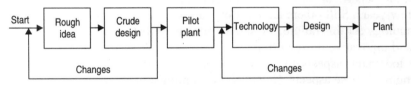

Figure 1.1 Technology and design.

We cannot leave out any of the factors, including intricate aspects, while developing a technology. For example, when experiments were done and a pilot plant was also operated, it was never noticed that a particular chemical reaction was significantly light catalyzed. This is because in Delhi (where this technology was developed) all the experimental work was done only in daylight. Pilot plant was also operated only during daytime. When converted into the real plant, one of the chemical reactions was found to stop during night-times and restart again during the day. Then lights were put inside the reactor and it started performing all twenty-four hours.

Ideally, any chemical used for a particular purpose should also finally breakdown into harmless components, once its duty is done, for the conservation of the environment. For example, DDT is a good pesticide but after killing the pests it continues to lie in the soil since it is not biodegradable, with the danger of it migrating into food products. To combat this menace, a company (Hindustan Insecticides Ltd.) converted DDT into Dicofol. Dicofol is readily biodegradable and is also very effective for a particular group of pests called phytophagus mites.

This book is also useful to set a tone for the future design philosophies in chemical process technology development. It is expected to be useful for all chemical and mechanical engineers in the process industry, project managers, aspiring plant designers, chemists, general students of science and technology and all who are generally interested in the development of chemical industry.

Introduction

1.2 PLANNING AND PREPARATION

A good planning leads to a good design, which in turn leads to a successful technology. To get a better insight into the working processes, planning aspects have to be discussed.

What are the preparations required for achieving a good chemical production unit? One very important preparation for the project team is to visit the site and make a careful field study. Similarly, to develop a new technology, a careful observation of pilot plant study is important. Points to be discussed with clients are carefully prepared before discussion with the clients. The most important thing is to acquire the correct domain knowledge. Get yourself trained to the software that you may have to use and be ready to write one or two computer programs independently, if need arises. Most important of all, plan all the activities. Even the industrial revolution was a result of good planning[1].

Whether to use simple techniques or opt for complicated ones is a matter of both judgment and the prevailing situation. Albert Einstein once gave a job to one of his assistants. The assistant asked, "Should I make it simple?" Einstein replied, "Make it simple. But not simpler." Where there is urgency and simplicity suffices to get a good answer, we can use simple techniques. Where it is not the case, we need to go in for complicated calculation procedures, programming, modelling, simulation and the like.

1.2.1 Approaches to Planning

Traditional approaches to planning are top-down approach, bottom-up approach, and combined or object-oriented approach. These approaches are explained below.

Top-down approach

In top-down approach (Figure 1.2) the goals, quotas and time requirements for activities are decided at the topmost level and passed downwards for further splitting up and elaborating on these goals, quotas and timeframes. The different activities are denoted by A_1, A_2, etc. These can be at any level.

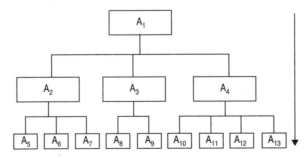

Figure 1.2 Top-down approach.

Bottom-up approach

In bottom-up approach (Figure 1.3), we start with the lower levels of activity for which data is readily available, set goals, quotas and timeframes, and move to higher levels until all activities are completed including the higher level activities.

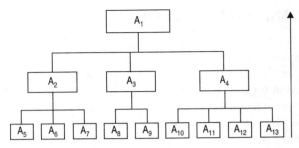

Figure 1.3 Bottom-up approach.

Combined or object-oriented approach

In actual design practice, a combination of top-down and bottom-up approaches were found more effective. It starts simultaneously at different points with a clear objective defined for each activity. It can also be called an object-oriented approach.

First, we identify the activities. Also, we identify those activities which can be grouped together as a class so that the same design methods can be used for all such activities coming under a class. We set goals, quotas and timeframes to each activity based on the class to which it belongs. However, we have to be cautious against simplification that can creep into the planning, with resultant problems. Figure 1.4 shows the combined or object-oriented approach.

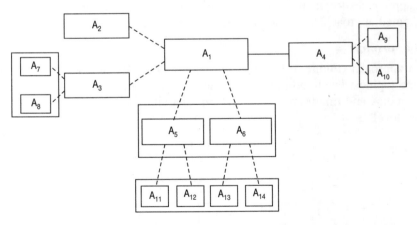

Figure 1.4 Object-oriented approach.

1.2.2 Time Schedule Charts

Two types of time schedule charts, bar chart and Program Evaluation Review Technique (PERT) chart, are popular in scheduling a project. The bar chart is simple and an example is given in the appendix 'Tables and Charts'. It contains horizontal lines of activities with start and finish times. A PERT chart is a chart in which the events and activities are connected together. Once we start a project which includes process design, we make a simple chart with events and activities connected together. This chart is used only for preliminary working. It facilitates better understanding of start and finish times of different activities and also helps in making manhour estimates. Such a chart is shown in Figure 1.5.

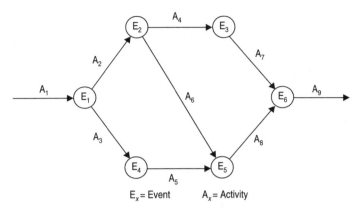

Figure 1.5 Project planning PERT chart.

1.2.3 Planning Software

At present, for planning all medium and large-scale projects it is normal to use some standard software available in the market. The most popular planning softwares available at present are 'PRIMAVERA' and 'MS Project'. These softwares have extensive applications in project management and execution.

Once the design engineers have completed their estimates, the planning engineer will review the same and arrive at the overall estimate figures (cost and manhours) using a standard planning software application.

1.2.4 Document Production Sequence

The production and release of process design documents follows a general order given below.

1. Process flow diagrams with heat and material balance
2. Piping specifications
3. Piping and instrumentation drawings
4. Configuration of layout
5. Equipment data sheets
6. Instrument data sheets
7. Hazardous area classification drawing
8. Cause and effect diagrams (or SAFE charts for offshore platform jobs)
9. Operating manual.

At the time of the starting of a project the following documents need to be finalized so that the design proceeds smoothly.

1. Document numbering procedure
2. Document control procedure

The above documents indicate how each text document or drawing is numbered, how revisions are incorporated, who is responsible for checking and approving the documents and other functions which ensure systematic progress as well as efficient storage and retrieval of documents.

1.2.5 Environmental Planning and Statutes

Planning and preparation of a project should also focus on environmental risks. It is commonsense to say that every company or corporate that starts a chemical manufacturing unit is liable to compensate for any damage or degradation of human health as well as environmental damage and degradation caused by it.

Various countries have various laws to prevent environmental degradation. After the Stockholm conference on Human Environment in June 1972, it was considered appropriate to have uniform laws across all regions so as to protect the health and safety of people as well as of flora and fauna. Many countries enacted laws and rules for the same. In the case of India, the Parliament enacted the following statutes.

1. The Water (prevention and control of pollution) Act, 1974
2. The Water (prevention and control of pollution cess) Act, 1977
3. The Air (prevention and control of pollution) Act, 1981
4. The Environmental Protection Act, 1986, consisting of:
    (a) Rules for manufacture, use, export and storage of hazardous organisms, genetically engineered organisms or cells
    (b) The Hazardous Waste (management and handling) Rules, 1989
5. The Chemical Accidents (emergency planning, preparedness, and response) Rules, 1986
6. The manufacture, storage and import of Hazardous Chemicals Rules, 1989
7. The Public Liability and Insurance Act, 1991
8. The Recycled Plastics Manufacture and Usage Rules, 1999
9. The Ozone Depleting Substance (regulation and control) Rules, 2000
10. The Batteries (management and handling) Rules, 2001.

Amendments along with new areas of concern are also included as per requirement with the aim of having a clean environment in the future. Planning of any industry should progress hand in glove with these statutes and rules.

Tabulation of exact figures of pollution, in a hundred per cent truthful manner, will be a true investment for a better future.

'Zero Effluent' is a very important concept even though only very few companies are attempting to take it up as a rule. A few plants producing chemicals/petrochemicals have claimed to have achieved the zero effluent level. While this could be a distant dream for many of the other manufacturers, steps in this direction will definitely promote pollution mitigation.

Another matter of concern is that while 'economy of scale' indicates the benefit of building bigger and bigger plants, the effects on the environment of the transportation of huge quantity of products through long distances is sometimes not taken into account. This has to be recognized by the manufacturers as an environmental liability and included in the Environmental Impact Assessment (EIA) for any project.

An example of a statute is illustrated below:

Stack height calculation for emission regulation

Minimum stack height for all plants is calculated by the formulae given below:

Based on emission of sulphur dioxide: $H = 14(Q)^{0.3}$ (Q in kg/h)
Based on emission of particulate: $H = 14(Q)^{0.27}$ (Q in tonnes/h)

Standard minimum stack height for all plants (except thermal power) = 30 m
    The higher of the above two calculated values is to be taken and used. If both the calculated values are less than 30 m, the minimum stack height should be specified as 30 m.

1.3 SIMULATION AND DESIGN TOOLS

Usually when engineers start a plant design, they will have to use several methods and programs. The following are some of the general tools used for the design of process plants. These can be divided into two categories, 'conventional methods' and 'process simulation software'. The other related multidisciplenary methods also need consideration.

1.3.1 Conventional Methods

The conventional methods comprise hand calculations using graphs and charts, mathematical modelling and user written programs in any of the programming languages such as FORTRAN, C, C++, for technical applications, and spreadsheets such as EXCEL sheets.

Hand calculations

Hand calculations, being the mother of all calculation methods, have a larger perspective (similar to that viewed by mind), and may be adopted whenever they are sufficient. They make use of various graphs and charts, many of them usually developed within the design organization.

Mathematical modelling and user written programs

User written programs give a good perspective of the procedures.
    Let us consider the simulation of steam reformers which produce hydrogen by the chemical reactions involving hydrocarbons and steam. The reforming reaction takes place inside tubes filled with nickel catalyst, the endothermic heat of reaction being supplied by burning fuel on the outside of the tubes.
    Ideally, the heat added to a reformer should be absorbed by the feed in such a way that a uniform temperature is maintained across the length of the tube to the extent possible. This ensures that tube thickness required is a minimum.
    The temperature changes as the reaction proceeds. Thus, it is important for us to find out the variation in temperature with length. A mathematical model can be built using reaction kinetics, thermodynamic equilibrium, radiation and convection, heat transfer equations to find out the heat flux and the tube skin temperature along the length of the tube. A computer program can be developed, based on the mathematical models, using FORTRAN/C++, which gives temperatures,

heat fluxes and composition at every centimetre along the tube. The program with some variations can be used to optimize the reformer designs and examine revamp options.

Spreadsheets

Spreadsheets such us Lotus, QuattroPro and Excel can be used to perform a limited amount of simulation. Typical examples where spreadsheets become very handy are trains of heat exchangers, convection section coils of fired heaters, etc. We can use heat transfer equations depending upon the circumstances. This is a very significant advantage over the bought-out heat exchanger programs. Hence, the time and effort required can be justified.

1.3.2 Process Simulation Software

Process simulation is the study of how any number of organic and inorganic chemicals and any number of hypothetical (pseudo) components you wish to generate, interact and behave inside any number of chemical equipment, pipelines or any other open or closed containers. Simulation softwares are used to design new processes and improve or de-bottleneck the existing ones, improve the purity of a product, discover routes to pollution reduction and energy saving, reduce capital investment, optimize and support operation, generate and store information.

Solution algorithms

There are basically three types of solution algorithms for process simulators. They are sequential modular, equation solving and simultaneous modular. In the sequential modular approach, the equations describing the performance of equipment units are grouped together and solved in modules, i.e. the process is solved equipment-piece by equipment-piece. In the equation-solving technique all the relationships for the process are written out together and then the resulting matrix of nonlinear simultaneous equations is solved to yield the solution. The simultaneous modular approach combines the modularizing of the equations relating to a specific equipment with the efficient solution algorithms of the simultaneous equation-solving technique.

Evolution of software

ASPEN PLUS was first developed by the United States Government department and later privatised. DESIGN2 FOR WINDOWS was developed by Chemshare Corporation, now known as Winsim Inc. HYSYS was developed independently by Hyprotech in Calgary, Canada. UNISIM was developed by Honeywell Inc., after it purchased HYSYS intellectual rights, but not exclusively. BJAC was developed independently. HTRI and HTFS were developed by the respective professional bodies of the same name. PIPELINE STUDIO—TGNET and TLNET—are specialized software developed by Energy Solutions International, UK, for single phase networks (gas and liquid) and include surge analysis. PIPESIM specializes in 'well fluids' and is for two and three phase flow. PIPENET is a useful program for the design of fire hydrants and other water spray systems and water networks.

Many of the following programs are being used in the examples given in this book, not necessarily because any one of these is most suited, but mainly to highlight the use of these programs in relation to other forms of designing.

ASPEN PLUS
Invensys Simsci's PRO II
Winsim's DESIGN II (formerly called Chemshare)
CHEMCAD
BioPro Designer and SuperPro Designer for pharmaceutical industry
HYSYS, simulation package for oil and gas and refining industry
UNISIM, simulation package for oil and gas and refining industry
PIPESIM for multiphase 'well fluid' lines
ASPEN-BJAC heat exchanger design package
HTRI heat exchanger design package
HTFS fired heater design package
PIPENET, pipeline simulation package for water and firewater
TGNET (PIPELINE STUDIO), gas pipe network systems package
TLNET (PIPELINE STUDIO), liquid pipe network systems package
FLARENET, design of flare and flare header systems.
OLGA for oil and gas flowlines, reservoir fluid, artificial lift and merging pipelines.

General steps in process simulation (using software)

1. Select all of the chemical components that are required in the process from the component database.
2. Select the thermodynamic models required for the simulation. These may be sometimes different for different pieces of equipment.
3. Select the topology of the flow sheet to be simulated by specifying the input and output streams for each piece of equipment.
4. Select the properties (temperature, pressure, flow rate, vapour fraction and composition) of the feed streams to the process.
5. Select the equipment specifications (parameters) for each piece of equipment in the process.
6. Select the way in which the results are to be displayed.
7. Select the convergence method and run the simulation.

Each of the above steps is now explained below in separate subsections.

Selection of components

The first step in setting up a simulation of a chemical process is to select the components to be used from the simulator data bank. All components including inerts, by-products, utilities and waste chemicals should be identified. If the components needed are not available in the data bank, then they can be added through several ways. These are specific to the process simulator. The VDI heat atlas[2] also gives a lot of data on components not readily available.

Selection of thermodynamic model

This is a very important part of the simulation. Some of the important aspects of thermodynamic model selection are the choice of groups of equations to predict heat flow, phase equilibria and viscosity. As an example, you can select BWRH, BWRK and BWRV (Benedict Web Rubin method, H for Enthalpy, K for Equilibrium constant and V for Viscosity) for inorganics and similarly for organics, PRH, PRK, PRV (Peng Robinson method, H for Enthalpy, K for Equilibrium constant and V for viscosity).

Whenever we model a system for simulation, the most important consideration is thermodynamics. Nature does not truly conform to the basic van der Waals theory of corresponding states. We need to understand the thermodynamic assumptions and ensure proper application[3].

Let us consider a distillation unit in which a mixture of cyclohexane, cyclohexanol and cyclohexanone is distilled to obtain pure cyclohexane as the desired top product. All the normally available thermodynamic options in the simulators failed to give realistic predictions. The only way to solve this system is to obtain the experimental vapour-liquid equilibrium data and regress it to obtain a data file which can be used along with the simulation program to obtain realistic values. The VLE (Vapour Liquid Equilibrium) data can be obtained from some websites on payment.

The thermodynamic models available on most simulators are based on empirical information which is in fact the data applicable to certain conditions. One cannot say how a model will behave, when extrapolated beyond the region in which data was available. Hence, the data should be as close as possible to the problem at hand. For example, in the cyclohexane distillation, the data used should be obtained at temperatures and pressures close to the actual working temperatures and pressures of the column.

The various thermodynamic options can be categorized as equations of state, activity coefficient methods and special methods. The following information gives the general chronological order of evolution of thermodynamic options[4]. The symbols used have their usual meanings.

1. For Non-polar Components —Equation of State

   Basic $pv = RT$

   van der Waals $\left(p + \dfrac{a}{v^2}\right)(v - b) = RT$

   Redlich Kwong $p = \dfrac{RT}{v - b} - \dfrac{a}{T^{1/2} v(v + b)}$

   Soave (SRK) Included acentric factor in addition to T_c and P_c (Acentric factor is a measure of non-sphericity of a molecule)

   Benedict Web Rubin (BWR) For light hydrocarbons, ammonia, etc.

   $$p = \dfrac{RT}{v} + \dfrac{B_0 RT - A_0 - \dfrac{C_0}{T^2}}{v^2} + \dfrac{bRT - a}{v^3} + \dfrac{a\alpha}{v^6} + \dfrac{C}{v^3 T^2}\left(\dfrac{1 + \gamma}{v^2}\right)\exp\left(-\dfrac{\gamma}{v^2}\right)$$

   Peng Robinson Improved cubic equation
   Lee Kesler Improved BWR with use of acentric factors, etc.
   Sanchez–Lacombe For polymers

2. Systems with Polar Components — Activity Coefficient

    Basic Raoult's Law $p = p_x$

    Activity Coefficient $p = \gamma p_x$

    Van Laars Equation

$$\ln Y_1 = \frac{a}{\left[1 + \left(\dfrac{x_1}{x_2}\right)\left(\dfrac{a}{b}\right)\right]^2}$$

$$\ln Y_2 = \frac{b}{\left[1 + \left(\dfrac{x_2}{x_1}\right)\left(\dfrac{b}{a}\right)\right]^2}$$

Margules Equation	Similar to above
Wilson ⎫ NRTL ⎭	Regress interactions Parameters from VLE data
UNIFAC/UNIQUAC	Use of group contribution method
Electrolyte NRTL	For electrolytes

3. For Special Mixtures — Special

SOUR	For sour water systems
Chao Seader ⎫ Grayson Stread ⎬ BRAUN K10 ⎭	For pseudo components and heavy oils

The concepts and derivations of the above equations form part of the subject of Chemical Engineering Thermodynamics. For simulation, we need only to select the correct option and apply it at the correct portion of flowsheet or equipment.

Normally, one option will work for one full flowsheet or a definite part of it or excluding a certain equipment. The selection will have to be mainly by reasoning and sometimes by validation with existing data.

Input the topology of the flowsheet

The most reliable way to input the topology of the process flow diagram is to make a sketch on paper and have this with us when constructing the flowsheet on the simulator. Certain conventions may be there in the numbering of equipment and streams which help the simulator to keep track of the topology and connectivity of the streams.

Selection of feed stream properties

All feed streams have to be specified in terms of composition, flow rate, vapour fraction, temperature and pressure. Vapour pressure can be specified instead of temperature or pressure. In addition, rough estimates (temperature, pressure, flow rate) of the recycle streams should also be specified.

Selection of equipment parameters

Process simulation is best done in two stages. The first stage involves specifying the minimum equipment parameters required for simulation and solving the material and energy balance. At the next stage, more data can be provided to obtain the desired design parameters.

Selection of output display options

There can be many options available to display the results of a simulation. The user manual has to be referred to get the different options.

Selection of convergence criteria and running a simulation

The most important convergence criteria are the number of iterations and tolerance. These criteria will often have default values set in the simulator. The user can change these values according to the nature of the problem. For equipment requiring iterative solutions, there will be user selectable convergence and tolerance criteria in the equipment module. There will also be convergence criteria for the whole flowsheet simulation. If the simulation has not converged, the results do not represent a valid solution and should not be used.

Checking the results

Process simulation problems should be tackled in a composite manner by using simulation software, inhouse programs, spreadsheets and hand calculations. Aspen Plus, Design II, Hysys process, Pro II, etc. are some of the torch bearers of process simulation packages available in the market and they still continue to be of immense service to process industry.

It may be noted that no software program has claimed itself to be bug-free. Also, many bugs may be present in user-written programs which are attached to the main software. Hence, it is important to check the simulation results fully before accepting them.

While executing a simulation in a computer, we usually look for favourable or preconceived results. This is like looking at the glitter and not the gold. But simulation is a case in which the path is more important than the result. The reliability of individual steps has to be continuously and painstakingly checked during a simulation. Due to the existence of many recycle loops in a typical chemical plant flow diagram, it so happens that when trying to correct a mistake in a succeeding step, the results of the preceding step get altered as given in Figure 1.6.

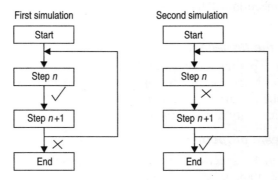

When trying to correct a mistake in step $n+1$, step n may produce some other undesirable value.

Figure 1.6 Flowchart.

The number of recycle loops in a typical chemical plant is very large and hence, it is easy to imagine that the amount of checking that would be required is very high.

Process simulation is both a science and an art. Making process simulations is similar to writing a story on a scientific framework, such that it conforms to thermodynamic and stoichiometric laws. En route, the possible number of pitfalls is very large. Since many things can go astray, it is essential to keep checking all the results after every simulation. The success of a simulation depends mainly on the scrupulous checking of results in each individual step.

General discipline to be observed while doing simulation includes the following.

1. Have a notebook handy.
2. Make a plan of simulation.
3. Make a chart with a list of answers you are looking for at the top.
4. Make a list of intermediate parameters you are looking for.
5. Make a list of parameters you are not looking for but which can change abruptly.
6. Every time you make a run of the simulation, keep entering values in the above chart.

Some other points given in literature are as follows[5].

1. Documenting the physical property assumptions, the data sources, and the range of applicability
2. Making a record of any properties which were not well defined
3. Components that should not be added, such as electrolytes for the equation of state methods
4. Commenting on properties that were not of interest and not validated
5. Keeping property estimation, regression and simulation files together
6. Using own judgement to evaluate simulation errors and suspicious results with a view to finding their source and making rectifications

1.3.3 Other Related Software (Multidisciplinary)

Computational fluid dynamics (CFD)

CFD is an elaborate simulation technique which uses numerical methods to solve fluid flow problems by finite element analysis in complex geometries. CFD programs are very useful to get an insight into flow that occurs within equipment. One example is Free Water Knockout (FWKO) tanks for separating free water from crude oil. New tanks can be designed with vortex technology wherein a slow moving vortex can be created in the tank to increase the residence time of oil and water and thereby achieve better separation. Existing tanks can also be retrofitted using this technology. CFD can be used to study mixing, pumps, turbines, hydroelectric power generation, pool fire modelling, fluidization, gasification of coal, and so on.

Plant design softwares (PDS and PDMS)

There are special multidisciplinary software for doing the complete three-dimensional (3D) engineering including process, piping, mechanical, civil, electrical and instrumentation on a single platform. These are mainly (PDS) Plant Design System and (PDMS) Plant Design Management System. PDS appears to be more integrated whereas PDMS gives more departmental freedom. These softwares do not affect process design to any great extent except that process and instrumentation diagrams (P&IDs) have to be entered with sufficient information as required by the software.

'SMART PLANT P&ID' by M/s Intergraph is a suitable P&ID tool in which information can be added which is directly interfaced with PDS.

HAZOP and safety

Hazard and Operability Study (HAZOP) is an important multidisciplinary tool to ensure safety of the designed plant. It is conducted during certain stages of design and at the completion of design.

Safety should be uppermost in the minds of process engineers right from the initial stages of design. Possibilities of gas and liquid leakages, vapour cloud formation, explosions, contaminants, runaway reactions, poisoning, effect on environment, relief, isolation, blowdown, quenching, controllability, redundancy, maintainability, etc. should be considered and discussed with experts at different stages of design. Different climatic and environmental peculiarities should also be kept in mind. HAZOP software is used during the discussion to ensure complete coverage.

During HAZOP discussions the process designer has to take initiative and explain in clear language the precautions which are already in place.

The design has to undergo at least two HAZOPs before going into the construction phase. Clarity is of utmost importance while resolving any outstanding HAZOP points. Ultimately HAZOP is only a committee's recommendation. If a new problem gets created (as already shown in Figure 1.6) due to the implementation of a HAZOP point, it should be taken up with the plant owner's representitives and resolved. While making the HAZOP reports, 'make it simple, but not simpler' principle should be strictly followed.

REVIEW QUESTIONS

1. What are the benefits of adopting an evolutionary approach to developing chemical process technologies?
2. Explain with the help of a figure, how a new technology is developed?
3. What is the importance of biodegradability?
4. Describe the three approaches to planning a project.
5. What are bar charts and PERT charts in planning? Which softwares combine the advantages of both?
6. Make a list of process design tools (conventional, simulation and multidisciplinary).
7. What is process simulation? What are the advantages of simulation programs?
8. Which organizations developed heat exchanger and fired heater design softwares?
9. What are the general steps for making a process simulation using software?
10. Explain what computational fluid dynamics means. What are its uses?
11. Explain the general discipline rules that should be observed while making a process simulation.
12. What is HAZOP? What are the safety aspects covered in HAZOP?
13. Make a list of process simulation software being used in the industry.

Chapter 2: Natural and Bioprocess Chemicals

Many chemicals are obtained directly from nature, undergoing only some physical transformations but without undergoing big or major chemical transformations. Vegetable extracts and oleo resins are good examples. Some important chemicals are discussed in this chapter with design/cost aspects included.

2.1 CARDANOL FROM CASHEW NUT SHELL LIQUID

Cashew nut shell liquid (CNSL) is obtained from the cashew nut shell during the thermal and mechanical dehulling and processing of raw cashew nuts. This liquid contains an important chemical called cardanol which was used in the past as a protective coating mainly for ships and boats.

At present cardanol is an important resin (a material that can be polymerized) used in the paint industry. Cardanol is obtained by distillation of CNSL. We now discuss this process as well as its design and cost estimation.

Properties of the CNSL (raw material)

Specific gravity at 25°C	between 0.95 and 0.97
Viscosity at 25°C	between 140 and 200 cP
Volatile matters, on heating beyond 200°C	between 3% and 5%

Properties of cardanol (product)

Specific gravity at 25°C	0.93
Boiling point at 10 mmHg	240°C
Freezing point	–20°C

Process description

A simple and small-scale batch operation plant is used. The capacity will be 500 litres of CNSL per batch. The properties of the liquid, the process of production, the design details and cost estimation are elaborated.

First, the volatile matter gets distilled. This will be about 5% of the total volume. Then the product distills at a temperature between 225°C and 240°C at 10 mmHg absolute pressure, yielding about 60–70% distillate which is mainly cardanol. What remains in the distillation vessel is a viscous liquid. This liquid known as residue has to be discharged quickly from the distillation still when it is hot, i.e. before it becomes sticky and difficult to drain.

The CSNL will get heat polymerized on being subjected to high temperatures for longer periods. In view of this, the distillation has to be completed within 60–80 min at the temperature range described above.

The CNSL liquid is taken in a batch distillation vessel, also called a reactor. See Figure 2.1. This is a 500-litre (working capacity) batch distillation apparatus. It consists of a jacketed vessel with an agitator. Hot oil is circulated in the shell of the vessel to provide heat for distillation. The vapours go into a vertical condenser of 4 m^2 heat transfer area. The condensed liquid goes into a vertical cooler with 1.3 m^2 heat transfer area.

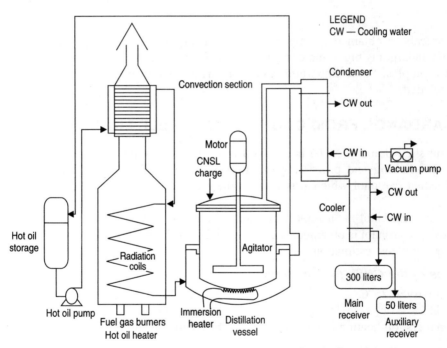

Figure 2.1 Cardanol distillation unit.

Vacuum is applied to the distillation vessel by means of a vacuum pump. Hot oil circulation is started and gradually the temperature of the liquid in the vessel is increased so as to prevent any degradation of the material. Initially, a small portion of the distillate is sent to the auxiliary receiver. When the temperature reaches 225°C, it is diverted to the main receiver vessel. The auxiliary receiver is drained to collect the lights. The distillation is continued and the distillate is sent to the product cooler for cooling it and then sent to the main receiver until the temperature suddenly shoots up to 240°C. At this point the distillate is again diverted to the auxiliary receiver. The heavies collected in it are drained.

The product is taken out from the main receiver and sent to a storage tank after ensuring the quality by chemical analysis in the laboratory.

The heat exchangers (condenser and cooler) can be kept vertical for compactness and easy draining. We can also use horizontal tubes which are better for the purpose of tube cleaning. There is no vent for the condenser since the vacuum is produced by a vacuum pump which will ensure a smooth flow of gases through the condenser. The vacuum produced in the distillation vessel is about 5 mmHg absolute. The product can flow either to the 50-litre auxiliary receiver (receiving lights or heavies or for testing) or to the 300-litre product vessel, i.e. main receiver.

The above process scheme is also referred to in Section 2.10 when discussing distillation of essential oils.

Design details

The distillation vessel should have an agitator with a stuffing box (or mechanical seal to prevent vapours from the vessel escaping to atmosphere from the gap between the agitator stem and agitator nozzle). The agitator speed is to be about 50 rpm. A jacket system for hot oil circulation for heating of the still is used. The still (distillation vessel) material of construction will be stainless steel AISI 316. The top product is condensed in a condenser with water on shell side. The condenser body is of carbon steel and tubes shall be SS316.

The cooler shall be made of carbon steel with SS304 tubes. Both receivers will be made of SS304. The size of the main receiver is 300 litres. The auxiliary receiver will have a size of 50 litres.

Calculation of size of distillation vessel

Volume of the liquid to be fed = 500 litres or 0.5 m^3

Take volume of straight portion including vapour space (50% more) = 0.75 m^3

Diameter of vessel (assumed) = 0.9 m

Required height of vessel = $0.75/(\pi/4)(0.9)^2 = 1.179 \approx 1.2$ m

Take straight height as 1200 mm

Calculation of heat (or power) requirement for hot oil (or electrical) heating

Vaporization rate	= 300 l/h
Specific gravity of CNSL	= 1.5
Latent heat of cardanol	= 21 × boiling point in K (for high boiling point liquids like cardanol the latent heat in cal/g·°C is taken as 2 times the boiling point in K)
	= 21 × 513 cal/g·°C
	= 10,773 cal/g·°C
Heat added	= 300 × 10773/1.5
	= 21,50,000/228
	= 9430 kcal/h

Take heat added as 9500 kcal/h
kW requirement = 9500/860 = 11 kW (∵ 860 kcal = 1 kWh)
Accounting for about 10% extra, kW requirement = 12 kW

Calculation of heat transfer area of condenser

Heat to be removed = 9500 kcal/h

Heat transfer coefficient = 50 kcal/h/m^2/°C
(based on previous experience—calculation not feasible since the product viscosity is very high.)

Log mean temperature difference ± 50°C

Heat transfer area = 9500/(50 × 50) = 3.8 m^2 ≈ 4 m^2

Hence, a 4 m^2 area condenser is selected.

Auxiliary facilities of still (distillation vessel)

1. Motor and reduction gear is mounted on still.
2. The still should have a level gauge, sight glass and temperature and pressure indicators.
3. A vacuum pump to produce 5 mm Hg absolute pressure is required.
4. Hot oil circulation is to be used for heating. And the hot oil itself shall be heated by fuel gas in the hot oil heater.

The points which are to be considered while designing this unit are described below.

The motor for reactor should be kept at a height from the vessel so as to avoid the motor from getting overheated. A suitable lead stuffing box or a dry mechanical seal should be provided to withstand a maximum temperature of 280°C.

A jacket should be provided up to the middle of the reactor with oil circulating arrangements.

A suitable method for heating, storing and circulating the thermic oil should be made.

Wherever the liquid or vapour from the reactor comes into contact, the material should be AISI SS316.

The condenser should have a slight slanting if it is horizontal. If a vertical condenser is selected, it will save space and its draining will be easy.

There can be some practical difficulties with the vacuum pump since choking can take place inside the pump under the cold condition. A water ejector in place of vacuum pump is a good proposition, in places where there is abundance of water.

Air pump can also be used for circulating the hot oil inside the jacket, but is usually not selected since the electric pump is more robust.

Quick mechanical design and cost estimation

A quick mechanical design for cost estimation purpose and cost estimation for the above plant is given in Section 10.1.1.

A typical plant and machinery cost summary (general off-site facilities, taxes, etc. not included) is given in Table 2.1.

The cost given is approximate, based on already suggested conditions, and for illustration purpose only. Also note that, in this case, the plant cost will be only a portion of the project cost.

What we have done is a detailed estimation. For a much quicker estimation of modern plants the book by Peters and Timmerhaus[1] can be referred.

Table 2.1 Cost summary

Item	Weight (kg)	Cost + Fabrication (₹)	Purchased items (₹)	Items as %	Total cost (₹)
Distillation vessel	600	226,000			226,000
Condenser	200	23,000			81,000
Cooler	100	14,000			59,000
Main receiver	70	7600			35,000
Auxiliary receiver	10	3000			7400
Vacuum pump			100,000		100,000
Hot oil pump			100,000		100,000
Hot oil heater			550,000		550,000
				Subtotal	**1,158,400**
Piping items				20	231,680
Insulation				10	115,840
Instruments				20	231,680
Electrical				20	231,680
Civil				20	231,680
				Total	**2,200,960 ≈ 22 lakhs**

2.2 FERMENTATION PRODUCTS (ALCOHOL, FOOD AND PHARMACEUTICALS)

2.2.1 Alcohol from Molasses

Fermentation is a microbiological process that is widely used for making chemicals such as ethanol, butanol, acetone, citric acid, lactic acid, medicinals, antibiotics, etc. from low cost hydrocarbon raw materials. Rectified spirit (Industrial Alcohol) is 94.5% ethyl alcohol and remainder water. It is used for making absolute alcohol used as a biofuel and in pharmaceuticals, and for making potable spirits as well as for making organic chemicals such as butanol, acetaldehyde and acetic acid. Therefore there are two grades produced in most countries, one for industrial uses (Grade 2) and another called 'silent spirit' (Grade 1) which is used for making potable alcohols.

Molasses, obtained from sugar mills, is the raw material used for the production of alcohol. Molasses contains about 50% total sugars. It is diluted by adding water to about 12% sugar content before it is fermented. The rectified spirit plant consists of two main sections, fermentation and distillation. The fermentation part is described here and the distillation in Section 2.2.11 after discussing other fermentation products.

Process description

During fermentation, yeast strains of the species *saccharomyces cerevisiae*, a living microorganism belonging to class fungi, convert sugar present in the molasses such as sucrose or glucose to alcohol. Chemically, this transformation from sucrose to alcohol can be approximated by the following equations:

Cane sugar to glucose:

$$C_{12}H_{22}O_{11} + H_2O \rightarrow 2C_6H_{12}O_6$$

Glucose to ethyl alcohol and carbon dioxide:

$$C_6H_{12}O_6 \rightarrow 2C_2H_5OH + 2CO_2$$

Heat is evolved during fermentation. The rise in temperature causes large losses from vaporization, also the quantity of by-product and that of aldehydes (lights) increases. The best temperature for fermentation is 30°C. The heat evolved during fermentation is 120 to 132 kcal/kg. Approximately 1% ethanol is lost at 35°C and 1.5 % at 40°C.

The fermentation time cycle is generally 48 hours for ordinary heat exchanger cooled fermentation plants. With special methods of yeast reuse, this cycle may be reduced. The fermentation tanks have a sloping bottom for easy pumping out of the sludge, and have a conical roof for collecting vapours.

Ammonium sulphate is added as nutrient for the yeast. The rate of addition is about 30 kg per 100,000 litres of rectified spirit. Sulphuric acid is also added at the rate of 0.05% based on the weight of molasses.

About 180 g of sugar on biochemical reaction gives 92 g of alcohol. Therefore, 1 tonne of sugar gives 511 kg of alcohol. The specific gravity of alcohol is 0.7934. Therefore, 511 kg of alcohol is equivalent to 551/0.7934 or 644 litres of alcohol.

During fermentation, other by-products like glycerine, succinic acid, etc. may also be formed from sugar unless it is totally free of contamination. Therefore, actually 94.5% total fermentable sugars are available for alcohol conversion. Thus, one tonne of sugar will give only 644 × 0.945 = 608.6 litres of alcohol, theoretically. Normally, 80 to 82% efficiencies are realized in the plant. Hence, one tonne of molasses containing 45% fermentable sugars gives an alcoholic yield of 230 to 250 litres.

For bringing out the above biochemical reaction, we require proper and careful handling of yeast, optimum parameters like pH and temperature control and substrate concentration, which results into an effective conversion of sugars to alcohol. For manufacture of yeast, separate equipment known as pure yeast culture apparatus is required. Initially, yeast is developed in the laboratory from the correct yeast culture.

In the laboratory, 10 ml of yeast is taken in a test tube and then it is transferred to a bigger flask of 500 ml containing the sterilized sugar solution. After due fermentation it is transferred to a 5-litre flask, again containing sterilized and diluted molasses. The pH of the molasses solution is adjusted in the range of 4.5 to 5.0 by the addition of acid. Necessary nutrients such as ammonium sulphate or urea, diammonium phosphate, etc. are also added. Each stage of development of yeast from 10 ml to 500 ml and from 500 ml to 5 litres and so on to 5000 litres is done in sterilized conditions.

The smaller equipment are also designed so as to facilitate boiling of molasses solution, in order to sterilize it and also cooling to bring it to the proper temperature of 33°C, letting in the yeast culture, etc. Keeping the fermentation medium free and away from any kind of infection is most important.

The next stage of yeast propagation (population growth) is done in open tanks as shown in Figure 2.2. The pre-fermenters require about 8 hours in order to build up the necessary

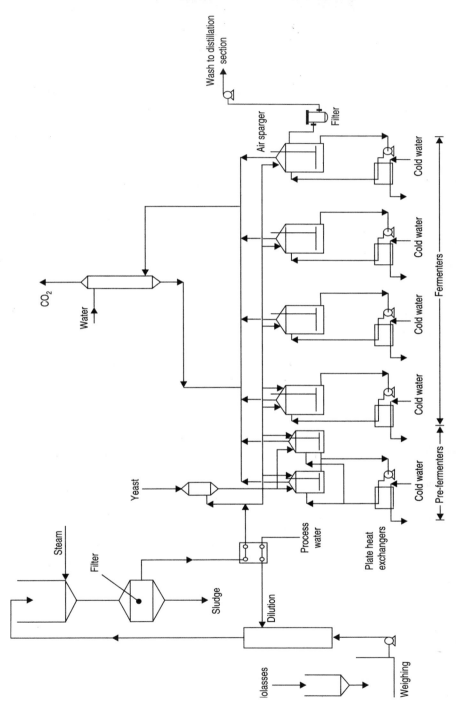

Figure 2.2 Fermentation of molasses.

concentration of yeast in them. Then the last pre-fermenter is emptied to a next empty fermenter, which is previously cleaned and kept ready. Dilute molasses solution is allowed to flow in this fermenter so as to fill it to its working capacity, say, about one lakh litres. The 'wash' (dilute alcohol) in each fermenter is cooled by cooling water in plate-type heat exchangers. The advantage of plate-type exchangers is that they can be cleaned and disinfected easily.

Readymade compressed yeast can also be used directly in the pre-fermenters depending on economic viability. Good quality yeast is available for use in distilleries. The yeast is manufactured under strict controlled conditions. This yeast is useful to obtain a good yield of alcohol by fermentation of molasses. In this case, the stages of yeast propagation described above for producing yeast from laboratory scale to pre-fermenter stage may be totally eliminated.

The fermentation of molasses in fermenters with about one lakh litres capacity take about 24 to 36 hours for completely exhausting the sugars in molasses.

As already mentioned all the sugars are not converted to alcohol during the process of fermentation because chemicals like glycerine are also produced by yeast during their metabolic process. Thereby, the efficiency of conversion of sugars to alcohol varies. The final product of fermentation is called 'wash' and it contains roughly 7% ethyl alcohol in water.

2.2.2 Bakers' Yeast

Bakers' yeast is the yeast that is produced in bulk, which is used in the production of bread. It increases the airiness and taste of bread. It is produced by the growth of microorganisms. The substrate used is either molasses or glucose. The fermentation is carried out under aerobic conditions and excess oxygen is provided. Partial oxidation will result in production of ethanol which is not acceptable and hinders yeast growth.

Basically, there are two ways to conduct any fermentation reaction. They are batch mode and continuous mode. Even though continuous mode has many advantages from a cost angle (smaller equipment, automatic control, etc.) there is a higher risk of mutations and contaminations during the growth. Bakers' yeast is produced in a semi-batch mode (or 'fed batch' mode as per biotech terminology). Fed batch operation is a compromise between batch and continuous operations.

For a metabolic process to occur, a source of carbon, energy, nitrogen, trace elements and sometimes vitamins are required. During the start-up of the fermentation, substrates and other additives are added stepwise. This gives a better control of the process.

The sugar substrate is added gradually to ensure continuous growth of biomass but at the same time preventing the production of alcohol. In this system, periodic withdrawal of a certain portion of the reacting volume is done. The cells have to be retained in the reactor and at the same time the toxic compounds have to be continuously purged. For this purpose, the liquid is passed through a membrane separator where the cells are collected and sent back to the reactor.

Process description

In a yeast plant shown in Figure 2.3, there are reactors of different capacities, in steps such as 300 litres, 2 cubic metres, 20 cubic metres and 100 cubic metres. A production cycle consists of making yeast from the smallest reactor in steps to reach the largest reactor which forms the final product. Typically, it takes many days to complete a cycle.

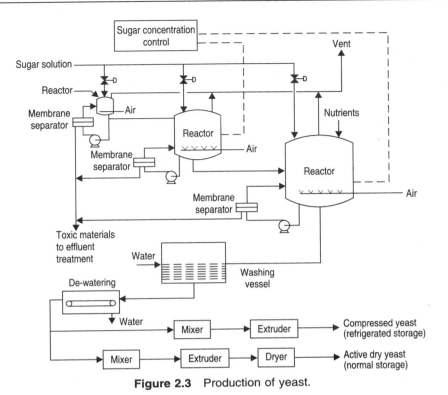

Figure 2.3 Production of yeast.

First, the reactors are sterilized with steam. Diluted molasses (it can be either cane sugar or beet sugar) is added along with yeast innoculants, proteins, salts and vitamins. It is also treated with acid to a pH of four and the precipitates are removed. It is sterilized at a temperature of about 150°C by steam injection for a few seconds. Ammonia is added to supply nitrogen and also adjust the pH.

The specific growth rate may vary as the biochemical reaction proceeds. At the end of the reaction the pH is raised to five. The yeast is filtered and de-watered. Emulsifiers are added and the moisture content is adjusted. The yeast is made into blocks and stored under refrigeration. This yeast is called compressed yeast.

The yeast can also be made into active 'dry yeast' by extrusion immediately after filtration. Then it is chopped into small pieces and dried. These pieces are capable of being stored at normal temperatures without degradation.

2.2.3 Citric Acid

Citric acid is a crystalline powder with the fomula $C_6H_8O_7$. It has the acid structure written as $C_3H_4OH(COOH)_3$. It decomposes into vapours on heating above 175°C.

Citric acid is an important organic acid widely used in the manufacture of carbonated beverages, foodstuffs, jams and jellies. It is also widely used in the biotechnological industry. Futher, it is also a constituent in medicines.

Citric acid is made by the submerged fermentation process by the fermentation of sucrose and dextrose present in sugar syrup, which is the raw material. The biochemical reactions are as follows.

Sucrose Fermentation:
$$C_{12}H_{22}O_{11} + H_2O + 3O_2 \rightarrow 2C_6H_8O_7 + 4H_2O$$

Dextrose Fermentation:
$$2C_6H_{12}O_6 + 3O_2 \rightarrow 2C_6H_8O_7 + 4H_2O$$

Process description

The various steps involved in the manufacture of citric acid are as follows. See Figure 2.4.

1. Purification of sugar syrup which includes vacuum filtration, cation exchange for removal of trace elements, pasteurization, cooling and pumping to fermenters.
2. Addition of inoculums into fermenters.
3. Addition of nutrients into fermenters.
4. Sparging sterile air into fermenters for citric acid production.
5. Purification of citric acid by separating off the mycelium from the broth, treatment with hydrated lime, forming calcium citrate.
6. Regeneration of citric acid by addition of sulphuric acid.
7. Purification by ion exchange, evaporation and crystallization, remelting and recrystallization under vacuum.
8. Classification and packing for sales.

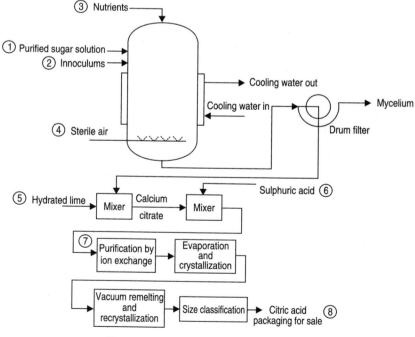

Figure 2.4 Production of citric acid.

2.2.4 Lactic Acid

Lactic acid is a liquid with the formula $C_3H_6O_3$. It has a weak acidic structure, and is written as $CH_3CHOHCOOH$. It is a solid, with a melting point of 53°C and a boiling point above 200°C. Since it is readily miscible in water, it is sold as a liquid solution in water. Lactic acid is present in the human body as lactates. They are used (burnt) by the muscles during prolonged exercises or workouts to produce energy for the body. Lactic acid is also used as a pharmaceutical.

Lactic acid is a well known acidulant for foods (bakery products, meat products, cheese, yoghurt, pickles, etc.) and beverages and soft drinks. Small amounts are used in making chemicals and plastics. Lactic acid can be converted to polylactic acid (PLA) which can be used for the production of biodegradable plastics.

Lactic acid gets produced naturally by fermentation of milk sugar by *streptococcus lactase* bacteria. It can also be produced from whey, a by-product of milk industry.

Process description

There are two processes by which glucose can be converted into lactic acid. These are heterolactic fermentation and homolactic fermentation. In homolactic fermentation, glucose splits into two similar molecules as follows:

$$C_6H_{12}O_6 \rightarrow 2CH_3CHOHCOOH$$

In heterolactic fermentation, glucose splits into three different molecules as follows:

$$C_6H_{12}O_6 \rightarrow CH_3CHOHCOOH + C_2H_5OH + CO_2$$

In the heterolactic reaction, only one part will get converted to lactic acid, the other being split into alcohol and carbon dioxide.

The homolactic fermentation process is shown in Figure 2.5.

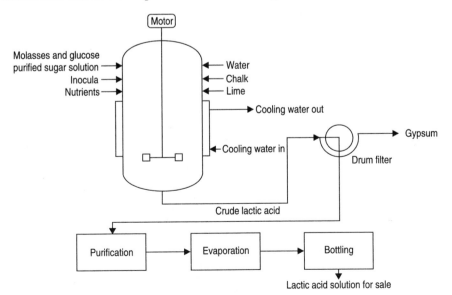

Figure 2.5 Production of lactic acid.

Lactic acid is commercially made by the fermentation of a mixture of molasses and/or glucose. Sucrose (molasses), glucose, water, chalk and lime are fed into a ferrmenter along with bacteria inoculum and nutrients. Crude lactic acid and gypsum are formed.

The gypsum is removed in a drum filter and the liquid obtained is the crude lactic acid. The crude lactic acid is purified by any of the methods such ion exchange, or newer methods such as esterification and hydrolysis, etc. It is then concentrated by vacuum evaporation. The concentrated liquid solution is filled in bottles, packed and sent for sales.

2.2.5 Fructose (as High Fructose Corn Syrup)

While fermentation is based on cell activity, it was found that enzymes that exist in the cells of bacteria can also be extracted and used without losing their activity. Glucose is converted into fructose by activity of enzymes.

The amount of sweetness in fructose is almost twice that of in glucose. This property was taken advantage of by sweetener manufacturers, resulting in heavy demand for fructose.

The conversion of glucose to fructose is carried out by the enzyme called *glucose isomerase*. With the biotechnological achievement of immobilization of the enzyme, shallow beds of the immobilized enzymes can convert glucose to fructose. The immobilized enzymes are supported on solid medium and are also more resistant to pH and temperature fluctuations. This technology has been used for the production of high fructose corn syrup (HFCS). See Figure 2.6.

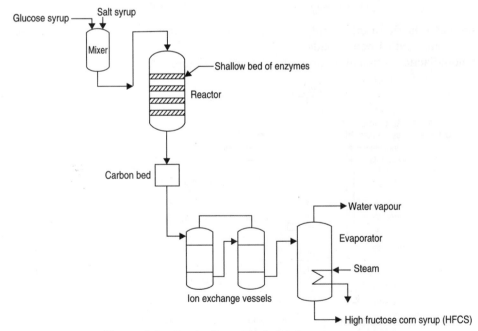

Figure 2.6 Production of high fructose corn syrup.

Process description

The process consists of first mixing the glucose syrup with metal ions such as cobalt, magnesium, etc. which provide catalyst activity. This is done in a mixer. Then it is passed through a reactor with four beds of immobilized enzyme carriers. After conversion, impurities and colouring matter are removed in an activated carbon bed and then passed through ion exchange vessels to recover metal ions and produce a pure solution. Then the pure liquid is sent to an evaporator where excess water is removed to produce concentrated high fructose corn syrup.

2.2.6 Penicillin

Pharmaceutical industry uses fermentation processes extensively for the production of antibiotics such as penicillin, streptomycin, vitamins, and other medical composition additives.

Antibiotics are naturally occurring substances that are capable of destroying harmful microorganisms that cause disease. It is known that certain curds have antibiotic action. The Chinese are known to have used certain antibiotics about 2500 years ago.

The raw material used for the production of antibiotics consists of a broth made of maize, soya bean or molasses. The organisms that produce the antibiotics are isolated and added to the broth typically in fermenters of about 30 to 150 cubic metres in volume. When the required quantity is produced which may take about 5 days, an extraction process is used to recover the antibiotic with solvents such as isobutyl ketone. After a further thorough refining process, the material is sent for making into capsules. Quality control holds the key to the manufacturing process.

Penicillin was first discovered in 1929 and the first commercial plant established in 1940. It is an important commercial product produced by an aerobic submerged fermentation. Later, many other products were made by the fermentation process. However, some of the older antibiotics have gone out of favour due to resistance build-up of disease bacteria as well as the after-affects of the treatment. Nowadays the older antibiotics are continued to be made, which are further converted to newer antibiotic products such as ampicillin.

Process description

See Figure 2.7. Penicillin is made by fermentation of a broth of maize to which penicillin mold is added along with starch and salts. Sterilized air is sparged through the reactor. After fermentation the product is treated with dilute phosphoric acid. Then it is purified by carbon decolourization, extraction with amyl acetate solvent using podbielniak liquid-liquid extractor separator, and re-extracted with salt solution. (The podbielniak extractor is centrifugal liquid-liquid separation equipment which uses centrifugal force instead of gravitational force to make the separation.) The solids are separated by crystallization, and then dried to form penicillin in powder form. It is then blended, packaged and sold.

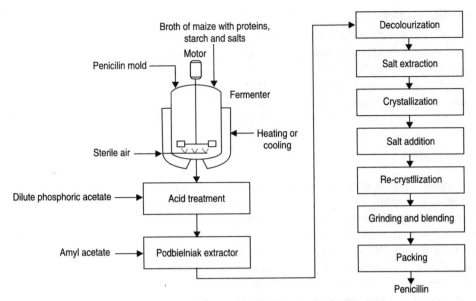

Figure 2.7 Production of penicillin—typical process.

2.2.7 Streptomycin

Two of the most commonly used antibiotics are streptomycin and ampicillin.

Process description

See Figure 2.8. Streptomycin is manufactured using its initial culture of *streptomyces griseus*. This is added into a sterilized fermentation media consisting of a broth of maize and mold along

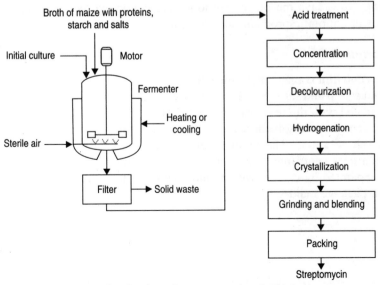

Figure 2.8 Production of streptomycin—typical process.

with proteins, carbohydrates, and minerals required for its growth. Mechanical agitation is done by passing air for a period of about two hours at favourable constant temperature and to the extent required. Then the cells and insoluble matter are filtered off. The filtrate is purified by acid treatment or ion exchange.

The finishing operations include concentration, decolourization, and hydrogenation to obtain a saleable product. Then it is crystallized, ground and bottled and packed for sales.

2.2.8 Ampicillin

Ampicillin is available in three different forms, namely, ampicillin trihydrate, ampicillin anhydrous and ampicillin sodium. Ampicillin trihydrate is the most widely used form of ampicillin. Its chemical name is D-alpha-aminobenzylpenicillin. There are many methods used for the production of ampicillin. Ampicillin can be made from penicillin as the starting material. First, the penicillin is converted into 6-amino penicillanic acid. The amino penicillanic acid can be condensed with D-alpha-phenyl glycine chloride in the presence of E. coli bacteria to form ampicillin trihydrate. In the earlier process, Danish salt was used instead of the chloride radical.

In a new process, ampicillin is made by an enzymatic process (a typical enzymatic process is described in the next section). The enzymatic production of ampicillin is done starting from 6-amino penicillanic acid and a functional derivative of D-alpha-aminophenylacetic acid.

2.2.9 L-ascorbic Acid (Vitamin C)

Vitamin C is chemically known as L-ascorbic acid. It is a very important component of human and animal food. It also serves as an antioxidant in the food industry. Huge amounts, about a lakh tonnes, is produced by the pharmaceutical industry every year.

Process description

See Figure 2.9. The production consists of many steps, some of them chemical and others microbial. The raw material used is D-sorbitol. The D-sorbitol is oxidized to L-sorbose by the action of the microbes called gluconobacter oxidants. This is a submerged fermentation process with aeration and vigorous agitation. The biochemical reaction is completed in about 24 hours. The L-sorbose produced is condensed with acetone to form sorbose diacetone. This is again fermented to produce 2-keto L-ascorbic acid, which is converted by lactonization to L-ascorbic acid or Vitamin C.

In the newer process developed in China, which is widely used, sorbose (or glucose) is converted directly into 2-keto ascorbic acid (or 2-keto gluconic acid) by genetically modified microbes. This is further converted by lactonization to L-ascorbic acid or Vitamin C, thus avoiding one step of fermentation.

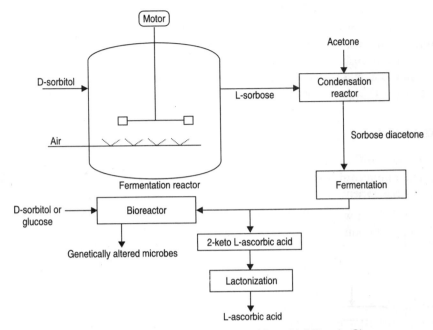

Figure 2.9 Production of L-ascorbic acid (Vitamin C).

2.2.10 Vitamin B₂ (Riboflavin $C_{17}H_{20}N_4O_6$)

Vitamin B_2, also known as riboflavin, is an easily absorbed micronutrient with a key role in maintaining human health. It is also obtained by fermentation. It is an orange yellow crystalline powder. It is added to bread and flour in large measures since it is a component useful for growth and aids the transfer of oxygen to tissues.

Riboflavin is produced by any of the following two methods.

1. As a by-product of the fermentation of molasses for the production of butanol and acetone.
2. By a synthetic process of condensation using oxylene, D-rebose and alloxan. Alloxan is condensed with boric acid to get riboflavin.

Many other vitamins, anti-viruses and anti-toxins, etc. are also obtained by biological fermentation processes. Products for immunization are also obtained by biological fermentation methods.

2.2.11 Industrial Alcohol (Rectified Spirit)

In Section 2.2.1, we have already seen how fermented 'wash' containing 7% ethyl alcohol in water is obtained. This 'wash' is converted into rectified spirit by distillation.

The important specifications of rectified spirit are as follows:
Specific gravity at 15.6°C to be less than 0.8171
Ethanol content: per cent by volume at 15.6°C, minimum: 94.68

Alkalinity: nil
Acidity, as acetic acid, maximum: 20 ppm for industrial grade.

(The acidity is 10 ppm for potable grade, known as 'silent spirit', which requires one more distillation column called 'silent spirit column' during distillation. The silent spirit is diluted with demineralized water and blended with additives to make potable liquors.)

The distillation process for obtaining rectified spirit is described below and the methodology of working out the cost of a typical plant is given in Chapter 10. Traditionally, all the items of equipment are made of deoxidized copper.

Process description

A distillation system as shown in Figure 2.10 is employed. This system consists of the following main items of equipment.

1. Wash column with degasifying-cum-heads column as top portion.
2. Rectifying column

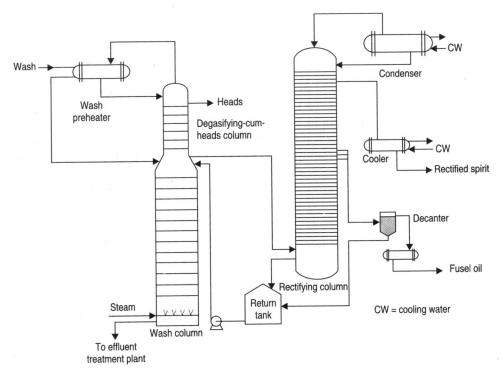

Figure 2.10 Rectified spirit plant.

The fermented wash first enters the wash preheater, which is a condenser for condensing alcohol vapours by using wash as a cooling medium. The objective of this wash preheater (also called beer preheater) is to recover the heat from the hot vapours of alcohol. Wash from the wash preheater goes to the degasifying column top plate. The purpose of this column is to get rid of foul gases dissolved in the wash. The wash from the degasifying column bottom goes to top plate of the wash column. The steam is admitted through the steam sparger situated at the bottom of the wash column. As the steam rises up, the wash descending from the top of the column gets heated, distilling the alcohol contained in it. When it reaches the bottom plate, it contains practically no ethyl alcohol.

The water going out is called spent wash, which is discharged through the drain pipe. The vapours coming from wash column now consist of approximately 50% alcohol and 50% water with impurities such as higher alcohols, aldehydes, acids, sulphur dioxide, etc.

One method to remove heads (aldehyde and low boiling impurities) is to draw the heads as a liquid draw from a plate at the top of the degasifying column as shown in the Figure 2.10.

Another method, more efficient but costlier, of removing aldehydes is to take a part of the vapours from the rectifying column to a separate heads column, where the low boiling impurities are separated from the spirit.

Heads will be removed and stored separately at the rate of 5 to 10% of the total production of alcohol depending on the extent of purity required.

Vapours from the wash column are led into the bottom of the rectifying column. This column consists of about 50 plates for rectification, which also helps the removal of bad smelling fusel oil, which is a mixture of higher alcohols. As the vapours rise to the top of the rectifying column, the concentration of alcohol goes on increasing and finally it reaches a concentration of 95% alcohol. The alcoholic vapours from the rectifying column are condensed in the main condenser using water as coolant and finally in the vent condenser (not shown in figure). The condensates of both the condensers go back to the top of the rectifying column as reflux, and uncondensed gases are let out through a vent pipe. The actual product, i.e. the rectified spirit is drawn from the third plate from the top and cooled in an alcohol cooler and taken out as a product.

The fusel oil, which is a mixture of higher alcohols, is drawn from the sixth to tenth plate (counting from bottom of rectifying column) as a side stream. It is cooled and let into a decanter where it is mixed with water. Fusel oil being immiscible with water remains at the top and is decanted through a funnel and a part is sent back to wash column for recovery of alcohol. Fusel oil is recovered at the rate of 0.2% of alcohol produced. The alcohol, both rectified spirit and heads are first led into separate receivers. The quantity and quality of alcohol produced is assessed daily in the receiver and it is finally transferred to storage vats or storage tanks.

Quick mechanical design and cost estimation

A quick mechanical design for cost estimation purpose and cost estimation for the above plant is given in Section 10.2.

A summary of the result of the above estimation is given below.

Cost summary

A typical plant and machinery cost summary (general off-site facilities, taxes and duties, etc. not included) is given in Table 2.2.

The plant cost worked out is approximate and useful for the illustration purpose only. The plant and machinery cost is only a portion of the project cost.

Table 2.2 Cost summary

Item	Weight (kg)	Cost (₹)	Fabrication and erection (%)	Items as (%)	Total cost (₹)
Wash column	6200	3,720,000	20		4,464,000
Rectifying column	6600	3,960,000	20		4,752,000
Wash preheater	2200	1,320,000	20		1,584,000
Condensers	1920	1,152,000	20		1382,400
Cooler	490	294,000	20		352,800
Reflux tank	100	60,000	20		72,000
Pumps etc.					900,000
				Subtotal	13,508,800
Piping items				20	2,701,440
Insulation etc.				15	2,026,080
Instruments				20	2,701,440
Electrical				20	2,701,440
Civil				20	2,701,440
Total cost					26,340,640 (say, 2.7 crores)

2.3 ABSOLUTE ALCOHOL WITH CASE STUDIES

Absolute alcohol is used in pharmaceutical industries. It is also used as an environment-friendly fuel for blending with petrol to reduce the consumption of mineral oils. It is considered environment-friendly or 'CO_2 neutral' provided the alcohol is originated from flora and fauna. This is true for most countries since alcohol from petrochemical sources is produced only in either developed or oil-rich countries.

Brazil and United States produce large quantities of absolute alcohol for fuel use commonly known as fuel alcohol. Brazil produces about 26.2 billion litres of fuel alcohol per year for use in automobiles. The main reason for this is the availability of excess sugarcane and the government requirement of mixing minimum 22% alcohol in petrol for automobiles[2]. All petrol engines are tuned to use this fuel.

Absolute alcohol is an important product required by industry. As per International Specification (SI) absolute alcohol is to be minimum 99.9% v/v pure alcohol. Ordinary alcohol as manufactured is rectified spirit, which is 94.68% v/v alcohol and the remainder water. It is not possible to remove the remaining water from rectified spirit by straight distillation as ethyl alcohol forms a constant boiling mixture with water at azeotropic concentration. Therefore, a special process for the removal of water is required for manufacturing absolute alcohol. In order to extract water from alcohol, it is necessary to use some dehydrate, which is capable of absorbing and thereby separating water from alcohol.

A simple dehydrate is unslaked lime. Industrial alcohol is taken in a reactor and quick lime is added to it and the mixture is left overnight for letting the chemical reaction to be completed.

It is then distilled in a fractionating column to get absolute alcohol. Water is retained by quick lime. This process is used for small-scale production of absolute alcohol by batch process only.

For large-scale manufacture of absolute alcohol, two basic processes are used: (i) molecular sieve method and (ii) azeotropic distillation method.

2.3.1 Molecular Sieve Method

This method is more appropriate for the production of alcohol used as fuel since even though the plant is costlier, the cost of production will be less for large quantities produced.

This plant carries out dehydration of ethanol and water mixture, by water being adsorbed into molecular sieve material made of zeolite balls. The dehydration unit operates with two adsorbers with alternate steps of adsorption and desorption. Adsorption occurs in the vapour phase and under pressure. Desorption regenerates molecular sieves. This step is performed under vacuum by Pressure Swing Adsorption (PSA) principle. Part of the dehydrated alcohol is used for the molecular sieve desorption. Alcoholic effluent from desorption is regenerated with a distillation column.

Process description

Refer to Figure 2.11. Rectified spirit, along with a recycle stream of rectified spirit vapours from the top of the rectifying column, is passed through a heat exchanger and superheater and taken to the molecular sieve units for dehydration. The vapour passes through a bed of molecular sieves and the water in the incoming vapour stream is adsorbed on the molecular sieve material.

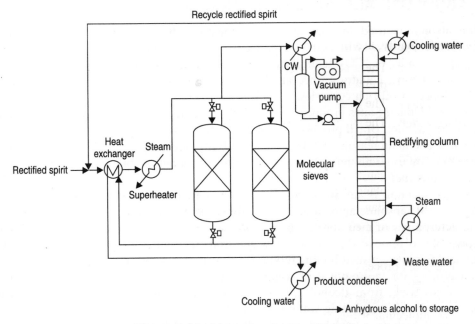

Figure 2.11 Molecular sieve process.

During regeneration hot anhydrous ethanol vapour from the molecular sieve units is passed through a heat exchanger to heat feed alcohol. The anhydrous ethanol (absolute alcohol) product is then further cooled down in the product cooler, to bring it close to the ambient temperature.

The two molecular sieve units operate sequentially so that one is under operation, adsorbing water from the vapour stream, while the other is being regenerated. The regeneration is accomplished by applying vacuum to the bed. The adsorbed water from the molecular sieve material desorbs and evaporates into vapour stream. This mixture of ethanol and water is condensed and cooled with cooling water in the molecular sieve regeneration condenser. Any uncondensed vapour and entrained liquid leaving the molecular sieve regeneration condenser enters the molecular sieve regeneration drum. The cooled regenerant liquid is weak in ethanol concentration, as it contains all the water desorbed from the molecular sieve beds. This alcohol water mixture is sent to the rectifying column for recovering ethanol. The waste water from the bottom of the column will contain only traces of alcohol.

In this process, the switch-over action is most important and can be provided through pneumatically operated or solenoid-based control valves. To provide continuous stable and efficient plant operation, a central programmed logic controller (PLC) or a distributed control system (DCS) based control system is to be used. The required control will be automatically carried out on the basis of the programmed algorithms.

An example with typical design basis

The basis used for a vapour phase molecular sieve based dehydration system is as follows. The same basis is also used for azeotropic distillation.

Absolute alcohol production, litres/day	60,000 litres/day
Product quality,	99.8 (minimum)% v/v
Feedstock	Rectified spirit
Feedstock quality, alcohol content	94.68 % v/v

Steam: The total steam requirement is about 2000 kg/h at 11.0 kg/cm^2 gauge pressure.

Electric power: The total connected load, for the unit will be about 28 kW.

Cooling water: The plant shall require cooling water systems for cooling the condensers. The cooling water requirement will be about 130 m^3/h at 30°C.

Instrument air: The plant shall require approximately 18 Nm^3/h of compressed air at a gauge pressure of about 7 kg/cm^2 for operation of control valves and other plant instrumentation. The air supplied shall meet the instrumentation air standards.

The advantages of molecular sieve technology for ethanol dehydration can be summarized as below.

1. The basic process is simple and automatic which reduces the labour requirement.
2. Since no other chemicals are used, there is no material handling problem, which might cause safety problems.
3. The molecular sieve desiccant material has a very long service life, with chances of failure occurring only due to fouling of the media or by mechanical destruction. A properly designed system will give a molecular sieve service life of 5 to 6 years.

2.3.2 Azeotropic Distillation Method

Most of the ethanol dehydration plants for the production of absolute alcohol, mainly for pharmaceutical use, are based on azeotropic distillation since the requirements of purity levels are stringent, but the quantities required are less. It is a mature and reliable technology capable of producing a very dry product. The application of PSA (Pressure Swing Adsorption) will not be economical for very small capacity plants.

Process description

A two-column azeotropic distillation system is proposed for dehydrating ethanol. Either cyclohexane or benzene can be used as an entrainer for the purpose. The total system consists of feed pumps, preheater, dehyderation column, recovery column, condenser, reboiler, decanter, intermediate tanks, transfer pump, etc.

Operation with fresh rectified spirit

Fresh rectified spirit is preheated in a feed preheater and fed to the dehydration column. See Figure 2.12. The column is fitted with a reboiler. A third component (cyclohexane or benzene) is introduced inside the system. The ternary azeotropic mixture coming out of the dehydration column is condensed in the condensers and collected in the decanter. This mixture containing entrainer, alcohol and water is heterogeneous in nature. It separates into two layers in the decanter. The organic layer is refluxed back to the dehydration column. The aqueous layer is sent to the recovery column for recovery of alcohol, and the entrainer is recycled.

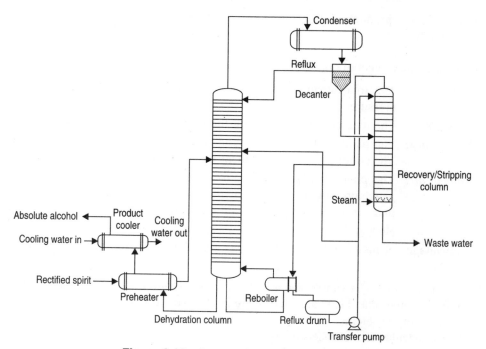

Figure 2.12 Azeotropic distillation process.

The recovery column is run under reflux and the top liquid draw is sent to the dehydration column. The bottom of this column is pure water and can be discharged to drain or used as usual. The absolute alcohol of desired specification is drawn out from the bottom of the dehydration column. It is cooled in two stages: first, in the feed preheater and second, in the product cooler. The cooled absolute alcohol is sent for storage.

There are two ways to give heat to the dehydration column reboiler. One is by introducing sufficient excess steam to the stripping column and the top vapours of stripping column giving heat to the reboiler. The other method is to give direct steam to the reboiler. In this case, stripping column should have a condenser. This method is shown in the example given below.

An example with design basis and calculations

The factors used to establish the design basis for a typical azeotropic distillation system for anhydrous ethanol are as follows. These factors are also similar to the molecular sieve process.

The objective is to produce absolute alcohol with a moisture content of less than 0.1% v/v.

Plant capacity : 60,000 litres/day of absolute alcohol
Feed : 68,600 litres/day of rectified spirit with the following compostion:
  Alcohol : 94.68% v/v
  Water : 5.32% v/v
Product rate : 60,000 litres/day of absolute alcohol with moisture content of less than 0.1% v/v.

Makeup entrainer reqirement: Benzene, as required for makeup.

For determining the number of theoretical plates required, a plate-to-plate calculation is performed for achieving design purity of the final product. The procedure given in the book, *Elements of Fractional Distillation* by Robinson and Gilliland (pp. 312 to 324)[3] may be used.

Assuming that the bottom tray outlet will be 99.9 % by volume (equal to 99.92% by weight) alcohol, the composition of all the components in the remaining trays is calculated using the tray-by-tray calculation starting from the bottom-most tray based on the relative volatilities of the components. Knowing the composition ratio of water to alcohol in the feed, the feed tray location is decided such that the above ratio coincides with the composition on the particular tray. Similarly, calculations are done for the enriching section of the column by taking the composition of the benzene layer from the separator as reflux. Alternatively, if full reflux is provided the product can be taken from two trays below the top, in which case the product will remain uncontaminated by any lighter components in the feed.

The calculation resulted in 25 theoretical trays for the dehydration column with feed at the 21st tray from the bottom or the fifth tray from the top. The second distillation column is designed as a normal stripping column for elimination of water from the system.

An effective modification to the above plant is to use double decantation of the top product of distillation instead of single decantation. This is applied where chilled water utility is available. The product of first decantation (which can be done in the tray itself with special arrangements) is again decanted after cooling, using chilled water at 19°C, which improves the quality of the product.

The mass balance diagram given in Figure 2.13 shows the mass balance obtained, based on the modified plant with two decantations instead of one.

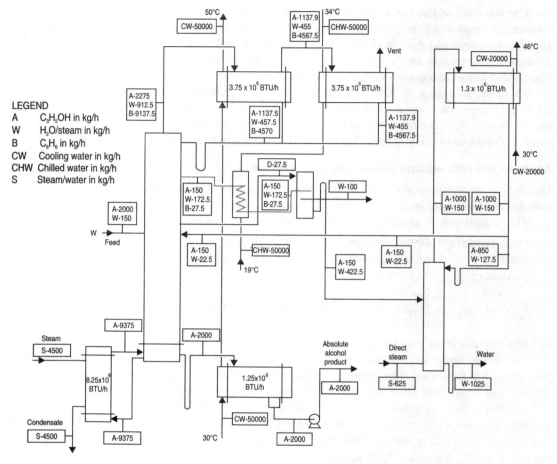

Figure 2.13 Mass balance—modified absolute alcohol plant.

Since the material handled is clean, sieve trays are sufficient and the design of the trays can be done as per the procedure given in the book on Distillation by Mathew Van Winkle[4] and for mechanical design Brownell and Young[5] may be followed.

The typical requirement of utilities is given below:

Saturated steam at gauge pressure 2.0 kg/cm^2 (through a Pressure Reducing Station):
2 kg/litres of product.

Cooling water supply and return at 35°C and 45°C : 250 m^3/h
Pneumatic air/Instrument air (dry, oil free at 7 kg/cm^2) : 10 Nm^3/h
Process water at 2 kg/cm^2 gauge pressure : 2.5 m^3/h

Power consumption (415 V, 3 phase, 50 Hz) including auxiliaries:
Connected load : 94 kW
Operating load : 55 kW

About 3% of cooling water flow is required as makeup for evaporative losses in the cooling tower.

Special features of the process: The process uses distillation with benzene (or cyclohexane). The ternary azeotropic mixture, which is formed at the top of dehydration column, allows the removal of water and thus the dehydration of alcohol. The azeotropic mixture is heterogeneous and the heavy phase, which is high in water content, is chilled and extracted by decantation. The regeneration column allows water extraction from the heavy phase as well as benzene (or cyclohexane) recycling. The water content in anhydrous alcohol is less than 1000 ppm.

2.4 PRODUCTS FROM SEA—AGAR AND AGAROSE

Products from the sea can be classified into inorganic and organic chemicals.

Inorganic chemicals: One of the oldest inorganic chemicals known to man was obtained from the sea. It is sodium chloride known as common salt. Solar evaporation of water is the most common method of its production. Many other compounds such as magnesium oxide, magnesium sulphate, potassium chloride, bromine, etc. are some of the other inorganic compounds obtained from the sea.

Organic chemicals: Many complex natural organic chemicals are obtained from the sea. Sea weeds are an important source of these products. Two of the fastest growing natural product industries in this area are agar and agarose, which are explained below in detail.

2.4.1 Agar

Unchecked growth of all types of industries may not be in the interest of mankind. It is the natural product that generates income and sustains a large population of the world and hence the importance of natural and sustainable technologies. Agar, extracted from seaweeds, is one such product beneficial to mankind.

Agar has shown itself to be notable and often indispensable in the gelling process, as an efficient thickening agent, a stabilizer for sugar-based mixtures, a calorie-free food additive, a natural surgical lubricant and a means of culturing. Its versatility can measure up to a wide variety of requirements especially in food and pharmaceutical products. Agar is especially suitable whenever the final products have to undergo extreme thermal treatments such as baking, sterilization, pasteurization, etc.

Chemical structure of agar and agarose

The chemical nature of agar varies according to the species of seaweed from which it is extracted, the environment where the seaweed grows and the method of preparation of the agar. Generally, however, agar consists of two fractions: agarose and agaropectin.

A typical structure of agarose is shown in Figure 2.14.

Agarose is a neutral, long-chain molecule formed by β-D-galactopyranose residues connected through C-1 and C-3 with 3,6-anhydro-L-galactose residues connected through C-2 and C-4. Both residues are repeated alternatively. However, depending on the origin of the raw material, some units of 3,6-anhydro-L-galactose are replaced by L-galactose. Also, D-galactose and L-galactose units can be in a methylated form. Polar residues, such as pyruvic and sulphuric acids, are also found in small quantities. The presence of agarose gives the gelling power to agar.

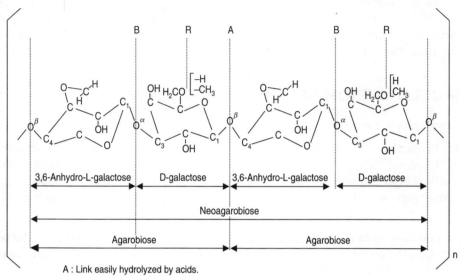

Figure 2.14 Chemical structure of agarose.

The basic structure of agaropectins consists of alternating D-galactose and L-galactose units. D-galactose can be substituted by D-galactose 4-sulphate, or by 4, 6-0-(1-carboxyethylidene)-D-galactose in certain terminal chain positions or even by D-galactose 2, 6-disulphate, while part of L-galactose can be replaced by 3, 6-anhydro-L-galactose. The different substitutions of the basic monosaccharide are responsible for an enormous number of possible chemical structures. Agaropectins have a low gelling power in water.

Properties of agar

Agar has very strong gelling power in aqueous solutions. Gels which are stronger than those of any other gel-forming agents are formed by agar, assuming that equal concentrations are used. Agar does not require any additives to produce gelation. It can be used over a wide range of pH values (varying from 5 to 8) and is able to withstand heat treatment, even at a temperature above 100°C, thus allowing for efficient sterilization.

A 1.5% aqueous solution of agar can gel at temperatures between 32°C and 43°C with a melting point of minimum 85°C. This is a unique property of agar. Transparent gels that are easily coloured can be obtained and their refractive index can be easily increased by adding sugar, glucose, glycerine, etc. This gives them an attractive brightness.

Agar produces gels without flavour and hence can be used without problems to gel food products with soft flavour. It assimilates and enhances the flavours of products and acts as a fragrance fixer. The gelling ability of agar is easiliy reversible. This allows it to be repeatedly gelled and melted without any loss of its properties. The gel is very stable and does not cause precipitates in the presence of cations.

Gelling mechanism of agar

The mechanism of gelation of agar is shown in Figure 2.15[6]. At temperatures above the melting point of the gel, thermal agitation overcomes the tendency to form helices and the polymer exists in solution as a random coil (Figure 2.15(a)). On cooling, a 3-dimensional network builds up in which double helices (agarose segments) form the junction points of the polymer chains, as shown in Figure 2.15(b). Further cooling leads to aggregation of these junction points linked by non-helical agaropectin segments to form a three-dimensional network as shown in Figure 2.15(c).

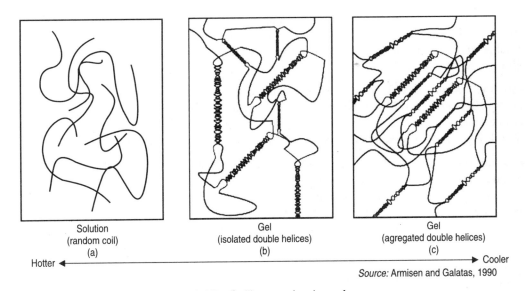

Source: Armisen and Galatas, 1990

Figure 2.15 Gelling mechanism of agar.

The presence of the sulphate at C_6 of the 1,4-linked-L-galactose residues, such as in the precursor of agarose, acts as a "kink". This prevents the double helix from forming. Closure of the ring to form the 3,6-anhydride and elimination of the C_6 sulphate group makes the chain straighten and leads to greater regularity in the polymer. This results in enhanced gel strength due to increased capability of forming a double helix.

Generally, the gel strength of the agar is influenced by the amount of agarose in it. An agar sample with higher agarose and lower sulphate content always displays higher gel strength. The ideal agarose should be free of sulphate groups. However, in practice there exists a trace (0.1–0.5%) of sulphate since no pure agarose molecules are present in nature.

Uses of agar

Agar has a wide variety of applications in the food, pharmaceutical, medical and other industries.

Food industry: Agar is used in edible jellies, salad dressings, icings, confections that set at room temperature. Agar is used in the food industry primarily as a stabilizing, thickening and gelling agent. It functions as a texture-improving and anti-stalling agent and acts as an emulsifier and stabilizer for fillings, icings and candies. Agar is sometimes mixed with flesh meats or meat products to enable them to withstand handling, transportation and preparation stresses. In another use, agar coatings are used to prevent constituents of fish, such as herring, from blackening the contents of cans.

Microbiology: The most widespread use of agar is as a culture medium for practically all pathogenic and non-pathogenic fungi and bacteria. This is because agar is not easy to metabolize and has firmness, elasticity, clarity and stability. Usually, a concentration of 1–2% of agar is used in microbiological preparations.

Tissue culture: An important application of agar is in tissue culture. There is a growing interest in tissue culture as a standard method for the propagation of orchids and other ornamental plants, vegetable and fruit plants. Here and in other agricultural products, agar is used as a culture medium.

Dentistry: Agar, mixed with other substances, is used to make accurate dental casts in dentistry. Between 15–18% of the impression material is made up of agar. Nowadays, however, agar encounters competition from alginate (a constituent of brown algae) as a dental impression material. Agar is also used as an impression material in tool making, criminology, etc.

Medicine and pharmacy: Agar is used as an ingredient in the preparation of capsules, in surgical lubricants, in the preparation of emulsions and as a suspending agent for barium sulphate in radiology. Agar is used as a laxative because when hydrated it provides a smooth non-irritant bulk in the digestive tract. Such agar is in the form of flakes which absorb 12–15 times their weight of fluids. It is also used as a disintegrating excipient in tablets. Sulphated agar has fat reducing activity. It is also believed to inhibit the aerobic oxidation of ascorbic acid.

Other applications: Agar is used for clarifying coffee, beer, wine, juices and Japanese shake. It is used in photographic films and paper and in solidified alcohol fuel, in dyed coatings for paper, textiles and metals, as a flash inhibitor for aluminium and for the action of nicotine as an insecticide in plant sprays. It is used in shoe polish mixed with shellac and wax. It is used as a setting inhibitor for deep well cement and as an adhesive in gloss finishing of paper products.

Production of agar

Agar is produced by a process that consists of the following fundamental phases.

1. Hot water extraction
2. Cooling and gelling
3. Purifying
4. Drying and milling

All these phases are subject to differing technologies, employed according to the type of seaweed extracted and the product's desired final characteristics.

Agar is produced in the form of flakes or powders of varying fineness. Colour differs according to the seaweed's origin and also upon the process employed during the production process.

Since agar is not a homogeneous material, the optimum conditions for extraction vary depending on the seaweed species used and the purity of the harvested seaweed. Furthermore, the quality of the final product is highly dependent on the quality of the raw material which, in turn, is dependent on the species of seaweed. The environmental conditions during growth of the seaweed, and the storage and handling conditions prior to processing also affect quality.

Seaweeds should therefore be carefully evaluated for their agar-yielding properties. This is done by taking many samples of seaweed and testing for moisture content, purity, agar yield and quality. The production of agar from seaweed involves five basic steps.

1. Cleaning and washing of seaweeds to remove impurities
2. Chemical pretreatment, if required, with acid or alkali to improve agar quality and yield
3. Extraction of agar
4. Filtration and gelation of the extract
5. Finally, washing, bleaching and dehydration of the gel.

Cleaning and washing: In order to obtain the purest possible extract of agar, seaweeds are generally hand selected and washed. They are placed in the sun and sprayed with fresh water until they are bleached. Foreign matter, such as other weeds, sand, stones and pieces of coral, is removed. Washing is carried out in open cement tanks, with or without mechanical agitation. The seaweeds are washed as many as four or five times.

Chemical pretreatment: Certain seaweeds, such as Gracilaria, contain considerable amounts of sulphated galactan (L-galactose 6-sulphate, the precursor of agarose) which varies with the species, growing season and location. In such cases, sulphate alkaline hydrolysis treatment is usually carried out to convert the L-galactose 6-sulphate to 3,6-anhydro-1-galactose so as to improve the quality of the agar product. This is usually done by diffusion with a dilute (0.1 mol) sodium hydroxide solution for one hour at a temperature of 80–90°C taking care not to extract the agar.

Sometimes an acid treatment is used to soften the seaweeds to prepare them for extraction. The weeds are immersed for 10–15 minutes in cement tanks containing the dilute acid. Usually, hydrochloric acid is used. No pretreatment is required for seaweeds of Gelidium species which usually contain agar of excellent quality, naturally with high gel strength and low sulphate content.

Extraction: Agar is extracted from the seaweeds using hot water. The seaweeds are pressure cooked at 1 kg/cm^2 guage pressure and at about 121°C for one hour to one and a half hours with an amount of water equal to 15–20 times their weight. An aluminium or stainless steel vessel is used for the process. Chemical bleaching, using sodium or calcium hypochlorite, may be done during or after extraction, as necessary.

Filteration, gelation, washing, bleaching and dehydration: Agar extracts of 0.8–1.5% are considered to be the optimal concentrations for subsequent dehydration. The freezing-thawing method, called syneresis, is usually employed at the industrial level for dehydration. The hot agar solution after extraction is filtered through a vacuum filter or filter press. The clear filtrate is then poured into shallow aluminium trays and cooled at room temperature to form a gel. The gels are cut into thick sticks or bars. The gel sticks or bars are freezed in freezers. Freezing should be done slowly to maximize the separation of agar. After freezing, the frozen gels are thawed, washed, centrifuged and dried in a drying chamber or under the sun.

Tables 2.3 and 2.4 give a good comparison of agar properties made from different species.

Table 2.3 Gel strength and physical properties of agar from various seaweeds[6]

Species	Yield (%)	Gel strength* (g/cm²)	Gelation temperature (°C)	Melting temperature (°C)
Gelidiella acerosa (1)	40	125	46	73
Gelidiella acerosa (2)	45	300	40	92
Gracilaria edulis	55	63	48	65
Gracilaria verrucosa	23	41	40	–
Gracilaria lichenoides	43	120	45	84
Gracilaria crassa	23	140	48	84
Gracilaria corticata	38	20	44	68
Gracilaria folifera	12	15	40	–

*Gel strength of 1.5% concentration agar in water at 28±2°C

Table 2.4 Chemical nature of agar taken from various seaweeds[6]

Species	Agar (%)	3,6 anhydro L-galactose	D-galactose (%)	6,0 methyl D-galactose	Sulphate (%)
Gelidium	20–22	33–55	47–58	0.1–2	1.5–4
Gracilaria	10–18	32–38	22–42	1–20	2.5–8

Production of agar powder

To produce agar as powder, the gel sticks are washed in water to diffuse out the soluble matter such as sulphated galactan portions, salts, pigments and other organic compounds. A bleaching agent may also be added. The washed gel sticks are then dehydrated in a hydraulic press. The dehydrated gels are dried and ground in a mill to give agar powder.

Small-scale agar manufacture

Agar production is commonly carried out at the village level, using basic equipment and technology in many countries, predominantly in South-East Asian countries like Phillipines. This process is described in Figure 2.16.

Seaweed is washed and dried repeatedly until it becomes lighter in colour. It is then soaked in water (one part seaweed to 20 parts water) for about 2–3 hours. The mixture is boiled for a long time until the seaweed becomes soft. The boiled seaweed is then ground in a blender after which the blended mixture is boiled again.

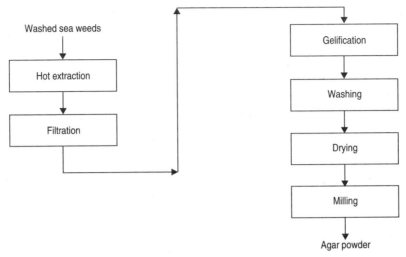

Figure 2.16 Small scale agar extraction.

The extract is filtered using a pressure filter or a high speed centrifuge. After filtration, the extract is poured into wooden or aluminium trays to gel. The gel is removed, cut into strips, passed through a filter press and dried in the sun to give crude strip agar.

To produce agar powder, the blocks of agar are crushed and the pieces washed. When the crushed pieces are melted, the agar is bleached and washed again before dehydration by centrifugation. The agar can be dehydrated slowly, over an extended period of time by means of a filter press. The agar is then dried and crushed into a powder by hand or sometimes by using a grinder.

As an alternative, a simple agar production process is also used. Seaweeds are washed with freshwater, sun-dried, and then re-soaked for 5–10 minutes. They are dried again until the colour becomes yellow and then bleached in dilute vinegar until the colour turns olive green. Drying is repeated until the seaweeds become light brown.

Extraction is done by boiling in dilute vinegar with constant stirring for 30–60 minutes. After boiling, the extract is strained through an ordinary cheese-cloth. It is then cooled and solidified. Cut into blocks or strips, the agar gel is packed into an ice box with dry ice or wrapped in cheese-cloth with ice and salt for about two days. The agar blocks are then thawed and dried under the sun.

In some countries crude agar is procured in bulk and re-processsed at centralized facilities into a purer form.

2.4.2 Agarose

Agarose is a purified form of agar after removing pectins with sulphate radicals from agar with the following approximate properties.

Density   : 1 g/cc at 20°C
Molecular weight   : 630 (approximate) × n (radicals)
Freezing point   : On cooling it solidifies at 35°C, on heating it melts at 90°C
Formula   : $C_{24}H_{28}O_{10} \times n$ (radicals)

Agarose is also a biologically inert material with controlled ionic properties being widely used in biomedicine and biotechnology. It exists as a water gel.

Uses of agarose

Agarose gel media are used in electrophoresis to separate and identify serum and spinal fluid protein, thus facilitating the diagnosis of diseases. In electrophoresis, agarose gels can be applied to a host of other biological mixtures including nucleic acids, lipoproteins, glycoprotiens, bacterial proteins, plant viruses, etc.

Columns of agarose gel (e.g. Sepharose, Bio-gel A) are used in chromatography for molecular weight based separations and for separation of artificial mixtures of proteins and viruses and those of ribosomes. Agarose is used in immunology in techniques such as agrose gel diffusion, radial immunodiffusion, immunoelectrophoresis, electroimmunoassay, counter electrophoresis, etc.

Agarose also plays an important role as a biologically inert carrier to which enzymes or cells are bound or introduced during gel formation or by diffusion and insolubilization. Such agarose beads are used as bioconverters to transform one chemical to another.

Manufacture of agarose

There are several methods for isolating agarose from the charged polysaccharide of agar. These are acetylation, sephadex chromatography, use of chitin and chitosan, rivanol, cetylpyridinium chloride, ammonium hydroxide, polyethylene glycol and ammonium sulphate. Four of these methods are given below, the last three being of commercial importance.

Acetylation: This method was developed by Araki in 1937. It involves the acetylation of agar with acetic anhydride in pyridine. The acetylated agarose is separated by chloroform extraction and then deacetylated by saponification to obtain agarose.

Cetylpyridinium chloride (CPC) method: A block diagram of the CPC method is shown in Figure 2.17.

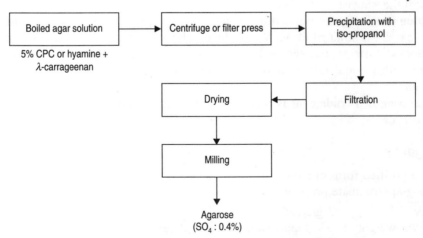

Figure 2.17 Preparation of agarose by the CPC method.

Cetylpyridinium chloride (CPC) is a cationic polymer which may react with highly sulphated galactan in agar, forming a precipitate. The addition of some λ-carrageenan, a highly sulphated polymer, may increase the bulk of the quarternary ammonium salt precipitate. This facilitates the removal of sulphated galactan by coprecipitation. It is again precipitated with isopropanol, filtered, dried and ground to obtain the agarose product.

DEAE-cellulose method: An adequate amount of anionic exchange resin DEAE-cellulose powder is added to the dissolved agar solution and stirred at a temperature of 80°C. The sulphated galactan in agar is absorbed by the anionic resin and removed by means of filtration. The filtrate is then allowed to gel and is frozen, thawed, dried and milled to produce agarose powder.

Polyethylene glycol method: Polyethylene glycol of 7–20% concentration and 2% NaCl are added to the boiled agar solution. The agarose is precipitated and isolated by centrifugation. After dehydration by acetone, dried agarose (sulphate content 0.23%) is obtained.

The above methods notwithstanding, the details of production are not generally revealed by the manufacturers[7].

Economics

Agarose is about twenty times costlier than agar and is used for high-end applications like DNA, RNA tests, etc. Coproduction of agarose along with agar will be highly economic and beneficial.

2.5 COAL AND COAL CHEMICALS

Coal is mined by both 'open cut' and 'underground' mining. In open-cut mining, the overburden of mud is completely removed whereas in underground mining, tunnels are cut to reach the coal deposits. Open cut is favourable only if the coal is available at lower depths below the ground.

Coal can be classified as anthracite coal, bituminous coal and lignite. While anthracite and bituminous coals are mined by underground mining, lignite, an inferior variety of coal, is mined by open-cut mining.

Apart from being a very important fuel, there are many other chemical products obtained from coal for use in chemical and metallurgical industries. Some of these are highlighted here.

1. Recovery of aromatics such as benzene, toluene, xylenes, phenol, cresol, naphthalene, dimethyl naphthalene, etc. These are also known as coal tar products.
2. Production of coke for iron and steel industry.
3. Production of methanol, ammonia and oxo-alcohols using synthesis gases produced by gasification of coal and further elaborate treatment. This use is declining in places with large scale availability of natural gas.

Coking of coal is a process by which the dissolved gases in the coal are removed by direct heating in inert atmosphere. The coal becomes porous due to gas escape and at the same time it becomes harder. Coke is much sought after by the iron and steel industry for use as a reducing agent. But petroleum coke has largely replaced the coke produced from coal.

2.5.1 Coal Gasification

From a historical perspective, gas obtained from coal (coal gas) was used as piped gas in towns and cities especially for street lighting puposes during the early part of the nineteenth century.

The gasification of coal produces a mixture of carbon monoxide and hydrogen which can be used for making chemicals such as hydrogen, methanol, ammonia, oxo-alcohols, etc. It is also a very good industrial fuel and also a clean way of power generation by the combined cycle method.

The main chemical reactions are the 'water gas reaction' and the 'water gas shift reaction' which are given below.

$$C + H_2O \rightarrow CO + H_2 \quad \text{(endothermic, heterogeneous)}$$
$$CO + H_2O \rightleftharpoons CO_2 + H_2 \quad \text{(exothermic, homogeneous)}$$

Other reactions taking place to a lesser extent are given below.

$$2C + O_2 \rightarrow 2CO \quad \text{(exothermic, heterogeneous)}$$
$$C + O_2 \rightarrow CO_2 \quad \text{(exothermic, heterogeneous)}$$
$$C + 2H_2 \rightarrow CH_4 \quad \text{(exothermic, heterogeneous)}$$
$$2CO + O_2 \rightarrow 2CO_2 \quad \text{(exothermic, homogeneous)}$$
$$C + CO_2 \rightleftharpoons 2CO \quad \text{(endothermic, heterogeneous)}$$

There are three main types of coal gasifiers as given below.

1. *Moving Bed Gasifier by Lurgi.* In this process, the gas cleaning process is very elaborate. Coking coals cannot be used in this gasifier.
2. *Winkler Generator.* This is a back mixing type fluidized bed reactor. It is ideal for brown coals and lignites as raw material.
3. *Entrained Flow Gasifier by Koppers Totzek.* This is suitable for combined cycle power generation.

Lurgi moving bed gasifier

This is a moving bed gasifier and consists of introducing coarse solids at the top of the gasifier while the oxygen and steam are introduced at the bottom countercurrently. The gasifier may be divided into four distinct zones, which are the top drying/preheating zone, the devolatilization zone, the gasification zone and the bottom combustion zone.

The gases that leave at the top contain H_2, CO, CO_2, H_2O, CH_4 and other hydrocarbons including oils and tars as well as other organic compounds and sulphur compounds such as H_2S, COS, some CS_2 and mercaptans, nitrogen compounds such as NH_3 and HCN. The temperature ranges from 980°C to 1040°C. Since this temperature is below the slagging temperature, only dry ash is removed at the bottom. The ash being removed from the bottom grate gets cooled by giving heat to the entering steam and oxygen.

Some of the disadvantages of this method are large steam requirement, large amount of by-products formed and difficulty of separation of these components. Another difficulty is that coals that become soft on heating (caking coals) are not suitable in this type of gasifier.

Winkler generator

Winkler generator shown in Figure 2.18, is used for the gasification of brown coals and lignite. The brown coal or lignite is fed to the bottom middle portion of the gasifier. The gas entering at the bottom, consisting of oxygen and steam, fluidizes the powdered coal mass.

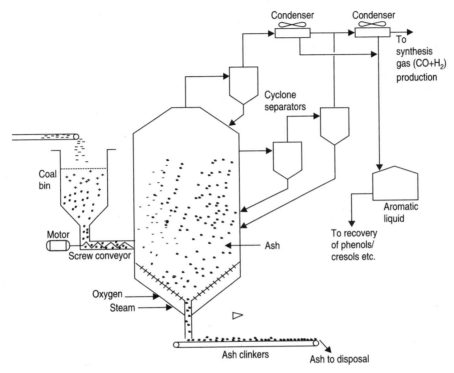

Figure 2.18 Fluidized bed Winkler gasifier.

The distinguishing feature of this generator is that the gases produced exit the gasifiers and enter the cyclone separators from where the entrained coal particles are sent back to the gasifier. The ash that is formed in the gasifier descends on the perforated grates at the bottom and falls on an ash conveyor and then taken out and disposed.

The gas produced from the Winkler generator is cooled by air coolers. An aromatic liquid condenses which is separated in separators (not shown in the figure) and then sent to storage. This liquid contains useful chemicals such as benzene, toluene, phenol, cresol, xylenol and catechol, which can be separated in a distillation plant and marketed.

The gas, after removal of liquids, is properly treated and mixed with nitrogen to produce synthesis gas ($N_2 + H_2$) for ammonia production or the synthesis gas ($CO + CO_2 + H_2$) for production of methanol. Even though the availability of cheap natural gas has replaced coal as a feedstock for the production of ammonia and methanol, the above system has the unique advantage of recovering many chemicals from brown coal or lignite.

The Winkler generator is similar to a circulating fluidized bed reactor. The term 'circulating' only indicates that most of the fines carried away by the exit gases are separated and recycled back to the generator.

For fluidization to occur in a bed, the velocity of the gas flowing through the bed from bottom to top should be greater than a value called "minimum fluidization velocity". This velocity is theoretically found out by equalizing the pressure drop across the bed with the weight of particles per unit area of cross section.

At the start of fluidization, the gas, in fact, passes through the bed in the form of bubbles which results in the slight expansion of the bed. In the fluidized bed reactor the pressure drop may decrease slightly when fluidization starts, and further increase in pressure drop is small, thus a lot of energy saving occurs in comparison to fixed or moving beds. The intimate contact between gas and solid particles also increases the efficiency of reaction. The theoretical aspects of fluidization are given in Unit Operations of Chemical Engineering by McCabe, Smith and Harriott[8].

Koppers Totzek gas generator

This is an entrained flow generator as shown in Figure 2.19. The flow in the generator is similar to plug flow with the coal particles travelling tangentially and coaxially along with the gases (oxygen and steam) and at the same time reacting at very high temperature. The product gases separate at the centre and move up and the residue liquid comes down to the bottom. The residence time in the generator is only a few seconds.

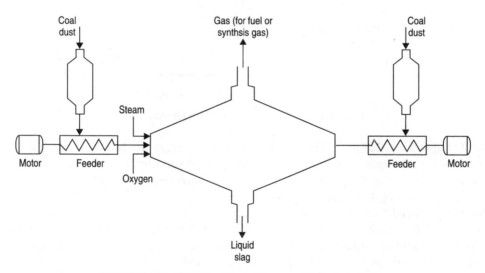

Figure 2.19 Entrained flow Koppers Totzek gasifier.

2.5.2 IGCC (Integrated Gasification Combined Cycle) Plant

As the name implies, in combined cycle power generation, power is produced through steam turbines as well as gas turbines. The chemical product obtained from this plant is pure sulphur by the Claus process. The Claus process is described in Section 4.6.

The pulverized coal mixed with oxygen from air separation unit goes into a gasifier/generator. See Figure 2.20. It can be a Koppers Totzek generator or any other efficient gas generator. The heat from the hot reacted gases is picked up by boiler feed water, producing steam which drives a turbine connected to a power generator for producing electricity.

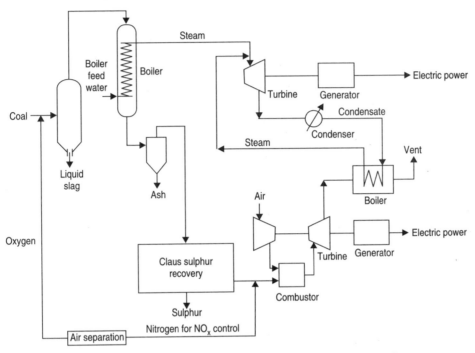

Figure 2.20 Combined cycle power generation using coal.

After leaving the boiler the gas goes to a cyclone separator where ash is removed and then the gas goes to a Claus sulphur recovery system for recovering sulphur (see Section 4.6.1 for details of the Claus process). The gas mixture then goes to a combustor where the gas is burned with compressed air and the flue gases sent to a turbine connected to a generator for producing electric power. Nitrogen is fed to the combustor to reduce NO_x generation by controlling temperature. The heat from the turbine exhaust is recovered in a boiler producing steam before the gas is vented or sent to natural underground CO_2 storages, this being a method of greenhouse gas reduction.

2.6 SUGAR AND STARCH

2.6.1 Sugar

The sugarcane, a bamboo-like plant, is considered to have originated in the island of New Guinea situated to the east of Indonesia. The plant and its use for the extraction of raw sugar spread to India and then throughout the world. Sugar is also made from other plant materials like beet sugar, but sugarcane still remains the most popular source of sugar.

Good quality sugarcane contains about 14% by weight sugar or sucrose. Sucrose has the molecular formula of $C_{12}H_{22}O_{11}$. A portion of it undergoes inversion to a mixture of glucose and fructose which takes place during its handling. The chemical reaction is shown below:

$$C_{12}H_{22}O_{11} \quad + \quad H_2O \rightarrow C_6H_{12}O_6 \quad + \quad C_6H_{12}O_6$$
(d-glucose + 66.6° polar)   (d-glucose + 52.8° polar)   (d-fructose −92.8° polar)

Polarization is the rotation of light, while it passes through a solution. For a substance in solution if the colour and path length are fixed and specific rotation of the substance is known, the observed rotation can be used to measure the concentration of sugar. A polarimeter uses this principle to determine the quantity of invert sugar in sugar syrup. The product of the above reaction is called invert sugar and it has a final polarization of −20°.

The process description for the manufacture of cane sugar is given in Figure 2.21.

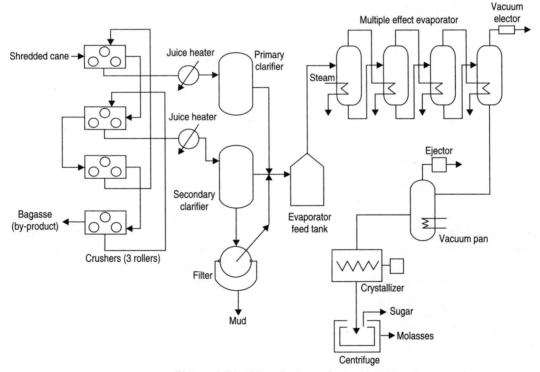

Figure 2.21 Manufacture of sugar.

The sugarcane is cut and shredded and fed to a series of four crushers. Each crusher has three grooved rollers which crush the cane between the rollers and squeeze out the juice. Water and recycled juice are added to macerate the cane-mass to facilitate juice removal. Water is mainly used for the last crusher. The juice obtained from the first and second crushers is then heated in steam heaters and fed to the primary and secondary clarifiers. The clarified liquids are filtered in drum filters and sent to the evaporation feed tank.

The bagasse obtained as by-product is used as fuel for power generation and excess power available is sold to local electricity boards.

The water in the juice is evaporated in a quadruple effect vacuum evaporator, steam being used as heating medium in the first evaporator. Other evaporators are heated by the vapours from the previous evaporator. The last evaporator is connected to a vacuum ejector. The thick sugar solution obtained is sent to vacuum pans which are steam-heated to produce a supersaturated liquid. This liquid is sent to the crystallization tank (crystallizer). In this tank, cooling and mild agitation take place. A slurry of liquid with crystals is formed. The crystals are removed in centrifuges. Two or three stages of vacuum evaporation, crystallization and centrifugation are done until a balance liquid is obtained which is called molasses. Molasses which contain around 50% fermentable sugars is sold to either alcohol plants to produce rectified spirit or to cattle feed manufacturers. It is also a raw material for making citric acid.

Raw sugar is the final product and contains 97.5% sugar. It can be sold as crystalline sugar. It can also be sent to a refining plant to produce refined sugar which is 99.9% sugar.

Manufacture of gur

In many countries, gur, which is the sugar with almost all the nutritious components intact such as iron, is made and used especially in villages. The sugarcane juice is evaporated to about 85% by weight. Then mucilagent (a plant-based coagulating agent) is added. All the impurities that rise to the top are removed and the balance liquid poured into moulds and solidified. This is sold as gur.

Manufacture of beet sugar

While sugarcane is grown in monsoon climates, sugar beet is grown in temperate climates. The manufacture of beet sugar is similar to cane sugar with some differences. The harvested beet is cleaned and sliced. The sugar is extracted from the slices by countercurrent extraction with hot water at about 75°C in diffusers which have ribbon screw conveyors to push out the resultant pulp, which is collected and sold as cattle feed. The resulting raw juice is treated with lime to precipitate undesirable impurities. Then it undergoes carbonation and sulphonation with CO_2 and SO_2 respectively to remove impurities and bleaching. The sugar solution is evaporated, centrifuged and again granulated to produce high quality white sugar.

2.6.2 Starch

Starch is always a major part of our daily food. It is contained in rice, wheat, cassava, corn and other cereals. It is a long chain of molecules (n = 40 to 100) with formula $(C_6H_{10}O_5)_n$. When rice is boiled in water, the water becomes a starch solution. This solution has been used for stiffening clothes. Starch has a number of uses in industry. Some of the industrial uses of starch include the following.

1. Food industry
2. Textile industry
3. Pharmaceuticals
4. Foundries

5. Air flotation
6. Leather tanning
7. Adhesives
8. Rayon
9. Tobacco industries

Process description

A cereal such as corn or maize is used as raw material for the production of starch, as shown in Figure 2.22.

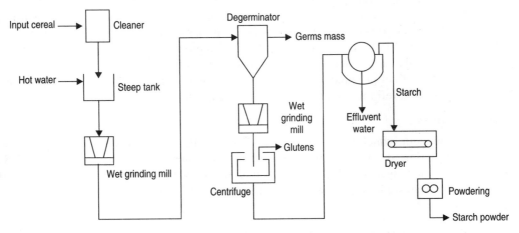

Figure 2.22 Production of starch.

First, the cereal is cleaned by using air purging and then by magnetic separation and all foreign materials including iron particles are removed.

The next operation is steeping. In this operation, the material is kept under hot water at about 55°C for a period of about two days. Wood or stainless steel vats are used for the purpose. Sulphur dioxide gas is passed through the liquid to prevent fermentation. The steep water removes salts, soluble organics and proteins from the material.

The starch portion is wet ground (the germs or protoplasm are torn off from the material) and the softened germs are floated off in the germ separator. Again it is wet ground and the fibre materials are removed as the light fraction of centrifuging.

The starch is then separated in drum filters as the filter cake and the material is passed through a drier. Then the dry starch is powdered and packed for sale.

2.7 PULP AND PAPER

2.7.1 Pulp

Pulp is made from wood and paper is made from pulp. The process of pulping is also a process used to make "viscose rayon" but this is covered under organic chemicals since many other organic processess are involved in the production of rayon.

Many processes are used for making pulp. The two major ones are:
1. Sulphate process, also called Kraft process
2. Sulphite process.

The sulphate process is more popular and more economically viable and hence described here.

Both batch and continuous processes exist. Like in other comparisons of continuous processes with batch processes, the continuous process has the following advantages.
1. For the same capacity the equipment is smaller.
2. Number of operators required is smaller.
3. Quality control is easier.

The continuous process is described below.

Wood consists of fibres joined together by lignin. The lignin has to be removed from the wood in order to get pulp. Both hard and soft woods (in other words, deciduous and coniferous woods) have been used for making paper but softwood is preferred since the fibres are longer, which makes the paper stronger.

Wood is cut into suitable size and debarked. The debarking is done by water jets at high pressure directed tangentially towards the barks. The debarked wood is chipped in chippers to small pieces of about one inch long, and fed into a bin located above the digester. See Figure 2.23. From the bin it is continuously fed to the top portion of the digester which is maintained at a pressure of about 10 kg/cm^2 and a temperature of about 145°C. Here it is mixed with recirculated cooking (or white) liquor, and chemical reaction occurs whereby digestion starts and vapours and light liquids are released.

The cooking (or white) liquor essentially consists of an approximately 15% solution of cooking chemicals. The approximate ratio of dissolved solids in the cooking liquor will be as follows.

    sodium hydroxide : 59%
    sodium sulphide   : 27%
    sodium carbonate : 14%

Wood consists of cellulose fibres held together by lignin. During cooking, the lignin gets dissolved in the cooking liquor releasing the cellulose fibres without any damage to form a good quality pulp. The pulp moves into the bottom section which is at about 175°C and pressure of 9 kg/cm^2 and can be mixed with additional cooking liquor. As it comes down, the reaction proceeds but is stopped at the bottom by mixing with cold recycled black liquor. The pulp gets transferred to the blow-down tank where the flash steam produced is used for many purposes such as heating and blowing.

From the blow-down tank the pulp goes into the washing section. The washing can be done in three to five steps, but it consumes more water. The new technique of washing is the use of "high density displacement washing" where the water use is minimized. The product is then sent to the bleaching section where chlorine dioxide is the preferable bleaching agent used for bleaching because it reduces the quantity of effluents. However, chlorine causes pollutants in the effluents. Hence, in order to reduce chlorine, bleaching has been largely replaced by hydrogen peroxide bleaching.

The pulp is then converted into coarse sheets, first by squeezing and pressing to remove liquid, and then lapping in lapping machines, and finally bundling and shipping to paper making plants.

56   **Chemical Process Technology and Simulation**

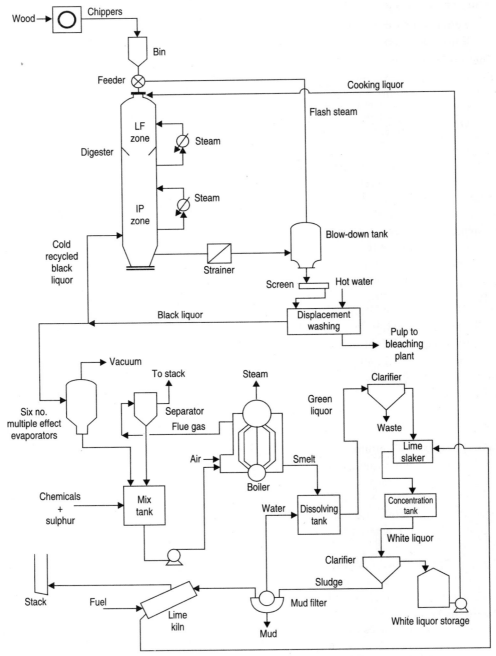

Figure 2.23   Sulphate process pulping plant.

Black liquor recovery

The liquid obtained from the washing section is black liquor which is thickened by evaporation in multiple effect evaporators. Then the liquid is burned in boilers producing steam with the following thermochemical reduction reaction taking place.

$$Na_2SO_4 + 2C \rightarrow Na_2S + 2CO_2 \quad (1)$$

The smelt obtained from the boiler is dissolved in water in a dissolving tank. Then it is sent to a clarifier and then to a slaker where slaked lime is added. Immediately the following recausticizing reaction (reversible) takes place in the aqueous phase with precipitation of calcium carbonate.

$$Na_2CO_3 + Ca(OH)_2 \rightleftharpoons 2NaOH + CaCO_3 \quad (2)$$

The slurry formed is separated in separators and solids are removed in a clarifier. The overflow is called white liquor and is sent to a white liquor storage tank. From here it is sent as cooking liquor to the top of the digester.

The sludge from the clarifier is sent to a mud filter and from where the calcium carbonate is sent to a lime kiln to convert it to calcium oxide by the following reaction.

$$CaCO_3 \rightarrow CaO + CO_2 \quad (3)$$

The quick lime obtained is sent to a lime slaker where it gets converted to calcium hydroxide by combining with water as shown below, which participates in the 'reaction 2' explained above.

$$CaO + H_2O \rightarrow Ca(OH)_2 \quad (4)$$

The slaked lime obtained as above is recycled.

2.7.2 Paper

Paper is made by compacting wood fibres in the pulp and by increasing the bonding between fibres, and also by making it to the required shape and size. This is done by either beating or refining.

Beating is done by cylindrical discs with attached blades rotating within a specially shaped metal tank filled with pulp. This is called a Hollander mill. Sizing materials are added to pulp, the most popular being $Al_2SO_4 \cdot 18H_2O$ also called papermaker's alum. Fillers such as titanium dioxide are also added.

Refining is considered a more effective method of making paper. In this method various machines are used. One such machine is a Conical Refiner also called Jordan engine. In this method, the pulp is passed through rotating cylinders with grooves.

Conventional wet process items of equipment are, 'Fourdrinier Machine' generally for thin papers, and 'Cylinder Machine' generally for thick papers. They operate by draining a dilute solution of pulp through moving screens so that the fibres build and bond on the screens as the water gets drained from the paper forming a continuous length of paper. Fourdrinier machine which is widely used is explained below in detail.

Fourdrinier machine

The Fourdrinier machine was first patented in France and later modified and patented in England. The process is explained in Figure 2.24.

58 Chemical Process Technology and Simulation

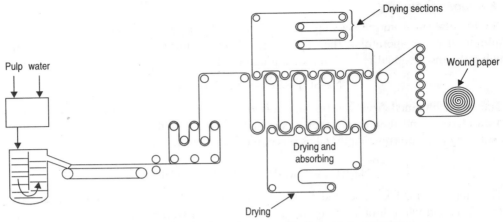

Figure 2.24 Production of paper with Fourdrinier machine.

Pulp and water are mixed in the ratio 1 : 200 and this mixture is fed to a moving wire-mesh endless belt which moves at a very high speed between 60 and 1800 metres per minute. There is a horizontal shaking of the screen provided so as to bond the fibres. The actual speed of the endless belt depends on the quality of paper required. Higher quality paper is produced at lower speeds. The water gets drained and the press rolls further remove water. Rubber straps are provided for edge formation. The material is then picked up by steam heated rolls and by felt belts pressing on the rolls, which further dry the paper from about 40% water content to 6% water content. The felt passes through a drying section and absorbing section. Then the paper is further dried by passing it between a set of rolls that are steam-heated. Lastly, the paper passes through calendaring rolls so as to increase its smoothness.

Sizing materials are applied as necessary. Fillers are also applied for brightness, flexibility, softness opacity and other qualities of paper. The newer Fourdrinier machines are computer-controlled so that correct speeds are maintained at various locations since stretching occurs as it is pulled. Quality of paper and dryness are achieved to perfection. The use of computer also helps quick changeover from one quality paper to another.

The strength of paper is also an important parameter. It is to be noted that conventional paper has a very low strength when it becomes wet. This is okay for some uses such as newspaper but some strength is required in wet condition too for some other uses. Amino-aldehyde resins are added to paper to increase its strength under wet condition.

Processes under development

The wet process of making paper uses a large quantity of water. The question arises whether air (instead of water) can be used for bonding the fibres into paper. This is being studied.

2.8 SOAPS AND DETERGENTS

2.8.1 Soaps

Soap is a salt formed by the chemical reaction of a fatty acid with sodium (or rarely potassium) hydroxide (NaOH). The hydrocarbon end connects with hydrocarbons in the dirt whereas the ionic sodium/potassium ends cause dissolution in water. So when we wash our hands with soap the dirt molecules are also drained along with the soap molecules. This induces the function of cleaning. The difference in the case of detergents is that the ionic end is either a sulphate or a sulphonate.

Tallow fat and vegetable oils are the main raw materials for making soap. Caustic soda (NaOH) is added to it for saponification. Glycerine is obtained as a by-product during soap manufacture. Glycerine is used to make alkyd resins, drugs, foams and as preservative for drugs, food products and cosmetics.

Manufacture of soap from fatty oil

Fat or fatty oil consists of glycerides. It is mixed with a catalyst and sent to the bottom of a tall vertical hydrolyser where it mixes with a countercurrent flow of hot water sent from the top of the vessel. See Figure 2.25. The following chemical reaction takes place.

$$(RCOO)_3C_3H_5 + 3H_2O \rightarrow 3RCOOH + C_3H_5(OH)_3$$
$$\text{Fat/Fatty oil} \quad \text{Water} \quad \text{Fatty acid} \quad \text{Glycerine}$$

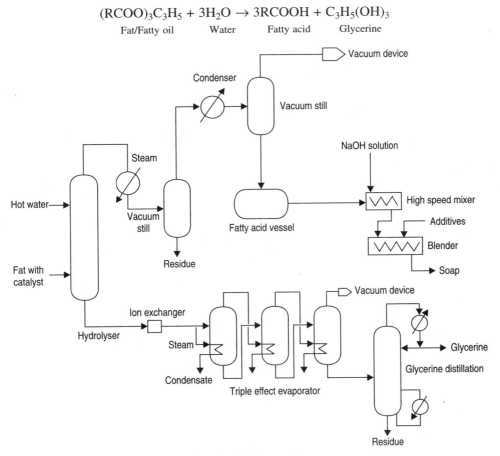

Figure 2.25 Soap production by continuous saponification.

The top product stream is the fatty acid. It is heated and fed to a vacuum evaporator. In the first still of the evaporator, the fatty acids are vaporized and the heavy residue is drained. In the second still, the light components are removed at the top and the bottom product is sent to the fatty acid vessel. The acid is then sent to a neutralizer where it is neutralized with caustic soda. The neutralization is carried out in a high speed mixer. The following chemical reaction takes place.

$$\underset{\text{Fatty acid}}{\text{RCOOH}} + \underset{\text{Caustic soda}}{\text{NaOH}} \rightarrow \underset{\text{Soap}}{\text{RCOONa}} + \underset{\text{Water}}{H_2O}$$

The mixture is then sent to a blender which is provided with a slow speed agitation. The reaction is completed in the blender wherein the other ingredients of soap (additives) are also added and thoroughly mixed.

The soap is either cast into bars for producing "bar soaps" or spray dried to give "soap powder". The bottoms of the hydrolyzer containing water and glycerine is passed through a triple effect evaporator to remove water from glycerine. Again higher boiling impurities are removed in glycerine distillations column to produce product glycerine at the top.

2.8.2 Detergents

Synthetic detergents have replaced soap especially for uses like washing clothes in washing machines. Detergents are also used for other miscellaneous cleaning requirements.

Water can be hard or soft. Hard water contains calcium and magnesium ions which react with soap to produce precipitated salts which hinder formation of lather. This is one of the disadvantages of soap especially in places where the water is hard. To alleviate this problem, detergent was introduced. The detergents lather even with hard water. Hence, detergents became popular especially with the advent of washing machines. Detergents are also called by the name "surfactants".

The most popular and largest volume detergent is linear alkyl benzene sulphonate (LABS). The main reason is that it is biodegradable. Experimentally it has been shown that straight chain hydrocarbons are always more biodegradable than the equivalent branched chain hydrocarbons.

Detergent is made using linear alkyl benzene (LAB), which is a petrochemical of formula C_nH_{2n+1}, and a straight chain alkyl radical (and therefore more easily biodegradable). The value of n, the number of carbon atoms, ranges from 10 to 14.

LAB is produced from 'normal paraffin,' which itself is produced by extracting a specific chain length of 'linear paraffins' from kerosene.

The process starting with kerosene is described below.

Production of linear alkyl benzene and its sulphonate (LAB, LABS)

The raw materials required are kerosene, benzene, sulphuric acid and oleum. The kerosene will be fractionated so that paraffins with chain length of 10 to 14 carbon atoms are recovered by removing lighter and heavier cuts. Further, oxygenates (alcohols ethers, etc.) are also removed. See Figure 2.26. The product is mixed with recycled olefins and sent to the alkylator. Benzene is also fed into the reactor. Benzene is alkylated in the presence of catalyst to form linear alkyl benzene (LAB).

Sulphonation/sulphation

The sulphonation of LAB with oleum occurs according to the following equation, forming linear alkyl benzene sulphonate (LABS).

$$C_6H_5-C_nH_{2n+1} + SO_3(+H_2SO_4) \rightarrow SO_3H-C_6H_4-C_nH_{2n+1}$$

The above chemical reaction occurs in heat exchanger reactors with cooling and circulation. The product is fed into another similar reactor with cooling and circulation. Fatty alcohols are introduced into this reactor where the following sulphation reaction takes place.

$$C_6H_5-C_nH_{2n+1}-CH_2OH + SO_3(+H_2SO_4) \rightleftharpoons OSO_3H-C_6H_4-C_nH_{2n+1}$$

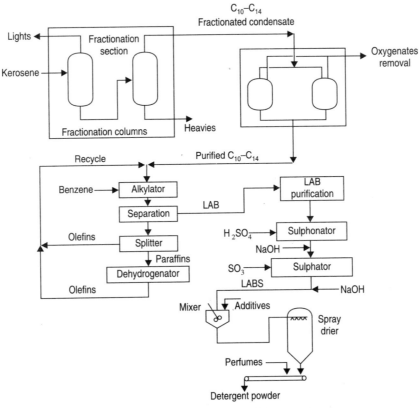

Figure 2.26 Production of detergent.

Again caustic soda is added and passed through the neutralization reactor under controlled temperature with cooling. This maintains fluidity of the detergent slurry. The above slurry is sent to storage and additives such as silicates and phosphates are added. Finally, the mixture is sent to a spray drier where the detergent (also called surfactant) is produced. Perfumes are added to the dried product before it is packed for sales.

2.9 RUBBER AND LEATHER

2.9.1 Rubber

Both natural and synthetic rubbers including the production of synthetic rubber monomers are covered in this chapter. Synthetic rubber monomers are, strictly speaking, petrochemical products.

62 Chemical Process Technology and Simulation

But as already indicated elsewhere, the final product is given primary importance and processes are described as per a sequence of different final products.

Rubber has a very wide range of uses in products such as tyres, footwear, sports utilities, protective clothing, and in areas such as defence, transport, mechanical industries and space science and technology.

Natural rubber

Natural rubber is obtained from the bark of the tree species "Hevea Brasiliensis", which is said to have originated either in West Indies or Brazil. However, at present natural rubber plantations are found mostly in the East Indies not only due to the favourable climate and availability of skilled and semi-skilled labour but also due to relative absence of diseases affecting trees in that region.

The name rubber arose because of its ability to rub out pencil marks. Today, rubber has very many uses out of which automobile tyres are the most predominant.

Rubber vulcanization

Rubber is a natural polymer material consisting of long chains of isoprene, butadiene and styrene, and similar radicals. Rubber in its natural form is sticky or tacky. Hence, it is difficult to use it as such for a wide variety of purposes. Vulcanization with sulphur removes the tackiness and rubber becomes a product with a fixed but sufficiently flexible form, making way for innumerable uses. The chemical products used in vulcanization are sulphur compounds including sulphur, sulphur monochloride, etc. Selenium, tellurium, and polysulphides are also used. When vulcanized, the material forms cross-linked structures. Additives such as accelerators and retarders are also used. The temperature at which the vulcanization is done is also important in obtaining high quality rubber.

Synthetic rubbers

There are many types of synthetic rubbers. Only a few are described here. It may be noted that the demand for natural rubber has been found to be growing steadily in spite of all the advancements in synthetic rubbers. Styrene butadiene rubber (SBR), polybutadiene rubber, butyl rubber, nitrile rubber, chloroprene rubber, silicone rubber are examples of synthetic rubbers.

Styrene butadiene rubber (SBR)

Styrene butadiene rubber (SBR) is similar to natural rubber in many of its properties and also in its structure. SBR is a product of the copolymerization of styrene and butadiene. Figure 2.27 shows the block diagram for the processes of production of styrene and butadiene.

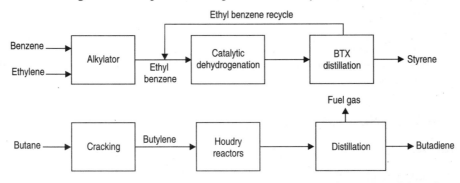

Figure 2.27 Block diagram for the processes of production of styrene and butadine.

Styrene reactions

Styrene ($C_6H_5CH=CH_2$) is obtained by alkylating benzene with ethylene, thereby obtaining ethyl benzene. Then the ethyl benzene is dehydrogenated using either of the following catalysts

1. Aluminium chloride with solid phosphoric acid
2. Silica alumina catalyst.

The reactions taking place are:

Alkylation $\quad C_6H_6 + C_2H_4 \rightarrow C_6H_5-CH_2-CH_3$ (ethyl benzene)

Dehydrogenation $C_6H_5-CH_2-CH_3 \rightarrow C_6H_5CH=CH_2 + H_2$ (styrene + hydrogen)

The second chemical reaction producing styrene is highly endothermic and requires a high temperature. The partial pressure of ethyl benzene should be low. Steam can be added both for supplying heat and lowering the partial pressure of ethyl benzene.

Styrene production

Benzene and ethylene are mixed and fed to an alkylator with a bed of catalyst consisting of aluminium chloride with solid phosphoric acid. See Figure 2.28. The effluents from the alkylator

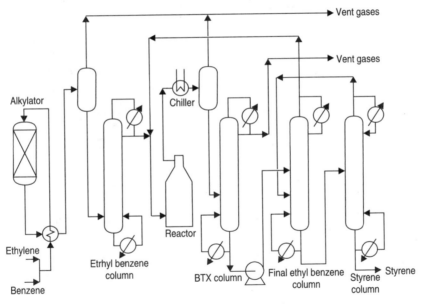

Figure 2.28 Styrene production.

containing ethyl benzene go to an ethyl benzene column, where ethyl benzene is separated at the top and mixed with recycled ethyl benzene from the final ethyl benzene column and fed to a dehydrogenation reactor. Styrene is formed along with hydrogen in this reactor. Heat for the endothermic reaction is supplied by superheated steam.

The reaction products are chilled using freon refrigeration. Uncondensed gases are separated and sent out as fuel. Condensed liquid is sent to a BTX column, where benzene and toluene are recoverd at the top and the bottoms are sent to a final ethyl benzene column. Ethyl benzene is separated at the top and recycled to a reactor. Styrene is obtained at the bottom. The styrene is further concentrated in a styrene column and sent to storage.

Butadiene reactions

Butadiene (CH_2=CH—CH=CH_2) is obtained during steam cracking of petroleum (gas or oil) for the manufacture of olefins. It can also be obtained by the Houdry process in which normal butane is cracked to produce n-butene and then by using oxidation dehydrogenation catalyst (bismuth molybdate) converted to butadiene. The chemical reactions are as follows:

Cracking
$$CH_3-CH_2-CH_2-CH_3 \rightarrow CH_3-CH_2-CH=CH_2$$

Oxidative dehydrogenation
$$CH_3-CH_2-CH=CH_2 \rightleftharpoons CH_2=CH-CH=CH_2 + H_2$$

Butadiene production

Butadiene as manufactured by Houdry process is already explained in Figure 2.27. The mixture of butane of butene along with some impurities from the steam cracker unit, is considered the raw material. See Figure 2.29. The charge is heated and sent to a set of Houdry reactors, where in the presence of oxygen, the butanes are converted to butenes and butenes to butadienes in two steps, as already explained as 'cracking' and 'oxidative dehydrogenation' reactions previously. The reactions are catalyzed by aluminium and chromium oxide catalysts. The difference between the ordinary dehydrogenation and oxidative dehydrogenation is that the oxygen in the air combines with the hydrogen produced in the reaction and pushes the reversible reaction to the right and hence almost 100% conversion of butane to butadiene is achieved. This is against 20% without the removal of hydrogen.

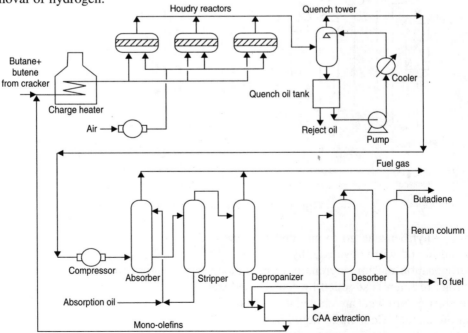

Figure 2.29 Butadiene production.

The product is cooled by quenching with quench oil and some of the impurities are removed by purging the reject oil. The absorber and stripper combination removes the light ends and the depropanizer removes the C_3 fractions which are all sent as fuel gas. However, in the crude butadiene obtained, there will be close-boiling hydrocarbons which cannot be removed by distillation. Hence, a polar reagent, cuprous ammonium acetate (CAA), markedly alters the solubility and volatility of butadiene by complex formation. Mono-olefins will not form complexes. The butadiene is absorbed and mono-olefins are returned to the charge heater.

In the desorber, butadiene is separated from the reagent. The reagent is recycled and butadiene is sent to a rerun column where heavies are separated and sent as fuel to the charge heater. The on-specification butadiene from the top of the column is sent to storage.

Styrene butadiene rubber (SBR) reaction

The styrene and butadiene are copolymerized in a copolymer reactor to obtain SBR. The chemical reaction taking place is:

$n(C_6H_5CH=CH_2) + n(CH_2=CH-CH=CH_2) \rightarrow (C_6H_5CH-CH_2-CH=CH-CH_2-CH)n$

SBR Polymerization, Coagulation and Drying

The polymerization, coagulation and drying is described for the manufacture of SBR rubber in Figure 2.30. The monomers are mixed with catalyst, extenders and dresinate soap in a mixer.

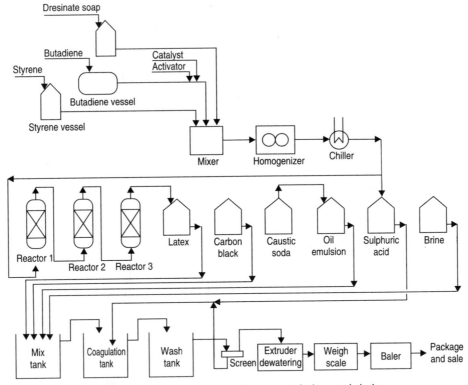

Figure 2.30 SBR polymerization, coagulation and drying.

The catalyst and activator increase the speed of polymerization, the extenders give more bulk to the rubber to be formed. In the case of rubber tyres, silica is also added to give rolling resistance. The mixture is homogenized and then chilled in a chiller. Then it is sent to a series of reactors with temperature control until polymerization is completed in the last reactor. The latex formed is sent to a storage tank and then to a mix tank.

In the mix tank the latex is mixed with carbon black, sulphuric acid and brine. The carbon black acts as a reinforcement in the rubber formed, which is mainly used for tyre making. Sulphuric acid and brine are used to coagulate the latex.

In emulsion polymerization, caustic soda is added to an oil and mixed to form an oil emulsion which is the medium used in the emulsion polymerization process. This oil emulsion is also added to the mix tank. From the mix tank it is sent to a coagulation tank and sulphuric acid added to control the chemical reaction.

The coagulated mass, obtained by either methods, is washed in a wash tank. It is dried by extrusion or pressing into moulds to form rubber sheets.

The sheets are weighed and packed into bales and dispatched to for sales.

Other types of synthetic rubber

Polybutadiene rubber: In this case, only butadiene is polymerized. It is blended with SBR and other rubbers in the production of tyres. It is also used for V-belts, conveyor belts, solid tyres, and anti-vibration mountings, etc.

Butyl rubber: Butyl rubber is a solution copolymer of isobutylene with a small proportion of isoprene. It is mainly used for making inner tubes of tyres.

Nitrile rubber: This is a copolymer of acrylonitrile and butadiene in which acrylonitrile proportion is 18 to 40%. It has high oil and heat resistance. It is used for the production of oil hoses, oil seals, gaskets, O-rings, brake linings, printing rolls and equipment linings in chemical plants. It is also used for making artificial leather and also as a lining for natural leather.

Chloroprene rubber: Chloroprene is similar to isoprene of natural rubber, but with a chlorine atom substituting for the methyl group. It is more resistant to oils and solvents. It is also used for mechanical purposes where heat resistance is required.

Silicone rubber: In this rubber, the polymer chain is not formed of carbon-based radicals but by silicon-based radicals. Silicone rubber is used for insulation capable of withstanding −100°C to 200°C and higher. Silicone rubber is used in aerospace industry, medical applications, conveyor belts, etc.

Polysulphide rubber: This rubber is resistant to all types of oils, aromatic compounds and chlorinated hydrocarbons and is used in hoses, sprays, sealants, etc.

Polyurethane rubber: This rubber is mostly used in solid tyres, upholstery, foam backed textiles, etc. Rigid foams provide heat insulation to buildings, and meet industrial requirements.

Vulcanization and compounding of rubber

Both synthetic and natural rubbers need vulcanization and compounding which is done by using intensive mixers and compounding machines along with characterisation and testing equipment.

The main polymers subjected to vulcanization are polyisoprene (natural or synthetic rubber) and styrene-butadiene rubber (SBR). These are mostly used for vehicle tyres.

Vulcanization essentially creates interlinks between polymer chains through sulphur atoms. An approximate or idealized equation of vulcanization of natural rubber with sulphur is shown below (for the reacted part only).

$$(C_8H_{14})_n + (8S)_n \rightarrow (C_8H_{14}S_8)_n$$

(The reactive sites—"cure sites"—are allylic hydrogen atoms. These C–H bonds are adjacent to carbon-carbon double bonds.)

The list of compounding chemicals include sulphur, accelerators, activators, retarders, antioxidants, antiozonates, fillers, etc.

As already mentioned, natural rubbers as well as synthetic rubbers (except some which do not need it), are subjected to vulcanization. During vulcanization, some of these C–H bonds are replaced by chains of sulphur atoms that link with a cure site of another polymer chain. The number of sulphur atoms in the crosslink influences the physical properties. Short crosslinks are related to better heat resistance. Crosslinks with more sulphur atoms (it ranges between one and eight) give the rubber good dynamic properties but with lesser heat resistance. Dynamic properties are important for flexing movements of the rubber article.

A typical vulcanization temperature for a car tyre (compression moulding) is ten minutes of vulcanization at 170°C. The rubber article will have to adopt the shape of the mould. For making door profiles for cars, hot air vulcanization or microwave heated vulcanization (both continuous processes) are generally used.

2.9.2 Leather

Leather is one of the oldest man-made materials. It was made by treating animal hides with vegetable oils and drying it in the sun. Most of the shoes and boots required for heavy duty work are still made from leather. This is in spite of the fact that modern materials like vinyl plastic and SBR shoes are more long lasting. Maybe natural leather shows the right amount of flexibility and hence it is more foot-friendly.

Manufacture of leather

The process of the manufacture of leather consists of the pre-treatment and tanning processes. The pre-treatment consists of trimming the hides, soaking in water and beaming to get a good "surface finish" for the leather.

The process of treating leather with tannin, or other equivalent materials, is called tanning of leather. There are many types of tanning but two important types are discussed here. They are vegetable tanning and chrome tanning. Pre-treatment and vegetable tanning are shown in Figure 2.31.

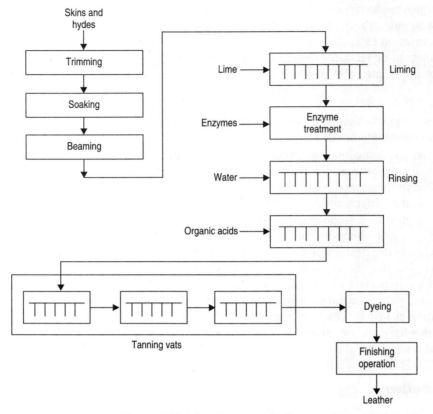

Figure 2.31 Leather manufacture.

Vegetable tanning

Tannins are glucosides of certain alcohols such as catechols and pyrogallols. These are contained in the barks of certain trees. These tannins are made to react with collagen fibres of the hides. The following are the general steps required.

1. Trimming and preparation of skins and hides
2. Soaking in water up to 24 hours
3. Beaming—usually done by hand
4. Liming—using hydrated lime for about 10 days
5. Scudding
6. Enzyme treatment
7. Rinsing
8. Treating with organic acids
9. Tanning with glucosides obtained from tree barks in tanning vats and layaway vats
10. Dyeing, stuffing, splitting, embossing and making into rolls and packing for sales.

The total time taken is about 2 to 4 months which is mainly because of the time taken in layaway vats.

Chrome tanning

Among all the steps given in vegetable tanning, all except steps eight and nine are common to both vegetable and chrome tanning.

The differences are given below:

Step 8. The pickling is done by mineral acids instead of organic acids.

Step 9. It consists of tanning with sodium dichromate ($Na_2Cr_2O_7$) and sodium chloride solution and reduction with sodium thiosulphate ($Na_2S_2O_3$), then setting with borax and rinsing with water.

The total time taken for chrome tanning is one to three weeks.

Other methods

There are several other methods of tanning. Some of these are given below.

1. Aldehyde—tanning using formaldehyde
2. Brain tanning—using emulsified oils similar to those compounds from animal brains.
3. Syntan—using sulphonated phenols and formaldehyde, aromatic polymers, etc.
4. Alum tanning—aluminium salts mixed with egg yolk, etc.
5. Chamois—using oils such as cod oil

Leather can also be produced without removing the hair and this type of leather is called hair-on leather.

2.10 OILS AND THEIR HYDROGENATION

2.10.1 Vegetable Oils

Vegetable oils have been produced from ancient times. They were used for lighting lamps and cooking in the beginning. The oil-containing seeds were crushed and dried and the oil was mechanically extracted. Animal power was used for the extraction which continues to some extent even now. While the mechanical process continues, more efficient methods like solvent extraction are also combined with it to increase the overall efficiency. It is important to notice that no part of the oil seeds is wasted since all the remaining solids are converted to either animal feed or fuel.

Vegetable oils are mainly used in cooking and preparation of food items. The non-food uses are for making soaps, toileteries, paints and varnishes.

Structure of fats and oils

Fats and oils are triglycerides of fatty acids with structure as given below.

$$\begin{array}{l} CH_2\!-\!O\!-\!CO\!-\!R_1 \\ | \\ CH\!-\!O\!-\!CO\!-\!R_2 \\ | \\ CH_2\!-\!O\!-\!CO\!-\!R_3 \end{array}$$

(where R_1, R_2 and R_3 are straight chain hydrocarbon radicals.)

The only difference between oils and fats is that oils are liquids at normal room temperature whereas fats are solids at normal room temperature.

They are also called lipids in biosciences. They are the storehouses of energy for plants and animals, for use when the need arises, as in winter due to shortage of food, or for other reasons like reproduction.

Oils and fats are hydrolysed by water in the presence of alkalies and acids to produce glycerol (glycerine) and fatty acids, as already illustrated in Section 2.8, dealing with 'Soaps amd Detergents'.

The major vegetable oils for edible purposes are groundnut oil, rapeseed oil, soyabean oil, cottonseed oil, sunflower oil, safflower oil, coconut oil, sesame oil, etc. The main non-edible oils are linseed oil and castor oil.

Processing of oils

The oils cannot be distilled without undergoing some decomposition and hence purification has to be done by other methods.

The length of carbon atoms in the fatty acid of various oils is different. In coconut oil, it is 12 to 14. For other vegetable oils it ranges from 16 to 18. For fish oil it is 24.

The oils are unsaturated whereas fats are saturated to a higher degree. The process of hydrogenation is to make oils saturated to a particular extent, because of which it is converted into shortenings (also called vanaspati) used in cooking.

Essential oils

Essential oils are a group of oils obtained in small quantities from rare and specific plants and are used for making perfumes and flavours. Essential oils are volatile whereas vegetable oils are non-volatile. The distillation of essential oils is similar to batch distillation covered in Section 2.1. Many essential oils such as citronellal, citronellol, gereniol, are produced in this manner. Vanillin is one of the most widely used flavours. It can be made from clove oil. Benzaldehyde is the simplest of essential oils and is used as a flavouring agent and in pharmaceuticals. It can be obtained from bitter almonds by the process of extraction.

Extraction of oils—physical and solvent extraction

Physical extraction of oil from oilseeds (mostly using animal power to squeeze out the oil) has been practised from ancient times. However, solvent extraction of vegetable oils using hexane or other solvents is now used in addition to physical extraction. The oilseeds are first crushed to increase their surface area and then fed to an extractor. There are different types of solvent extractors. One type is the extractor that consists of a chain of buckets which carry the seed flakes and move either horizontally or vertically with solvent entering from one side. The extract is then filtered and sent to a vacuum stripping section where the solvent is recovered.

The stripped oil is then treated with alkali to remove free fatty acids. Then the oil is bleached by passing through Fullers Earth or other clay adsorbent beds. The oil is filtered using vacuum drum filters to obtain refined oil.

Animal oils and fats

In contrast to vegetable oils and fats which contain predominantly unsaturated fatty acids, animal oils and fats contain more saturated fatty acids.

Animal fats are obtained during the dressing of slaughtered animals. These are also called slaughtered fats. Fats obtained during the preparation of meat for sales are known as butcher fats.

Lard is an animal fat obtained by melting pig and other animal fats. It is used in many food products such as margarine. Apart from food uses, they are used to make industrial food products such as fatty acids, soaps and animal feeds.

2.10.2 Hydrogenation of Vegetable Oils

Hydrogenation is a catalytic reaction of unsaturated fatty glycerides by which they are converted into saturated glycerides as illustrated by the following equation.

$$(C_{17}H_{31}COO)_3C_2H_5 + H_2 \rightarrow (C_{17}H_{33}COO)_3C_2H_5 \quad \text{(Exothermic)}$$

Nickel in the activated state was originally used as catalyst. However, the catalyst can be either of the following.

1. A reduced nickel on an inert catalyst support such as diatomaceous earth
2. Kieselghur
3. A high surface area sponge catalyst obtained by the chemical reaction of nickel-aluminium alloy with sodium hydroxide.

The product being an edible substance with medium demand and varying raw materials, the batch operation is found to be more feasible than the continuous operation. Temperatures and pressures are also varied based on the input and the output quality requirements.

Batches of up to 30 tonnes can be loaded into a reactor. The reactor has a steam heating jacket and is fitted with an agitator and a hydrogen gas sparger just below the agitator. See Figure 2.32. Once the reactor is filled with the required quantity of oil, steam is introduced into the jacket and the liquid is heated to expel any dissolved gases to a safe vent or flare.

The catalyst slurry is introduced into the vessel. The catalyst concentration varies between 5 to 15 kilograms per tonne of oil. The steam given to the heating jacket is turned off and hydrogen is introduced through the sparger. The hydrogenation reaction takes place. The quantum of hydrogenation is checked and on reaching the desired level the hydrogen supply is turned off. The slurry is cooled and filtered. The catalyst is sent to the catalyst slurry tank for re-slurrying and reuse.

The hydrogenated oil goes to the vacuum stripper column. The oil is de-odoured in this column. Steam is admitted at the bottom. Each of the trays is heated by steam or hot oil (heating medium utility). The odorous materials and other impurities are stripped off by steam which is pulled out by the ejector at the top and sent to a condenser fitted with barometric seal.

The oil from the bottom of the column is then decolourized in active carbon beds (not shown in the figure). Then, additives like vitamins are mixed with the product in a vessel as required, along with refrigeration or slow cooling so that a granular structure is obtained. It is pillow packed (or any other type of packing) and sent for sales.

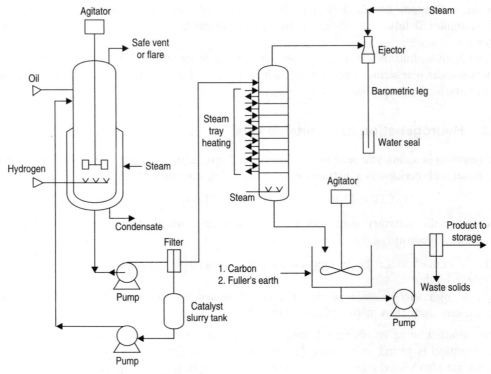

Figure 2.32 Hydrogenation of vegetable oil.

2.11 BIODIESEL FROM JATROPHA SEED OIL

Biodiesel is a fuel with properties resembling petroleum diesel (high speed diesel obtained from refineries) and which can be used instead of diesel or can be mixed with diesel. Alkyl esters obtained by the esterification reaction of alcohols with fatty acids present in vegetable oils are found to have the required properties. The advantage of using biodiesel is that it is carbon neutral and leads to reduction in global warming.

Jatropha is a wild and robust tree which grows in all tropical and semi-tropical regions of the world. It produces seeds which contain about 65% vegetable oil. Being inedible, this oil is a cheap and an easily available raw material for making biodiesel.

Even straight vegetable oil can be used on a diesel engine. But due to poor atomization of the fuel and incomplete combustion, coke deposition on fuel injectors can occur. The best method of overcoming this problem is the trans-esterification of the oil.

The trans-esterification reaction occurs as given below.

If R represents the ester radical of the fatty acid in the glyceride, the equation can be written as given below.

$$CH_2OCOOR \cdot CHOCOOR \cdot CH_2OCOOR + 3CH_3OH = 3CH_3COOR + CH_2OH \cdot CHOH \cdot CH_2OH$$

Fatty acid glyceride + Methanol = Fatty acid Ester (biodiesel) + Glycerine

In short, what we are doing is converting the three ester radicals connected to a single tri-alcohol radical to three separate ester radicals each connected to different methane radicals, to obtain the product that resembles diesel. Instead of methanol, ethanol can also be used if available at lower cost, which is unlikely.

The glycerine produced has a density higher than that of the biodiesel and it can be easily removed using separators. Excess methanol is recovered and reused.

2.11.1 Batch Process of Producing Biodiesel

In the trans-esterification reaction, the fatty acid triglyceride molecule in the oil is converted by methanol to produce fatty acid esters and glycerine. Sodium or potassium hydroxide is used as a catalyst in the chemical reaction. The glycerine produced is removed by decantation. The oil containing the fatty acid esters has exactly the same ignition properties as diesel and can be used as such. The overall benefits of trans-esterification are lowered viscosity, complete removal of glycerides, lowered boiling point and lowered pour point. It can replace diesel in almost all its uses. Batch production operation is found suitable for small-scale manufacture of biodiesel especially for use in large farmlands.

By-product glycerol production: In the above process, glycerine (also called, glycerol actual chemical name is 1, 2, 3-propanetriol) is obtained as a by-product. The main uses of glycerine are in the soap industry and also in the food and pharmaceutical industries. It is converted to propylene glycol and then converted to epichlorhydrin for making epoxy resins. Epoxy resins offer a steady market for glycerin.

2.11.2 Continuous Process Technologies

Some of the promising new continuous process technologies are the following:
1. Henkel technology
2. Continuous deglycerolization technology
3. Continuous process using supercritical technology

Henkel technology

In this continuous trans-esterification process, the glycerin is separated by separators fitted with coalescer packs. The ester after trans-esterification is purified by distillation. The ester has light colour, high purity and less glycerides. The glycerine obtained is of good quality.

Continuous deglycerolization (CD) technology[9]

This technology, known as CD technology, follows low pressure continuous trans-esterfication conducted in long reactor columns. Closed-loop chemical reaction with methanol and centrifugal separation are carried out. Closed-loop water extraction of glycerin is also carried out. The ester is then washed and purified to obtain the product with very low impurity content.

Continuous process using supercritical technology

This is a new technology. The critical point of a fluid on its phase diagram is the point where the vapour-liquid coexistence curve terminates. A fluid is said to be supercritical when its temperature and pressure exceed the temperature and pressure at the critical point on the phase diagram. The fluid above this point has been called either a 'dense gas' or a 'supercritical fluid'.

The basic principle is that propane at supercritical conditions becomes an excellent solvent and dissolves the feedstock (oil and methanol) so that the molecules of the reactants are in close proximity of each other and therefore react readily without a distinct catalyst[10]. This feature is exploited in the supercritical technology. Propane is generally called the cosolvent. Propane has a critical temperature of 96.8°C and critical pressure of 42.34 kg/cm^2.

This continuous process is suitable for large-scale manufacture of biodiesel. A more important advantage is that it can use waste oils (such as those obtained from plant or animal origin) as part of raw material thereby justifying the investment cost. Further, the chemical reaction being at a high temperature, there are no dirty by-products to be disposed.

A conceptual flow scheme to produce biodiesel from jatropha seed oil is given in Figure 2.33. Jatropha oil is preheated by flash recycle stream fed to a feed and recycle mixer. Fresh methanol, recycled methanol and recycled liquid propane are also sent to this feed and to the recycle mixer. The output of the mixer is further heated to a supercritical state by pumping through a feed effluent heat exchanger. It enters the tubular reactor at a pressure of about 130 kg/cm^2 at a slow speed. Hot oil is circulated outside the tubes to maintain a temperature of 300°C inside the tubes. The residence time in the reactor is about 5 minutes to achieve full conversion.

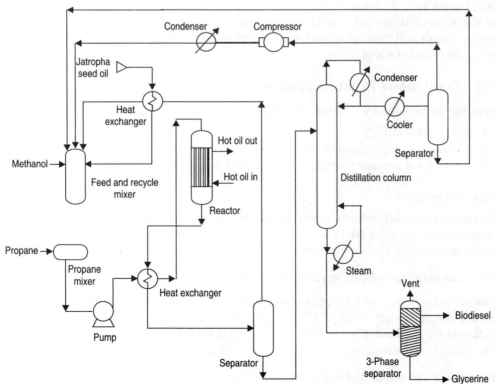

Figure 2.33 Biodiesel from jatropha oil (conceptual).

The product is taken out, cooled and flashed in a flash vessel. The gas from the vessel is recycled with a compressor or blower to the feed and recycle mixer after getting cooled in the heat exchanger.

The liquid goes to a distillation column where it is distilled to produce a mixture of biodiesel and glycerine at the bottom. This is taken to a settler where glycerine being heavier is taken out at the bottom and the product biodiesel is taken out from the top.

The top product of the column is cooled and sent to a gas separator. Here the propane gas comes out at the top which is compressed and cooled and put back into the feed and recycle mixer. The bottom liquid portion is also pumped back into the mixer.

REVIEW QUESTIONS

1. What been the past uses and what are the present uses of cashew nut shell liquid?
2. Describe a distillation unit for producing cardanol fom cashew nut shell liquid.
3. Describe the process with a flow diagram of fermentation of molasses to produce ethyl alcohol.
4. Can enzymes be used for the manufacture of chemicals? Give an example and describe the process.
5. Describe two methods used for removing aldehydes from alcohol while producing rectified spirit (industrial alcohol). Can you suggest a third method?
6. What is fusel oil? How are higher alcohols and acids removed from alcohol?
7. Which method of producing absolute alcohol is better if it is used for making pharmaceuticals?
8. What is pressure swing adsorption? Describe a process using it.
9. What are the steps for producing agar from seaweed?
10. Show the structure of an agarose molecule.
11. Explain the CPC method of production of agarose.
12. What is water gas reaction and water gas shift reaction? Explain what a fluidized bed is and what the entrained flow type coal gasifiers are.
13. Explain the process of manufacture of sugar or starch with a sketch.
14. Describe the sulphate process of making pulp.
15. What is vulcanization of rubbers? How is leather made?
16. What are the various types of synthetic rubbers? State their uses.
17. Explain with a figure the process of hydrogenation of oil.
18. Apart from agar and agarose, what are the other chemical products that can be produced with raw materials obtained from sea? (for further reading).
19. How is biodiesel made?
20. What is supercritical technology?

Chapter 3 ORGANIC CHEMICALS

3.1 METHANOL, PHENOL AND ACETONE

The first organic chemical 'methanol' is described in greater detail because the same will be useful for understanding 'hydrogen and ammonia' described in Chapter 4 and the modelling methodology presented in Chapter 8.

3.1.1 Methanol

Methanol is the first alcohol in the alcohol series. It was originally called wood alcohol since it was obtained during the destructive distillation of wood.

Some important properties of methanol are given below:

Chemical formula : CH_3OH
Molecular weight : 32
Density at 25°C : 0.791g/cc
Viscosity at 25°C : 0.55 cP
Other   : It is dangerously poisonous if consumed.

Methanol has a variety of chemical uses, including as a building block for plastics, paints, manufacture of formaldehyde and other chemicals. It is also used in the production of biodiesel.

Methanol has also been picturized as an all-purpose fuel of the future and also as a clean fuel in the futuristic cars. It is found that the quantity of methanol produced is showing a steady upward trend and consequently the cost of production is also decreasing.

Briefly, methanol is made by the reaction of carbon oxides with hydrogen. A mixture of CO, CO_2 and H_2 gases in stochiometric ratio is called methanol synthesis gas. Generally speaking, a capacity of about 600 tonnes per day can be assumed as an economically viable proposal and such a plant is described below.

Process description

The production of methanol synthesis gas (also called reformed gas) for methanol synthesis is based on steam reforming of natural gas or naphtha delivering a synthesis gas at about 20 kg/cm^2 and 875°C. The manufacture of methanol from synthesis gas is carried out at about 60 kg/cm^2 and 250°C.

The process plant is grouped into the following sections for convenience in presenting the process description.

1. Hydrodesulphurization section. Here the sulphur present in natural gas is removed.
2. Steam reforming and heat recovery for production of steam. Here the methane in natural gas is conveted to CO and H$_2$ by the main reactions.

$$CH_4 + H_2O \rightarrow CO + 3H_2$$
$$CO + H_2O \rightleftharpoons CO_2 + H_2$$

Huge amount of heat produced during reforming is recovered by producing steam.

3. Synthesis gas section. Methanol is produced in the reactor by the main chemical reactions.

$$CO + H_2 \rightarrow CH_3OH$$
$$CO_2 + 2H_2 \rightarrow CH_3OH + H_2O$$

4. Distillation section to produce pure methanol.

An overall block diagram of a methanol plant is shown in Figure 3.1.

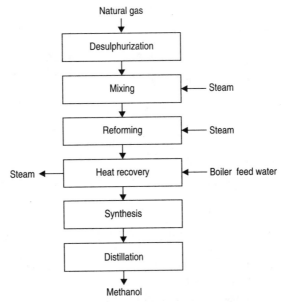

Figure 3.1 Overall block diagram of a methanol plant.

A detailed description of the individual stages in Figure 3.1 is given in the following subsections.

Hydrodesulphurization and reforming sections

The natural gas is compressed in a natural gas compressor to about 28 kg/cm^2. It is then delivered to a feed heater. This feed contains organic sulphur and H$_2$S. These are removed by two steps of catalytic reactions. The feed gas is mixed with recycled hydrogen from the synthesis loop section. The ratio is adjusted to maintain two mole % of hydrogen in the mixture.

A sketch of the steam hydrocarbon reformer with desulphurizer is shown in Figure 3.2. It consists of reformer tubes, as shown on the right-hand side of the figure, filled with a catalyst. Shown on the left side is the heat recovery section, where the heat from the combustion gases is recovered through the various heat exchanger tube bundles. At the left end is an induced draft (ID) fan which discharges flue gases to the atmosphere.

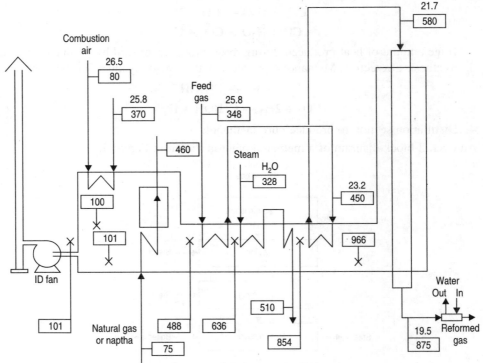

Note: Pressures are given outside the boxes (kg/cm^2). Temperatures are given inside the boxes (°C).

Figure 3.2 Pressures and temperatures at the enhanced capacity reformer section.

The feed heater is located in the reformer convection section and separated into a pre-heating part and a vaporizing part. After pre-heating up to 370°C, the feedstock flows to the hydrodesulphurizer to decompose organic sulphur to H_2S on nickel-molybdenum catalyst as the first step and to remove all sulphur by the reaction in a packed bed of ZnO, forming ZnS as the second step. The required H_2S content in the outlet stream of ZnO bed is less than 0.25 ppm in the case of natural gas so as to limit the sulphur content to 0.06 ppm as H_2S in the synthesis make-up gas.

Reforming and heat recovery section

The effluent natural gas is mixed with process steam in an amount equivalent to a steam–carbon ratio of 3:1.

While discussing ethyl alcohol in Section 2.2 we saw that fusel oil was also produced as a side reaction. In the methanol reactor too, some fusel oil is produced as a side reaction (a minor by-product of methanol manufacture which contains propanol, iso-butanol and water). This fusel oil is also sent to the reforming section and mixed with the reformer feed.

The gas–fusel, oil–steam mixture is pre-heated to about 575°C in the mixed gas heater located in the reformer convection section and distributed to catalyst tubes suspended in the radiant section of the reformer. This reformer feed gas passes down in the reformer tubes and makes contact with the nickel reforming catalyst.

The reformer operates with firing of fuel gas between the rows of tubes to achieve the process gas temperature of about 875°C at the outlet of the catalyst tubes. Under this condition, the gas will contain some unconverted methane. The pressure at the outlet of catalyst tubes is about 20 kg/cm^2. For more design details of the radiant section of reformers and a comparison of different reformers, see the section on Hydrogen and Ammonia in Sections 4.4 and 4.5.

The reformer is designed to attain maximum thermal efficiency from the gases by recovering heat in the convection section.

The convection heat is used for the following services:

1. To preheat the feed natural gas (or naphtha) with steam
2. To superheat high pressure steam
3. To preheat natural gas (or naphtha)
4. To preheat the combustion air.

The flue gas leaves the base of the furnace at approximately 970°C and the heat in this gas is utilized for steam raising, steam superheating and reactants heating before being discharged to the chimney stack by the induced draft (ID) fan.

The flue gas first enters the mixed gas heater which heats the mixture of process gas and steam entering the reformer. The flue gas then passes to the steam superheater, superheating the steam from the steam drum.

The flue gas is then divided into two streams—one of them enters the feedstock heater and the other passes through the combustion air heater, in which the combustion air is preheated to about 460°C.

The endothermic reforming reaction occurs in the tubes over the nickel reforming catalyst, and the heat of reaction is provided by firing burners arranged in the furnace.

For firing the reformer furnace the following fuels can be used.

1. Naphtha and/or natural gas
2. Purge gas from the synthesis loop section and
3. Waste gas from the distillation section.

An example of maximum heat recovery from reformed gas as practised is given in Figure 3.3.

The reformed gas is collected in the outlet header and transferred to the heat recovery area through the transfer line with water jacket. First, the reformed gas enters the reformed gas waste heat boiler, which is of the horizontal fire tube type combined with the steam drum, and is used to produce high pressure steam at about 105 kg/cm^2.

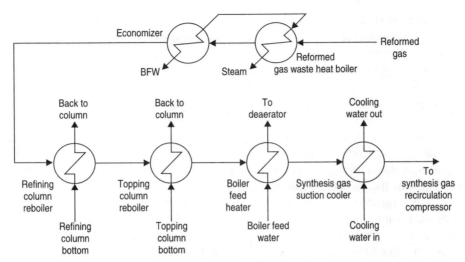

Figure 3.3 Block flow diagram of reformed gas heat recovery system.

The reformed gas leaves the reformed gas waste heat boiler at about 480°C and enters the process steam superheater where high pressure saturated steam is heated up to about 360°C.

The reformed gas enters the reformed gas economizer in which the heat is used to preheat the boiler feed water (BFW) before entering the steam drum. Then the reformed gas passes through the following series of heat exchangers where it is cooled to temperatures as given below.

Heat exchanger	Entering temp., °C
1. Product column reboiler	164
2. Lights column reboiler	152
3. Boiler feed heater	147
4. Synthesis gas compressor suction cooler	135

In both of the product column reboiler and the lights column reboiler, the use of reformed gas for heating in the reboiler considerably enhances the thermal efficiency of the reforming section since this heat would otherwise be discarded in the synthesis-gas compressor suction cooler.

In the boiler feed heater, the reformed gas heats the feed water to the deaerator and then passes to the process condensate knockout drum and the condensate is separated and cooled in the process condensate cooler.

The reformed gas is finally cooled in the synthesis-gas compressor suction cooler and passes to the synthesis gas compressor suction knockout drum and the gas then passes on to the suction of the synthesis gas compressor. Process condensate in the knockout drums is reused as boiler feed water.

Synthesis and reactor gas cycle

The synthesis gas is compressed to about 60 kg/cm^2 by the synthesis gas compressor driven by steam turbine. This compressor has a low pressure stage and a high pressure stage for fresh gas. It also has a recirculation stage to take care of recirculation.

The synthesis gas compressed in the low pressure stage is cooled in the synthesis gas compressor intercooler and the condensate is separated in the synthesis gas compressor interstage knockout drum. Then the synthesis gas passes through the high pressure stage and enters the synthesis loop at the discharge of the recirculation stage.

Essentially the chemical reactions that take place in the production of methanol are the following:

$$3H_2 + CO_2 \rightarrow CH_3OH + H_2O + \Delta H \quad (kcal/kgmol)$$

$$2H_2 + CO \rightarrow CH_3OH + \Delta H \quad (kcal/kgmol)$$

The synthesis gas is mixed with the recirculating gas. It gets heated in the exchangers to about 240°C in the synthesis gas-gas exchangers and enters the methanol reactor.

The methanol synthesis reaction is exothermic. It takes place at low temperature (210°C–290°C) over a highly active copper base catalyst. (High temperature converters were used in earlier times.)

The reaction heat produced is recovered by heating and vaporizing the boiler water. After vaporizing, the mixture of saturated steam (medium pressure steam approximately at 30 kg/cm^2) and boiler water enters the built-in steam drum in which steam is separated from water mist. This generated steam is used to the tune of about 50% of the required reforming steam. The make-up boiler water is supplied from a portion of the boiler feed water pump.

The reacted gases at 265°C containing about 4.5% methanol leave the methanol reactor. This reacted gas passes through the gas-gas exchangers to heat feed gas up to about 240°C.

An additional methanol reactor is shown in Figure 3.4. The purpose of this reactor is explained in Section 3.1.2 (Methanol Plant—A Case Study).

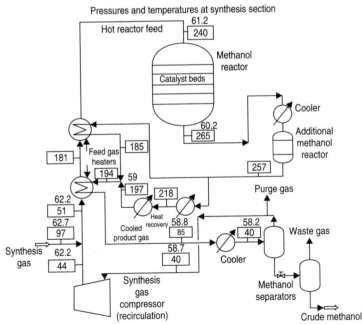

Note: Pressures are given outside the boxes (kg/cm^2). Temperatures are given inside the boxes (°C).

Figure 3.4 Synthesis loop section with an additional methanol reactor.

Finally, the gas is cooled by cooling water, and the liquids, methanol and water get condensed. The gas obtained is passed forward to the synthesis gas compressor for recirculation. The liquid obtained contains dissolved gases which are separated in the crude methanol separator after pressure reduction. The crude methanol is sent to the methanol distillation section.

Before entering the synthesis gas compressor, a portion of the gas is purged from the main stream, this purge gas serving to remove from the loop, the excess hydrogen, methane and nitrogen which were present in the reformed gas, thus controlling the concentration of these gases in the loop. This purge gas is used in the reformer as furnace fuel.

The synthesis loop contains the start-up heater which serves to preheat circulating gases at start-up for catalyst reduction, and is also used during plant emergencies to maintain the temperature level in the methanol reactor ready for recommencement of normal operation.

Distillation section

See Figure 3.5. Crude methanol is converted into product methanol by distillation. Crude methanol enters the lights column after being preheated through the crude methanol heater.

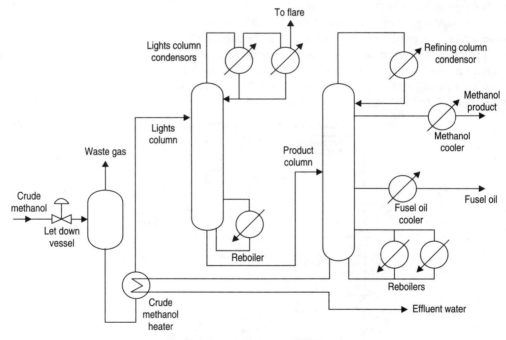

Figure 3.5 Methanol distillation.

The lights column containing a number of carbon steel sieve trays removes light ends. The light ends are physically dissolved gases, CO, H_2, CO_2 and CH_4 together with other impurities formed in the methanol synthesis reaction such as dimethyl ether and methyl formate. The reboiler heat for the column is provided by reformed gas as described in the reformed gas heat recovery section. The condensed methanol is taken separately from the outlet of the two condensers to the lights column reflux drum (not shown in the figure), thus decreasing the methanol losses in the non-condensing gases passing from the second condenser. The liquid methanol collecting in the reflux drum goes back to the lights column via the lights column reflux pump.

The gases passing from the second condenser consist mainly of non-condensible gases CO, H_2, CH_4 and CO_2, together with dimethyl ether, methyl formate and a small quantity of methanol. This is sent to flare.

The topped crude is pumped by the bottoms pump (pump not shown in the figure) to the product column. The product column is of similar design to the lights column containing mild steel sieve trays.

The refined methanol product is taken from the product column, on the fourth tray from the top, through the product methanol cooler and sent to the product methanol test tank to analyse the product specification. Fusel oil is removed in the side stream near the bottom of the product column.

The overheads from the product column are condensed in the product column condenser and collected in the product column reflux drum (drums not shown in the figure), and then recycled to the top plate of the product column by the product column reflux pump.

The reboiler heat for the product column is provided by two reboilers:

1. The reformed gas heated reboiler
2. Low pressure steam reboiler. The steam condensate is recycled to the deaerator after recovery of its heat.

The liquid passing from the bottom of the product column consists essentially of water and is cooled in the crude methanol heater.

The fusel oil purge from the product column passes through the fusel oil cooler and enters the fusel oil tank. This consists mostly of fusel oil (isobutyl alcohol and ethanol), the balance being water.

Steam generation including deaeration section

High pressure steam is generated in the steam drum which is placed above the reformed gas waste heat boiler. The generated steam is heated up to about 500°C in the process steam superheaters and the steam superheater. The superheater consists of a low temperature and a high temperature superheater.

The superheated steam enters the steam turbine drive of the synthesis gas compressor.

The medium pressure steam (25 kg/cm^2) from the extraction stage of steam turbine is utilized as about 50% of reforming steam. The balance 50% of reforming steam is supplied from the built-in steam drum in the methanol reactor.

The low pressure steam from the back pressure stage of steam turbine flows to the low pressure steam header and almost all of low pressure steam is used in the product column steam reboiler. The steam condensate is recovered and fed to deaerator.

The make-up boiler feed water from the water reatment plant is heated in the BFW heater and is fed to the deaerator. After deaeration with hydrazine and low pressure steam, the boiler feed water at about 125°C is discharged to the steam drum using the boiler feed water pump, through the reformed gas economizer.

Waste water facilities

The following waste waters (aqueous liquid effluent) are produced in the plant:

1. Process condensate from the heat recovery section
2. Process water from the distillation section
3. Boiler blowdown discharged to the battery limit
4. All drained water containing methanol is recovered to the methanol sump and re-distillated as a part of crude methanol in the distillation section.

Other facilities required

The other facilities required for the operation of the plant are:

1. Catalyst reduction facility
2. Cooling water facility
3. Flare and vent system
4. Emergency power generator

3.1.2 Methanol Plant—A Case Study

Capacity is critical to the productivity and profitability of any plant. In fact, it turns out to be the very feature for the existence of the plant in a competitive environment.

The intention here is to increase the capacity of an operating methanol plant based on natural gas as raw materal by 20% from 600 TPD to 720 TPD.

The procedure given below is followed for successfully increasing the production rate. The software "Design 2 for Windows" of Winsim Inc. can be used for the simulation of the plants at two different capacities.

As already seen, the methanol plant consists of three sections: reforming section, methanol synthesis section and distillation section.

Reforming section

For increasing the capacity of the reforming section, the following options can be considered.

1. Adding a pre-reformer
2. Changing the catalyst in the reformer to one of low pressure drop and high heat transfer rate type of catalysts with higher active surface.
3. Putting a secondary reformer fired by oxygen. This alternative is for large increases in capacity. Hence, this alternative is not considered here.
4. If excess cushion is available on pressure drop, then the throughput can be increased by increasing the pressure. But since no extra cushion is available, this is not considered in the present case.

The first option is described. A pre-reformer partially completes the steam reforming reactions upstream of the main steam reformer at a much lower temperature, using a highly active catalyst. Adding a pre-reformer, in effect, shifts a part of the reforming duty from the radiation section to the convection section. It also enables the reforming to be performed at lower "steam:carbon" ratios. The product from the pre-reformer is again heated in the mixed gas heater to a higher temperature, thereby reducing the heat duty of the reformer. The combination of the above two effects enables the reformer to be operated at a higher load. Some modifications of the convection coils are inevitable. The ID fan and FD fan will have to be run at higher loads depending upon the air and flue gas quantities.

Adiabatic pre-reforming process: In the pre-reformer all higher hydrocarbons will be completely converted into a mixture of carbon oxides, hydrogen and methane. The endothermic reaction is followed by the exothermic methanation and shift reactions, adjusting the chemical equilibrium between carbon oxides, methane, hydrogen and water according to the following reactions:

$$C_nH_m + nH_2O \rightarrow nCO + (n + m/2)H_2 \quad \Delta H = +ve \quad (i)$$

$$CO + 3H_2 \rightarrow CH_4 + H_2O \quad \Delta H = -206 \text{ kJ/mol} \quad (ii)$$

$$CO + H_2O \rightleftharpoons CO_2 + H_2 \quad \Delta H = -41 \text{ kJ/mol} \quad (iii)$$

For a typical natural gas feed the overall process is endothermic, resulting in a temperature drop. For heavier feed stocks such as naphtha, the overall process is exothermic or thermoneutral.

Since the operating temperature of the pre-reforming catalyst is low, the equilibrium for chemisorption of sulphur is favourable. Sulphur is quantitatively removed in the pre-reforming catalyst, giving a sulphur-free feed to the primary reformer.

By the addition of pre-reformer the life of the primary reformer catalyst is increased considerably.

Details of revamp: The results of the reformer section simulated for the production of 20% more synthesis gas, with the preferred option, are given. The steam: carbon ratio considered is 3.0.

The pre-reformer is loaded with a pre-reforming catalyst. The catalyst will be able to operate on a range of feedstocks from natural gas to naphtha with a final boiling point up to 200°C and an aromatics content up to 30%. The catalyst volume has been calculated for an expected catalyst lifetime of two years and a total conversion of all higher hydrocarbons. This will allow for a preheating of the process gas from the pre-reformer to maximum 650°C using the existing mixed gas heater.

The operating conditions for natural gas feed at the exit of the pre-reformer case are given below.

Composition	mol% dry
H_2	22.8
N_2	1.42
CO	0.17
CO_2	11.43
Ar	0.0
CH_4	64.18
Temperature, °C	450
Pressure, kg/cm^2	23.9
Pressure drop, kg/cm^2	0.14

The catalyst volume will be about 10 m^3 and the catalyst bed will have a diameter of three metres.

Higher active area catalyst: The most important variable to be kept under control is the skin temperature of the reformer tubes. The temperature should not exceed the limits based on the material of the tubes. The model calculation details are given in Chapter 9.

The maximum skin temperature of the reformer as per this calculation is expected to be 901°C after the revamp, with the new reformer catalyst. The high activity of the catalyst will accelerate the endothermic reforming reaction which is responsible for the lowering of the tube wall temperature. The other special features of the new generation shaped catalysts are:

1. Higher surface area
2. Higher thermal stability
3. Lower pressure drop
4. Good mechanical strength.

The temperature profile along the length of the reformer tube, after loading the new catalyst mix of high active area, is given in Table 3.1. The values are for a typical case of 7 hole catalysts. The length of the reformer tube is considered to be 13 metres.

Table 3.1 Temperature profile of the reformer tube

Axial distance (m)	Catalyst temp. (°C)	Skin temp. (°C)
0	615	615
1.0	564.5	817.4
2.0	626.33	867.9
3.0	675.83	887.9
4.0	716.26	900.1
5.0	748.34	900.1
6.0	774.51	900.1
7.0	790.32	900.2
8.0	796.94	900.2
9.0	815.24	900.2
10.0	830.88	900.2
11.0	844.19	900.2
12.0	855.41	900.2
13.0	865.11	900.2

The pressure drop across the catalyst also decreases with this catalyst change.
The pressure drop across the reformer tubes will be only about 3 kg/cm^2.

Synthesis and distillation sections

Gas compressor: An increase in capacity of the synthesis gas compressor can be achieved by providing chillers at the suction of the compressors so that the gas will be cooled down from 40°C to around 15°C for feeding chilled gas to the compressor. Since a similar chiller can also be used along with the methanol condenser, a common chilling system along with two chillers is considered.

Distillation: No alteration is required since columns were designed for higher capacity.
The new items of equipment considered for revamp are as follows:

1. Prereformer
2. Adiabatic reactor downstream of main reactor
3. New chilling system with chillers for synthesis gas compressors.

A model capital investment and financial analysis is shown in the next subsection purely for illustration purposes.

Illustrative capital investment and financial analysis

The financial calculations for converting a 600 TPD methanol plant to 720 TPD are given here.

Capital investment: The basis of estimation of capital investment for the augmentation of production capacity is given in Table 3.2. The table consists of rates for duties, taxes, insurance, freight charges and financial charges.

The capital investment consists of the following:

1. Cost of equipment
2. Cost of catalyst
3. Cost of piping, electrics, instrumentation and civil
4. Fee for detailed engineering.
5. Cost for erection, hook up and commissioning.

Table 3.2 Basis of economic analysis—rates for capital cost estimation

Item	Cost
Taxes and duties	
Customs duty	50%
Excise duty	10%
Sales tax	4% of FoR price
(FoR—free on road)	
Inland freight and insurance	
Domestic equipment	4% of FoR price
(FoR—free on road)	
Imported items	2% of FoB price
(FoB—free on board)	
Ocean freight and insurance	11% of FoB price
(FoB—free on board)	
Interest on long-term loans	15%
Contingencies	
Civil works	3%
Others	5%
Maintenance	
Civil works	1%
Others	2%
Depreciation	10%
Insurance premium	0.5%
Cost of natural gas, ₹/Nm^3	9
Cost of power, ₹/kWh	3

The total project cost thus works out to ₹ 10.31 crores. The details are given in Table 3.3. The cost of spares is not considered, this being a revamping of the existing plant.

A lumpsum amount of ₹ 32 crores is taken towards the cost of the modification of the synthesis section.

One month period is envisaged for hook-up and commissioning of the plant. The cost for production loss is not considered because it is proposed to do the hook-up during the catalyst change-over period/annual maintenance.

Table 3.3 Cost of revamping (All figures in rupees lakhs)

	Item	Indian currency	Foreign currency	Total
A.	Pre-reformer catalyst	–	675	675
	Primary reformer catalyst	–	330	330
	Equipment	2440	684	3124
	Piping (5% of equipment)	156.6		156.6
	Instrumentation (2% of equipment)	62.5		62.5
	Electrical (2% of equipment)	62.5		62.5
B.	Sub total (Ex-works)	2721.6	1689	4410.6
	Excise duty @ 10% of Indian currency	272.2		272.2
C.	FoB/FoR cost	2993.8	1689	4682.8
D.	Ocean freight and insurance @ 11% of foreign currency		185.8	185.8
E.	Customs duty at 50% of CIF cost	937.4		937.4
F.	CST at 4% of FoR	120		120
G.	Inland freight and handling	195.6		195.6
H.	Detailed engineering	320		320
I.	Civil works (3% of equipment)	132		132
J.	Erection and commissioning	525		525
H.	Works contract tax	12		12
I.	Erected cost of front-end revamp	5235.8	1874.8	7110.6
J.	Lumpsum cost for synthesis section revamp			3200.0
	Total FoR Revamp			**10310.6**

Cost of production

The annual cost of production for incremental production is worked out and given in Table 3.4. The cost of production is comprised of both fixed cost and variable cost.

The following assumptions have been made while working out the fixed cost:

The debt/equity ratio of capital investment is 1:1.
The interest on long term loans is at the rate 15% per annum.
The maintenance cost is 2% of plant and machinery.
The depreciation is 10% of plant and machinery.
The insurance cost is 0.5% of capital.
The number of stream days is taken as 330.

The cost of raw materials and utilities is taken as given below:

Cost of natural gas : ₹ 9 per Nm^3
Cost of power : ₹ 3 kWh

Since this is a revamp project, no cost is taken towards manpower and general overheads.

Table 3.4 Cost of production and profitability for incremental production

Sales

A.	Extra production = (120 × 330)		=	39600.0 tonnes per year
B.	Methanol price		=	₹ 29850.0 per tonne
	Additional sales turnover = (A × B)		=	₹ 11820.6 lakhs

Variable cost

	Natural gas		=	₹ 7350.0 per tonne
	Power		=	₹ 510.0 per tonne
C.	Total Variable Cost		=	₹ 7860.0 per tonne
D.	Total variable cost = C × A		=	₹ 3112.56 lakhs
	Contribution A × B − D		=	₹ 8708.04 lakhs

Fixed Cost

	Project cost		=	₹ 10310.6 lakhs
E.	Depreciation @10%		=	₹ 1031.0 lakhs
F.	Interest on long-term loans @15% for 50% loan		=	₹ 773.3 lakhs
G.	Maintenance @2%		=	₹ 54 lakhs
H.	Insurance premium @0.5%		=	₹ 52 lakhs
I.	Total fixed cost (E + F + G + H)		=	₹ 1914.3 lakhs
J.	Cost of production for extra qty (D + I)		=	₹ 5026.9 lakhs
K.	Profit before tax (A × B − J)		=	₹ 5283.7 lakhs
	Payback period (before depreciation) = Project cost/(K+E)		=	1.63 years
	Payback period (after depreciation) = Project cost/K		=	1.95 years

Profitability analysis

The annual incremental production consequent to this modification is 39,600 tonnes per annum.

The annual cost of production for this incremental quantity is ₹ 50.3 crores and the sales turnover from this incremental quantity is ₹ 118.2 crores.

The payback period before depreciation is 1.63 years and that after depreciation is 1.95 years, which means that this modification is economically viable.

3.1.3 Phenol and Acetone Coproduction

Phenol, earlier known as carbolic acid, and used as an antisceptic in very dilute solutions is a compound with formula C_6H_5OH. It is a solid at room temperature and has the following important properties.

Molecular weight : 94.11
Specific gravity : 1.07
Melting point : 42°C
Boiling point : 181.4°C

Acetone, formula $CH_3-CO-CH_3$, is a sweet smelling ketone gas with the following important properties.

Molecular weight : 58.08
Specific gravity : 0.79
Melting point : –95.1°C
Boiling point : 56.5°C

Phenol is an important industrial chemical mainly used for making phenol formaldehyde resins and other resins. Phenolic resins are used in electrical and structural components, automobile brake liners, glues, adhesives and moulds. Acetone is an important industrial solvent used for making various products. One such application is discussed in Section 3.2.

Phenol is manufactured with either benzene or toluene as the starting material. The toluene route is followed mainly in Europe whereas the benzene route is followed in America. In the toluene route, the toluene is first converted to benzoic acid and then to phenol. The manufacture of phenol and acetone starting from benzene is described here. The benzene route coproduces phenol and acetone and is more economical since the acetone produced also has a very good demand.

Phenol is a weak acid just like any other alcohol but it has highly recognized acidic properties because it forms strong phenoxide ions (C_6H_5O-) in the presence of water. Even though it is naturally occurring in dilute solutions, it belongs to a group of compounds that are considered to be toxic in higher concentrations. It can be absorbed into the body through the skin, by inhalation, or by swallowing. Hence, a lot of precaution is required in its handling.

Phenol and acetone are chemicals with many industrial uses. Hence, coproduction of phenol and acetone by the cumene hydrogen peroxide route is the most important method for producing phenol.

Cumene is produced in many refineries in many parts of the world by alkylation of benzene with propylene using solid phosphoric acid as catalyst in a cumene reactor, and purification by distillation. Alternatively, the cumene can be manufactured at the phenol site itself in the above-mentioned manner with both raw materials purchased from the refinery.

Coproduction of phenol and acetone

The process consists of mixing benzene with propylene in a correct ratio, heated and routed through the cumene reactor.

$$C_6H_6 + CH_3CHCH_2 \rightarrow C_6H_5CH(CH_3)_2$$

The reactor effluent then passes through a series of fractionation columns for the recovery of unreacted benzene. The cumene product is separated by distillation and stored.

See Figure 3.6. Phenol and acetone are made by splitting cumene hydrogen peroxide which is obtained by oxidation of cumene.

The first step of synthesis is oxidation wherein cumene is oxidized to Cumene Hydrogen Peroxide (CHP) in two oxidizers with compressed air. Oxygen is derived from air and the reaction is carried out in an alkaline environment.

$$C_6H_5CH(CH_3)_2 + O_2 \rightarrow C_6H_5C(CH_3)_2OOH$$

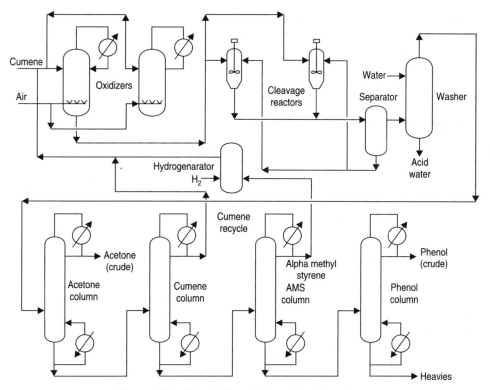

Figure 3.6 Production of phenol and acetone.

Cumene hydrogen peroxide is sent to cleavage reactors which are placed in protective safety bunkers. The cleavage is catalysed by dilute sulphuric acid. The chemical reaction is carried out at a temperature of about 65°C as given below:

$$C_6H_5C(CH_3)_2OOH \rightarrow C_6H_5OH + CH_3COCH_3$$

The reaction products are separated into oil and water layers; the water layer containing sulphuric acid is recycled. The oil layer consists of 14% phenol and 8% acetone as products and cumene, alpha methyl styrene and impurities like acetophenone.

The products are washed and sent to a distillation section with a series of distillation columns. The acetone column will remove acetone and sent to storage after purification.

The cumene column removes cumene which is recycled and then the Alpha Methyl Styrene (AMS) is removed in AMS column. The AMS is converted back to cumene by hydrogenation in a reactor with nickel catalyst and a temperature of about 100°C and recycled.

Crude phenol is produced from the phenol column and sent to phenol purification column to produce on-specification phenol.

Safety

The critical part of the above process is the oxidation reaction. A runaway reaction can cause major accidents. However, facilities for containment are provided in the design itself.

If the temperature shoots up, the oxygen flow is cut off immediately and water spray is given on the surface of the reactor. If the temperature does not come down in spite of it, the entire contents of the reactor are dumped into a large blowdown vessel with water in it so that it cools down immediately and no further chemical reaction takes place. The blowdown tank is provided in a corner of the plant itself.

3.1.4 Solvent (Acetone) Recovery

Acetone is a very important organic solvent and used in making coatings and depositions. In the process, it evaporates and is carried over by air. This section illustrates a typical solvent recovery plant using the adsorption process.

Acetone, also called dimethyl ketone, is a non-toxic organic solvent with a pleasant smell and with many uses. Mechanics, painters and fibreglass workers are common users of acetone. It is a good solvent for plastics and synthetic fibres.

Acetone and phenol are coproduced by the reaction of cumene hydrogen peroxide (CHP) with concentrated sulphuric acid in cleavage reactors. Under controlled conditions of temperature and acidity the CHP is cleaved to phenol and acetone as given below:

$$C_6H_5C(CH_3)_2OOH \rightarrow C_6H_5OH + CH_3COCH_3$$

Today, most of the acetone and phenol is produced by this method. The process is already described in detail in Section 3.1.3.

For use as solvent it is imperative that acetone is recovered after use. Impregnated fabrics manufacturing, produces air laden with acetone. The acetone evaporates during the manufacture, mixes with air and is blown out with a blower. The acetone is recovered from the mixture by the adsoption process. General description and special considerations for acetone are also given below.

The acetone in the acetone laden air is passed through fixed bed adsorbers where the acetone gets adsorbed. The adsorbers are usually horizontal cylindrical pressure vessels with a horizontal bed of activated carbon located at the centre. The size of each adsorber and the amount of activated carbon depends on the volume of vapour laden air to be handled, the concentration of the solvent, the adsorptive capacity of the activated carbon for the solvent to be recovered and the length of the adsorbing period used. Superficial air velocities of 50 to 100 feet per minute and carbon bed depths of 12–36 inches are frequently used.

Corrosion

Activated carbon acts as a mild catalyst in the hydrolysis or decomposition of solvents in the presence of steam. Therefore, any corrosion tests used as a basis for the selection of materials of construction should be made in the presence of activated carbon. Simple hydrocarbons like acetone can usually be made in plain stainless steel equipment. Other solvents may require the use of copper, everdur, monel or stainless steels depending on the severity and type of corrosion[1]. Direct prolonged contact of activated carbon with metals, in the presence of water or other electrically conductive liquids, results in corrosive attacks on the metal. With granular carbons in unprotected steel filter shells, such corrosion is rapid and destructive. Even with corrosion resistant steels, there is evidence that corrosive attacks occur, though not destructive.

Protection can be secured by coating the metal in contact with the activated carbon with an inert insulating material.

Regeneration

The steaming of acetone for full removal would involve too great a consumption of steam. It is better to keep a residual amount to reduce steam consumption. After steaming, the carbon is cooled and dried by passing a current of air. This air is very often provided by taking a percentage of collected air and passing it through the adsorbent during a short period. After this, the vessel is ready for the normal operation to be resumed.

Utilities

The carbon adsorbs substantially all the vapour from the air regardless of concentration, humidity and temperature. After the adsorption, steaming is done to recover the solvent.

The steaming is done for about 15 to 60 minutes. Rayon, transparent wrapping and lacquer coated or impregnated fabrics industries have typical consumption figures as given by Ray, A.B.[2]

Steam : 3–5 lb/lb of solvent recovery
(Additional steam will be required for distillation.)
Power : 0.04–0.08 kWh/lb of solvent
Water : 7 gal/lb of solvent
Acetone recovery (redistillation)
Water : 9.5 gal/lb of solvent (at 70°F)

The figure that is generally taken as the suitable concentration for industrial purposes is 0.4 to 1 lb of solvent per 1000 cubic feet of air[3]. However, higher concentrations are also found viable.

Importance is given to high recovery efficiency, low operative expenses, reliability of operation, hard dense pellets (low ash) which have maximum adsorptive capacity and minimum resistance to air flow. The carbon is installed in beds typically 24 inches deep contained in horizontal cylindrical adsorbers[4].

First, vapour laden air is filtered and cooled. A typical operating cycle is such that the vapour laden air is passed to two adsorbers in series at all times. A third adsorber is being steamed for removal of adsorbed solvent, and the fourth adsorber is being cooled by air recirculated from one of the two working on line. The completely solvent free air from the second adsorber working on line is discarded to the atmosphere. The cycle is so arranged that, of the two adsorbers working on line, the air for recirculation is taken from the more fully charged adsorber with low solvent content and is passed to the atmosphere. The adsorber receiving the recirculated air contains activated carbon which has just been regenerated by steam and is capable of removing any solvent present in the recirculation stream.

A case study

An installation employing four adsorber units is to be designed to handle 4000 Nm^3 of air containing 0.72 mol% of acetone per hour.

Process description

Air containing acetone vapours is passed through two of the adsorbers in series and the effluent air is let off to the atmosphere. See Figure 3.7. The first adsorber adsorbs all the acetone in the beginning. Towards the end, the second adsorber takes the residual acetone not adsorbed by the first adsorber. The first adsorber is saturated in 4 hours and the same operation is continued with second and third adsorbers, and so on.

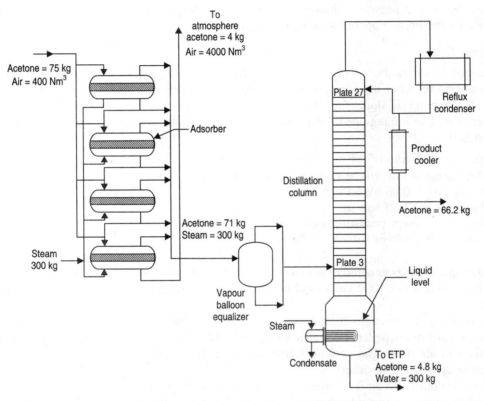

Figure 3.7 Flow sheet for acetone recovery.

The saturated active carbon bed is stripped by passing steam through it in a direction opposite to the adsorption. After stripping, the adsorbers are cooled and dried by blowing air through them.

The steam acetone mixture is taken to a vapour baloon equalizer to reduce pressure fluctuations during changeover of adsorbers, and then fed to the distillation column at the third plate from the bottom. Heat to the column is provided by a stab-in type reboiler at the bottom of the column. The top vapours are condensed and refluxed back to the column, a portion of the liquid is taken out as product. The product acetone is cooled and sent to storage.

Adsorber design

Amount of acetone in feed
= 0.72%
= 4000 × 0.0072 Nm³/h
= 28.8 Nm³/h
= 28.8/22.414 kgmol/h
= 1.285 kgmol/h
= 1.285 × 58
= 74.52 kg/h (say, 75 kg/h)

As per the information from active carbon vendors the weight of acetone adsorbed per unit weight of activated carbon will be 0.45(kg/kg). Out of a total of 75 kg/h of acetone an amount of 71 kg/h will be adsorbed while passing through a bed of 0.5 metre thickness. Most of the balance acetone, say, 4 kg is assumed to be adsorbed in the second bed. Hence, on saturation each bed would have adsorbed 71 kg from the main operation and about 4 kg from the previous operation giving:

Weight of acetone to be adsorbed per adsorber in a cycle	= 74 × 4
	= 296 kg
Weight of carbon to be used per adsorber per cycle	= 296/0.45
	= 658 kg
Bulk density of activated carbon	= 500 kg/m³
Volume of carbon to be used	= 658/500
	= 1.316 m³
Thickness of carbon bed	= 0.5 m
Cross sectional area of adsorber	= 1.316/0.5
	= 2.632 m²
Area of a bed (length 2.2 metres and width 1.2 metres)	= 2.2 × 1.2
	= 2.64 m²

Hence, an active carbon bed of 0.5 m depth, 1.2 m average width and 2.4 m length is selected.

Distillation column design

Feed is 370 kg/h containing 74 kg/h acetone.
    Distillate is to be 96.8 mol% acetone.
    Bottoms—acetone to be less than 0.5% by weight.
    Minimum reflux ratio:

Reflux ratio	*No. of stages*	*Feed stage*
Min 1:1	∞	—
1.5:1	17	15th from top
2:1	14	12th from top
5:1	11	8th from top
Selected 1.5:1		

Number of theoretical trays: The equilibrium curve is first drawn. The minimum reflux ratio is calculated from the figure where the operating line touches the equilibrium curve which was found to be 1. The actual reflux ratio to be fixed above the minimum and the actual top section operating line is drawn with reflux ratio of 1.5. Then the bottom section operating line is drawn from the bottom composition point (X_w) to the feed composition. Next the top section operating line is drawn from the point X_0 on the $x = y$ line to the point where it meets the bottom operating line. The number of theoretical plates is determined by stepping off the plates horizontally and vertically from top to the bottom. This is shown in Figure 3.8.

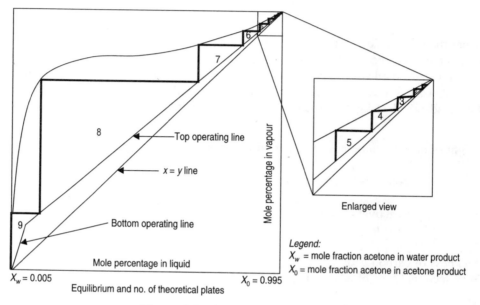

Figure 3.8 Acetone water system.

The actual number of plates is obtained by dividing the actual plates by the tray efficiency. [Note that this exercise is strictly not required to be done since the same can be done using one of the Process Simulation Softwares which contain the equilibrium data. However, this exercise gives a better perspective of the operation. This is important because any computer program can have bugs.]

Number of theoretical plates
   Rectifying section = 8
   Stripping section = 1

Number of actual plates
   Rectifying section = 24
   Stripping section = 3

Tray efficiency is taken as 0.33 since the trays have to function under alternating loads.

The total number of actual trays will be 27.

Cost estimation

Similar to the calculations done in the previous section, cost estimates are done here too. A summary of results is given below in Table 3.5. The major material of construction shall be Stainless Steel Grade 304.

Organic Chemicals

Table 3.5 Cost estimation for acetone recovery plant

Item	Weight (kg)	Material cost (₹)	Fabrication cost (₹)	Purchased items (₹)	% Profit (₹)	Total cost (₹)
Four adsorbers	2600	655,000	620,000	131,000		1,406,000
Distillation column	500	254,100	231,000		50,820	535,920
Preheater				1,50,000	30,000	180,000
Main condenser	350	246,400	154,000	–	49,280	449,680
Reflux condenser	185	130,900	100,000	–	27,000	257,900
Cooler	122	85,000	68,000		17,000	17,000
Pump				180,000	36,000	216,000
Blower				25,000	50,000	300,000
Receivers		200,000	120,000		40,000	360,000
Instruments				210,000	40,000	250,000
Piping				900,000	180,000	10,80,000
Civil & Electrical						400,000
Total cost						**₹ 5,605,500**

The cost of the bare plant comes to about ₹ 56 lakhs.

3.2 LPG AND PROPYLENE FROM CRACKED LPG

3.2.1 Liquified Petroleum Gas

Liquified Petroleum Gas (LPG) is a mixture of light hydrocarbons, especially propane (C_3) and butane/butylene (C_4), in such a ratio so as to have a ceiling on vapour pressure such that the pressure will not exceed the operating pressure of LPG cylinders under worst atmospheric temperature conditions in the country or region where it is to be used. Hence, the composition of LPG varies widely between countries. In a cold country like the USA, even pure propane can be used as LPG, whereas in a hot country like India propane in LPG should be below 50%.

The LPG for domestic use is supplied in the form of vapour-liquid mixtures in cylinders of convenient size.

Production of LPG

LPG is obtained from various sources mainly:

1. Gas processing units
2. Refinery gases

Each of the above processes with examples is described in the following subsections.

3.2.2 LPG Production from Gas Processing Units

In this case, natural gas is the starting material. Here also there can be many methods. The manufacture of LPG based on chilling and distillation is described below. A propane refrigeration system is used for chilling. See Figure 3.9.

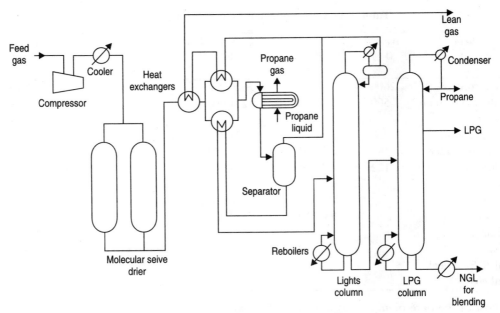

Figure 3.9 LPG production by chilling.

Feed gas compressed to a pressure of 17.5 kg/cm^2 and at temperature of 42°C is dried in molecular sieves of diameter 4Å to a moisture content of less than 1 ppm and filtered to remove foreign particles. The molecular sieves are regenerated by recycled lean gas (not shown in the figure).

The feed gas is then cooled by the recycled lean gas. Then it is split into two equal streams. One portion is cooled by lean gas and the other is cooled by feed to the lights column. The two streams are remixed and then chilled in a chiller, using propane refrigeration, to a temperature of −37°C. The chilling produces liquid which is separated in a separator and the liquid produced exchanges cold to one of the incoming streams, and being heated enters the lights column.

Feed enters the lights column at a temperature of around 35°C. In the lights column, C_1 and C_2 are removed at the top. The reflux condenser again uses chilled propane as cooling medium at −40°C. It is fitted with 40 numbers valve trays. The bottoms of the lights column enter the LPG column. The column has 53 valve trays. The top products of LPG column are propane and LPG produced as per requirement, propane being drawn from the top and LPG from an intermediate stage. The bottoms of this column are sent as NGL (natural gas liquid) to be mixed with gasoline.

Increasing the production of LPG—A case study

Suppose we need to increase the production of LPG using the same plant as described above. There are many alternatives possible for increasing the capacity. Some of them are as follows.

1. Enhanced pressurization and chilling. This will be very expensive.
2. Recycling of natural gas liquid (NGL).
3. Use of adsorption units.

Alternative 2 is the least expensive modification because capital investment and utility consumption are both less. Hence, this process is described below.

In Figure 3.10, you will notice that a new recycle pump has been added. The recycled natural gas liquid is mixed with the feed before the feed splits into two parts, going into the feed gas/lean gas exchanger and the feed gas condenser exchanger as shown in Figure 3.10. The ratio of natural gas that goes back as recycle to the quantity that is taken out as NGL product should be within the limits that the columns can handle with a new set of higher capacity valve trays.

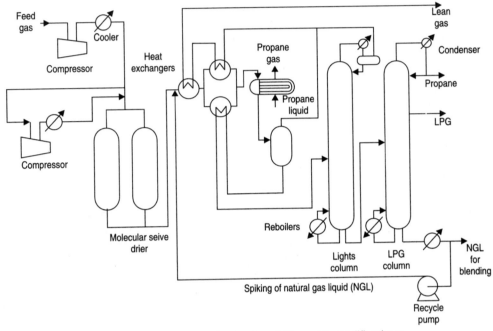

Figure 3.10 LPG production by chilling with modifications.

The production volumes of propane and LPG after modification are as given in Table 3.6, based on simulation with Winsim's 'Design 2 for Windows'. This shows a considerable increase in the production of LPG and propane which are high cost items under local conditions.

Table 3.6 Production of propane and LPG after modification

Stream no.	Feed	Propane	LPG
Flow, kgmol/h	557.77	7	18.06
Flow, kg/h	11811.3	309.24	945.15
Temperature, °C	42	42.8	45
Pressure, kg/cm² A	18.5	15	15.11
Composition (mol %)			
N_2	0.89		
Methane	74.5		
Ethane	17.61	0.02	
Propane	3.94	99.36	46.23
I-butane	0.68	0.47	18.32
N-butane	1.22	0.15	30.54
I-pentane	0.56		3.27
N-pentane	0.6		1.64
Total	**100**	**100**	**100**

The recycling will result in changes to the equipment as given below.
1. Additional exchanger bundles for feed to lean gas as well as condensate exchangers.
2. New pump for spiking the natural gas liquid.
3. Column trays to be replaced for both columns.
4. Increase in capacity of column condenser.
5. Compressors drive to be replaced.
6. Some changes will be required in piping.

The above results can be obtained by using any of the Simulation software.

3.2.3 LPG Production from Refinery Gases

The production of LPG is mainly based upon the principle of separation. The C_3–C_4 fractions can be separated from the rest of fractions in many ways, however, at present the following methods are used in industry.

1. Distillation at low temperature
2. Absorption and desorption
3. Compression and expansion
4. Combined methods.

Of all these methods, the first two are the most popular ones in the industry[5].

Distillation is based upon the boiling points or relative volatility, while absorption is intimately connected with the selective absorption capacity of oils for certain fractions.

The absorption method is presented here as depicted in Figure 3.11.

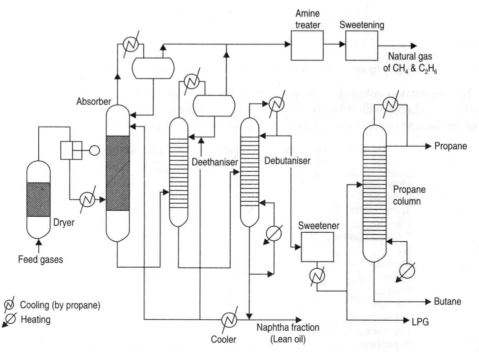

Figure 3.11 LPG production by absorption technique.

Off-gases from various units of the refinery such as Crude Distillation, Isomerization, Alkylation, Catalytic Cracker, Stabilizer, etc. are combined and first given a water wash to remove suspended particles (not shown in the figure). These feed gases are then passed through a dryer to remove water vapour, compressed to about 10 kg/cm^2, cooled and sent to the absorber. In the absorber, cold lean oil is used as absorbent which absorbs the heavier C_3 and C_4 fractions. The other light ends are obtained at the top of the column.

The rich oil containing C_3 and C_4 fractions is sent to a deethaniser column where any C_2 present is separated. Then, the oil is sent to a debutaniser column where all the C_3 and C_4 are taken out at the top and the lean oil that comes out at the bottom is partly recycled to the absorption column. The top product of the debutaniser is LPG but any sulphur compounds which may be present are removed in a sweetener (desulphurizing section). If pure propane is required, then a propane column is used.

The bottom product contains solvent and other hydrocarbons of grades higher than the C_4 fraction. When natural gasoline or virgin gasoline is the feed, C_5 and C_6 components will also accompany and these fractions are absorbed by lean oil. Some portion of this lean oil is cooled and sent into deethaniser and the remainder to absorber. Recovered C_5-C_6 fraction is blended with gasoline.

3.2.4 Propylene from Cracked LPG—A Case Study

LPG is obtained as a cracked product in cases where refineries use catalytic or other similar cracking processes. In these cases, propylene and butylenes will also be present along with propane and butane and this stream is called cracked LPG.

Propylene and butylenes are value added products since they can be polymerized to make plastics and also make other useful polymers and chemicals. In one such case, it is required to separate propylene from cracked LPG to be sold as a value added product.

100,000 TPA of LPG containing 28.8 mol % propylene forms the feed to the plant unit. The propylene plant is to be designed to produce 80% propylene as top product. The total C_3 (propane + propylene) in the bottom is maintained around 20 mol %. The recovery of top product propylene depends upon the propylene content in the feed and some cushion needs to be added to process the lesser propylene content as well. The plant will be designed to have a turndown capacity of 50%.

Process description

Cracked LPG containing 28.8 mol % propylene (min. 23%), the remainder being propane, butane and butylenes, is pumped to the propylene unit by means of LPG feed pump. See Figure 3.12. First, the LPG is preheated in a feed bottom exchanger using the heat from the bottom product of the propylene splitter column. The preheated LPG is fed to the middle of the propylene splitter column.

In the propylene splitter column the propylene in the LPG gets rectified against the action of reflux and 80% propylene is obtained at the top. The top vapours are condensed in an overhead condenser and collected in an overhead receiver. A portion of the liquid is returned to the column as reflux by means of reflux pump and the other portion from the outlet of the pump is sent to LPG storage.

Heat to the propylene splitter column is supplied by steam through a reboiler. The bottom product from the column is cooled first in feed bottom exchanger and then in product cooler.

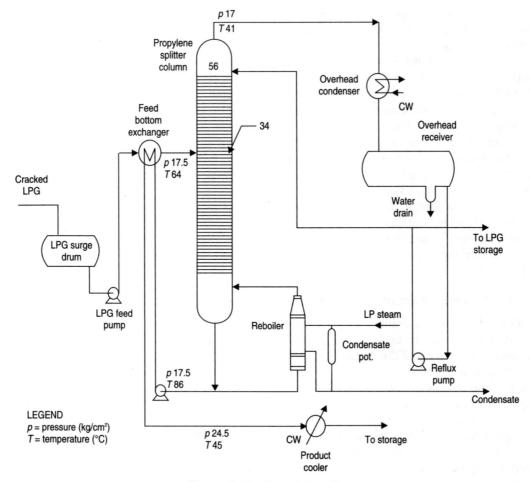

Figure 3.12 Propylene unit.

Automatic control of quality can be applied either at the top or at the bottom of the column by putting in line either of the TICs. LIC is used to vary the heat transfer area of the reboiler. PIC is controlled by the level of condensate in the overhead receiver. The bottom level of the column is controlled by the bottom product flow rate. All instruments are connected to the distributed control system of the refinery.

Based on simulations using "Design 2 for Windows" or other equivalent software, Table 3.7 shows the flow rates, the temperature and pressure conditions, and the composition for feed, propylene and balance LPG respectively.

Valve trays are selected for the column so that the plant can operate smoothly even at low flow rates (high turndown), which may occur during the connected refinery operations. See Figure 3.13.

Table 3.7 Flow rates, temperature and pressure conditions and composition for feed propylene and balance LPG

Stream no.	Feed	Propylene	Balance LPG
Flow, kgmol/h	183.8	48	135.8
Flow, kg/h	9399.66	2033.23	7365.87
Temperature, °C	30	41	35
Pressure, kg/cm^2	9	17	23
Composition, mol%			
Ethane	0.2	0.772	0.0
Propylene	28.8	80.71	10.473
Propane	11.6	18.516	9.158
I-Butane	16.6	0.002	22.459
N-Butane	4.5		6.089
1-Butene	8.7		11.77
I-Butene	10.2		13.802
Trans-2-butene	10.4		14.075
Cis-2-butene	8.4		11.366
Pentane	0.6	0.0	0.809
Total	**100**	**100**	**100**

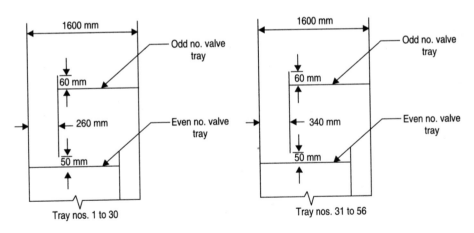

Figure 3.13 Valve trays.

The specifications for valve trays shown in Table 3.8 are for 100% capacity. The trays should have turndown of 50%. The maximum pressure drop for column = 0.5 kg/cm^2

Table 3.8 Specification for valve trays

Trays	Vapour rate (kgmol/h)	Vapour rate (kg/h)	Vapour density (kg/m^3)	Liquid rate (kgmol/h)	Liquid rate (kg/h)	Liquid density (kg/m^3)	Surface tension (dy/cm)
1–30	618.42	26255.71	36.54	550.37	23390	479.7	4.717
31–56	564.42	27665.71	39.48	754	38861.42	481.9	4.382

Other plant specifications: Valve trays parametres for different flow cases (1 and 2) are shown in Table 3.9.

Table 3.9 Valve tray specifications for flow cases 1 and 2

Case	Trays	Oprating pressure (kg/cm^2)	Operating temperature (°C)	Vapour rate (kgmol/h)	Vapour rate (kg/h)	Vapour density (kg/m^3)	Liquid rate (kgmol/h)	Liquid rate (kg/h)	Liquid density (kg/m^3)	Surface tension (dy/cm)
1	1–10	17.04	41.91	432.9	18379	36.54	385.26	16373	479.7	4.717
	11–20	17.12	42.54	434.1	18481	36.85	386.21	16463	477.4	4.677
	21–26	17.2	43.91	432.3	18492	37.05	381.27	16417	477.2	4.641
	27–28	17.23	47.54	423.3	18349	37.09	365.01	16165	780.6	4.654
	29–30	17.25	51.72	413.0	18194	37.13	350.74	16003	483.9	4.669
	31–32	17.27	58.03	398.7	18033	37.33	544.49	25867	487.1	4.671
	33–36	17.3	58.37	409.0	18524	37.46	545.0	25923	486.5	4.652
	37–48	17.39	61.10	407.7	18702	37.93	541.65	26410	484.8	4.582
	49–56	17.46	73.96	395.1	19366	39.48	527.8	27203	481.9	4.382
2	1–10	17.04	41.92	505.4	21457	36.55	469.85	19971	479.6	4.716
	11–20	17.12	42.63	507.1	21604	36.89	471.284	20104	476.8	4.66
	21–26	17.2	43.9	505.8	21655	37.12	66.99	20099	475.8	4.624
	27–28	17.23	47.11	496.9	21523	37.17	449.36	19827	478.6	4.63
	29–30	17.25	51.07	485.4	21345	37.21	431.97	19614	481.9	4.647
	31–32	17.27	57.49	486	21131	37.39	624.54	29563	485.5	4.654
	33–36	17.3	57.85	477.2	21581	37.52	625.14	29631	484.9	4.633
	37–48	17.39	60.66	475.5	21790	37.99	620.94	29879	483.2	4.561
	49–56	17.46	73.72	459.5	22510	39.52	603.1	31040	480.9	4.37

The equipment for the propylene unit is listed below:

1. Propylene column
2. Overhead receiver
3. Condensate pot.
4. Feed bottom exchanger
5. Overhead condenser
6. Reboiler
7. Product cooler
8. LPG feed pump
9. Reflux pump
10. LPG surge drum
11. Bottom product pump

The utilities required for the propylene unit are:
1. Steam for reboiler (at 5.5 kg/cm^2) = 3500 kg/h
2. Cooling water for condenser = 290 m^3/h
3. Other utilities comprise nitrogen, steam, water and plant air connections at suitable points.

Propylene and other light hydrocarbons are highly flammable. Extreme care should be taken in the handling of flammable chemicals. Providing well designed and suitable safety valves are extremely important. The outlet of safety valves should go to a flare stack for burning the gases. An erected standby safety valve is provided for all the safety valves for doing maintainance without plant shutdown. Isolation valves are provided such that one of them will be always under operation.

Typical maximum flow rates of LPG at discharge of the safety valves are given in Table 3.10. Note that one of the valves (RV 3 A/B) is to be designed for a two-phase flow.

Table 3.10 Typical specification for relief values

Valve	Flow (kg/h)	Temperature (°C)	Set pressure (kg/cm^2)
RV 1 A/B	11,000	42	30
RV 2 A/B	20,804	64	20
RV 3 A/B Gas	15,400	40	6
(Two phase) liquid	27,788		
RV 4 A/B	5,200	40	15

3.3 MAN-MADE FIBRES

3.3.1 Viscose Rayon

Synthetic fibre manufacture has been one of the main drivers of chemical industry. While there are a large number of products, only some of them (viscose rayon, nylons, polyester and acrylonitrile) are described here. Viscose rayon is one of the oldest man-made fibres but it still has a large market. The improvements required in the industry are mainly better availability of timber for pulp manufacture and better pollution control measures.

The manufacture of pulp is not discussed since it is similar to the description of pulp manufacture given under Section 2.7, 'Pulp and Paper'.

The pulp is immersed in a solution of 17 to 20% sodium hydroxide at a temperature of 18 to 25°C. This is called steeping. See Figure 3.14. In this process, the cellulose is converted to alkali-cellulose.

$$(C_6H_{10}O_5)_n + nNaOH \rightarrow (C_6H_9O_4ONa)_n + nH_2O$$

Further the material is pressed to make it dense. The material is then shredded to make it fluffy. The fluffy particles are called crumbs and they have increased surface area. Ageing is then done (batchwise or continuously) in stainless steel vats, at a temperature of 24°C, for a period of 1 to 2 days to depolymerize it to the required degree. Depolymerization is done to reduce the molecular weight of pulp to between half and one-third.

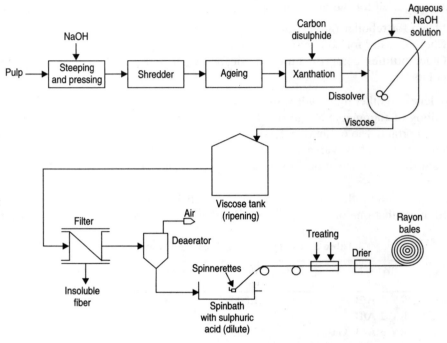

Figure 3.14 Production of viscose rayon.

Xanthation

The product (depolymerized pulp) is then reacted with carbon disulphide. The following chemical reaction takes place:

$$(C_6H_9O_4ONa)_n + nCS_2 \rightarrow (C_6H_9O_4-O-C(=S)-SNa)_n$$

The product obtained is called cellulose xanthate. Side reactions occur and hence the product has an orange colour. It does not dissolve in water. Hence, it is dissolved in dilute caustic soda. The solution so obtained is called viscose.

It is then kept in viscose tanks for ripening. The xanthate redistributes to cellulosic hydroxyls and free carbon disulphide. The loss of carbon disulphide reduces the solubility of cellulose and facilitates filament formation. The viscose is filtered and deaerated by applying vacuum.

Spinning

The viscose is sent to the spinning section where the viscose is introduced into a dilute sulphuric acid bath through spinnerettes which converts it into thread. The overall reaction (including ripening and spinning) can be represented as below:

$$(C_6H_9O_4-SC-SNa)_n + (n/2)H_2SO_4 \rightarrow (C_6H_{10}O_5)_n + nCS_2 + (n/2)Na_2SO_4$$

The thread of regenerated cellulose polymer is drawn, treated and dried. It is wound and converted into bales of rayon and sent to sales.

3.3.2 Nylons

The nylons (nylon 6 and nylon 6,6) are very strong and durable synthetic fibres as well as important industrial and engineering materials. In the western countries, both are mainly used at present for making carpets. It is recently found that nylon 6 is recyclable, almost 99%, since it can be converted back or depolymerized to caprolactam, which is its precursor[6,7]. Nylon 6,6 cannot be recycled by converting back to its precursors. Hence, nylon 6 is a relatively greener product when compared to nylon 6,6. The nylon 6 recycling process developed by Honeywell/Allied Signal and DSM Chemicals has resulted in a joint venture called 'Evergreen Nylon Recycling' to recover and reuse caprolactam.

Nylon 6

Caprolactam, the lactam of caproic acid with formula, $(CH_2)_5CONH$, is an organic compound produced in the form of white flakes. Nylon 6 is made by the polymerization of caprolactam.

In the western countries, making carpets is a major use of the nylons. Apart from it, nylon 6 is mainly used in making tyre cord. It is also used as thread in bristles for toothbrushes, surgical sutures, and strings for acoustic and classical musical instruments, including guitars, violins, violas, and cellos. It is also used in the manufacture of a large variety of threads, ropes, filaments, fishing nets, tyre cords, as well as hosiery and knitted garments. Both nylons are also used to make lightweight guns and pistols. 'Nylon films' are used for high quality/strength packaging.

Ammonium sulphate, a by-product of caprolactam manufacture, is a good fertilizer since it contains the major nutrient 'nitrogen' along with micronutrient 'sulphur'. This fertilizer is particularly effective for rice cultivation.

Production of caprolactam and nylon 6

Caprolactam $(CH_2)_5$ CO NH, is in fact, a cyclic amide (lactam) of caproic acid. About 20,00,000 tonnes are produced in the world every year. Its use is less in hot and humid countries. About 3% of the total production is in India.

Refer to Figure 3.15. The first step is the hydrogenation of benzene to produce cyclohexane. Benzene is hydrogenated to cyclohexane by the following chemical reaction.

$$C_6H_6 + 3H_2 \rightarrow C_6H_{12} + 49.25 \text{ kcal/kgmole}$$

The above reaction takes place in the vapour phase at a temperature ranging from 350 to 400°C and at a pressure of 30 kg/cm^2 at the inlet of the reactor. The heat produced is absorbed in a tubular reactor by means of hot oil flowing outside the tubes, so that outlet temperature is about 225°C. The catalyst is placed within the tubes. Catalyst consists of platinum with aluminium oxide support material. It will be in the form of cylindrical tablets. Excess hydrogen is required to complete the reaction. The sulphur impurity is converted to H_2S and is removed using a zinc oxide (ZnO) bed by the reaction given below.

$$ZnO + H_2S \rightarrow ZnS + H_2O$$

The cyclohexane is oxidized to cyclohexanone by the reaction shown below.

$$C_6H_{12} + O_2 \rightarrow C_6H_{10} = O + H_2O$$

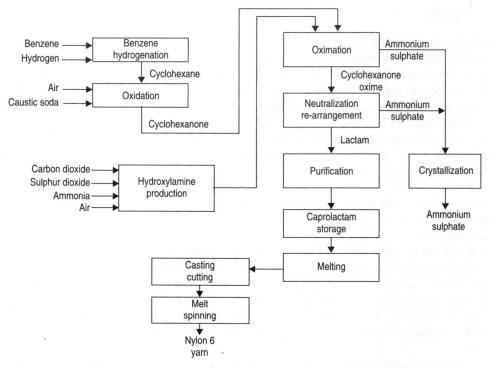

Figure 3.15 Manufacture of caprolactam and nylon 6.

The most important process steps of manufacture are oxime formation and rearrangement which are given below.

First, cyclohexanone is converted into cyclohexanone oxime in a reactor by reacting with hydroxylamine sulphate.

Cyclohexanone + Hydroxylamine Sulphate → Cyclohexanone Oxime + Ammonium Sulphate

$$C_6H_{10} = O + NH_2OH \rightarrow C_6H_{10} - O - NH + (NH_4)_2SO_4$$

Then the cyclohexanone oxime is treated with sulphuric acid in a reactor to catalyse the Beckman Rearrangement which produces caprolactam.

$$C_6H_{10} - O - NH \text{ (Cyclohexane Oxime)} \rightarrow NH - C_6H_{10} = O \text{ (Caprolactam)}$$

The exact structure of caprolactam is shown below.

Economics of caprolactam: The investment required for setting up a plant for producing 50000 TPA caprolactam, if basic materials like benzene, ammonia, etc. are available at the site, will be approximately ₹ 1500 crores. Profitability will largely depend upon the availability of cheap raw materials. The intermediate product cyclohexanone can also be sold to pesticide manufacturers as a constituent of pesticide formulation.

Nylon 6 production: The caprolactam is converted into nylon 6 by the following 'ring opening type' polymerization reaction by opening the heptagonal ring, previous to the nitrogen junction.

$$n \text{ (caprolactam)} = [-CO-CH_2-CH_2-CH_2-CH_2-CH_2-NH-]_n \text{ (n is chain length)}$$

The caprolactam is melted to a liquid and it is subjected to hydraulitic catalysed polymerization. It can be done in one or two stages. The reactors can be batch type or continuous catalytic reactor type. Continuous process is better for larger capacity plants. Titanium dioxide and other additives are fed as aqueous suspensions.

The steps in the production of nylon 6 yarn are almost similar to the steps in the production of nylon 6,6 yarn from the 'melting vessel' onwards, as shown in Figure 3.16, for manufacture of nylon 6,6. It consists of melting of caprolactam (it melts at 68°C and is soluble in water), addition of pigment, acid promoter, etc. passing through a casting wheel to produce ribbon which is cut into chips, feeding to melt-spin vessel, passing through spinnerettes to obtain yarn which is cooled, stretched and drawn and packed and sent to stores and/or shipped for sales.

Nylon 6,6

It has higher temperature and friction resistance than nylon 6. It is largely used in carpet manufacture. It is also suitable for the manufacture of tough engineering materials.

Nylon 6,6 has two C_6 rings but is mostly similar to nylon 6 and is made from the chemical reaction of hexamethylene diamine and adipic acid which produces hexamethylene diammonium adipate, commonly known as 'nylon salt'. (The hexamethylene diamine is a compound made from either butadiene or acrylonitrile. Adipic acid is made by combining cyclohexanone and cyclohexanol in the presence of nitric acid.)

The nylon salt is then polymerised to nylon 6,6.

The overall reaction for producing the 'nylon salt' and and its conversion to nylon 6,6 is approximately a combination of the two chemical reactions given below. The actual transition is slow and maybe slightly more complicated.

$$\text{Adipic acid + Hexamethylene diamine = Nylon salt}$$

$$HOOC(CH_2)_4COOH + H_2N(CH_2)_6NH_2 = [-OOC(CH_2)_4COO \cdot HN(CH_2)_6NH_3]$$

$$\text{Nylon Salt = Nylon 6,6 (Polyhexamethylene adipamide) + Water}$$

$$n[-OOC(CH_2)_4CO \cdot HN(CH_2)_6NH_3] = n[-OC(CH_2)_4CO \cdot HN(CH_2)_6NH-] + nH_2O$$

Production of nylon 6,6

Adipic acid and hexamethylene diamine in equimolar ratio are fed into a nylon salt reactor. See Figure 3.16. Acetic acid and water are added to control the polymerization reaction.

110 Chemical Process Technology and Simulation

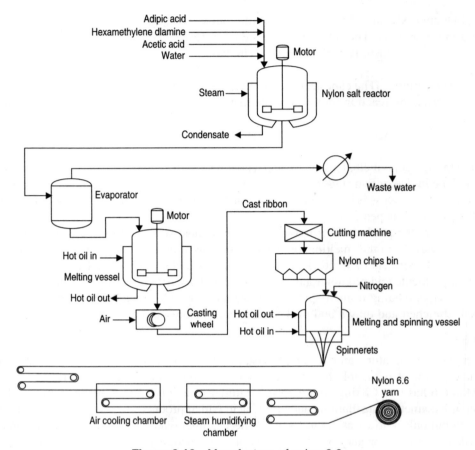

Figure 3.16 Manufacture of nylon 6,6.

The reactor is provided with an agitator and a jacket through which the mixture is heated using steam. The nylon salt produced is evaporated in an evaporator where water is removed by evaporation. Then it is sent to a second reactor for completion of polymerization. Here it is heated by hot oil (dowtherm) to a temperature of 263°C. The nylon 6,6 melt is sent to a casting wheel where it is drawn out as a ribbon of 3 mm diameter and further cut into pieces of 30 mm length in the cutting machine and fed to a bin.

From the bins the chips are sent to a melting and spinning vessel which is heated by hot oil. As they fall on the surface of a plate fixed on the vessel, they melt. Nitrogen under pressure is introduced into the vessel from the top. The melt is passed through spinnerettes and the yarn produced is drawn out and sent to a dryer and then passed through a steam humidifying chamber to make the yarns in equilibrium with water vapour. Then the yarn produced is stretched and drawn by passing through differential speed roller systems. The yarn obtained is packed and sent to stores and/or shipped to consumers.

3.3.3 Polyester

Polyethylene terephthalate, popularly called polyester, is one of the most popular synthetic fibres and is used for clothing and many similar other purposes. It is the strongest among all natural and synthetic fibres. Since it is bright in colour, it is comparatively difficult to dye.

There are two processes for making the ester.

Process 1 (PTA Route)

The first method is the condensation of purified terephthalic acid (PTA) with ethylene glycol as given below which gives only water as a by-product.

Purified terephthalic acid (PTA) + ethylene glycol → polyethylene terephthalate + water

$$n COOH-C_6H_4-COOH + nOH-CH_2-CH_2-OH \rightarrow$$
$$(-O-CH_2-CH_2-COO-C_6H_4-CO-)_n + (2n-1)H_2O$$

Process 2 (DMT Route)

The second method is the condensation of di-methyl terephthalate (DMT) with ethylene glycol which will give methanol as a by-product as shown below.

dimethyl terephthalate (DMT) + ethylene glycol → polyethylene terephthalate + methanol

$$n COOCH_3-C_6H_4-COOCH_3 + nOH-CH_2-CH_2-OH \rightarrow$$
$$(-O-CH_2-CH_2-COO-C_6H_4-CO-)_n + (2n-1)CH_3OH$$

Process comparison

In the manufacture of raw materials for polymerization, monomer purity is extremely important. Earlier, purification of terephthalic acid was found to be difficult and hence the DMT route was preferred. Later when the purification process was perfected, purified terephthalic acid (PTA) became available and this process became more popular.

Production of polyester by PTA route

A polyester plant based on PTA route, generally as per 'AMOCO' process, is very briefly described below. See Figure 3.17.

Terephthalic acid (TPA) is first produced by reacting paraxylene with oxygen in the presence of a catalyst consisting of cobalt acetate promoted by bromine. Acetic acid is used as a solvent to the catalyst. The following overall reaction takes place:

$$CH_3-C_6H_4-CH_3 + 2O_2 \rightarrow COOH-C_6H_4-COOH + 2H_2O \quad \Delta H = -ve$$

Paraxylene, acetic acid and catalyst are admitted into the reactor. Air is admitted at the bottom of the reactor. The vapours produced at the top are condensed and the exothermic heat of reaction is used for producing steam for the distillation section. A scrubber is provided for scrubbing unreacted air before venting. The reacted mixture in the form of a slurry is sent to the crystallization vessel provided with a condenser. The product from the bottom of surge vessel contains TPA crystals and liquid. This is filtered and the fitrate, a mixture of acetic acid, paraxylene and catalyst is sent to a distillation section from where the acetic acid, paraxylene and catalyst are recycled. The TPA crystals obtained from the filter are dried in a dryer. The final product obtained is 99.5% terephthalic acid (TPA).

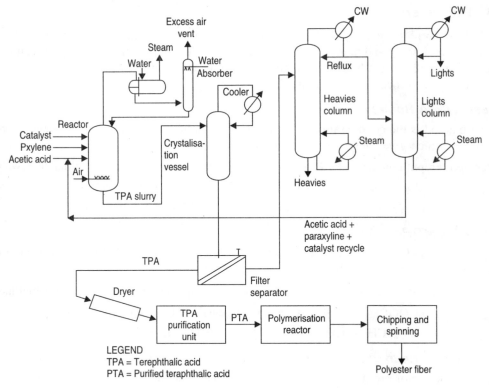

Figure 3.17 Polyester production (simplified).

The TPA is then purified by means of a re-crystallization process such that the maximum impurity content in the product is limited to 25 ppm, which is called purified terephthalic acid (PTA). This is sent to polymerization reactor vessels to obtain polymerized terephthalic acid which is called polyester.

The ribbons of polymer are drawn out, chipped, spun into fibre, baled and packed for sales.

3.3.4 Acrylonitrile

Acrylonitrile is an important monomer for the manufacture of synthetic fibres and plastics. It is a pungent smelling liquid and has the chemical formula $CH_2 = CH-C \equiv N$, consisting of a vinyl group and a nitrile group.

Acrylonitrile is soluble in most organic solvents and water. It is a liquid with a boiling point of 78.5°C and a specific gravity of 0.807. It is flammable and poisonous.

The main uses of acrylonitrile are in the production of synthetic fibres, mainly acrylic fibre, plastics consisting of styrene acrilonitrile (SAN), acrylonitrile butadiene styrene (ABS) and synthetic rubbers like acylonitrile butadiene rubber (ABR). It is used for making 'carbon fibre' which is a low weight, high strength, high stiffness material used in making sports equipment and sports gear. A small part of acrylonitrile is used as a fumigant.

Methods of production

There were many processes using different raw materials such as acetylene, ethylene cynohydrin and acetaldehyde. However, a new process based on propylene became the most popular process after its discovery which is explained below.

Propylene-ammonia air-oxidation process (Sohio process)

Sohio is a petroleum refining and petrochemical company founded by John D. Rockefeller. A research team of this company found out a short-cut for producing acrylonitrile from propylene in the year 1955. This resulted in a sudden fall in the price of acrylonitrile.

The production of acrylonitrile from propylene is one of the few chemical reactions in which three reactants combine together to form a product on the surface of a bismuth phosphomolybdate solid catalyst. The three raw materials are propylene, ammonia and oxygen.

Apart from the product, acetonitrile, a by-product, is also obtained in this process which is a good solvent, similar to acetone and is used in solvent extraction processes.

Process description

A mixture of propylene, ammonia and air is sent to a fluidized bed reactor (see Figure 3.18) in which a molybdenum–bismuth catalyst (bismuth phosphomolybdate) in the shape of microspheroids is present. The flow of gases into the reactor fluidizes the catalyst. The ammonia, propylene and oxygen combine at a pressure of 1.5 to 3 kg/cm^2 and at a temperature of 400 to 500°C to form acrylonitrile, in the presence of fluidized catalyst particles, as per the following chemical reaction.

$$CH_3-CH=CH_2 + NH_3 + 1.5O_2 \rightarrow CH_2=CH-C\equiv N + 3H_2O$$

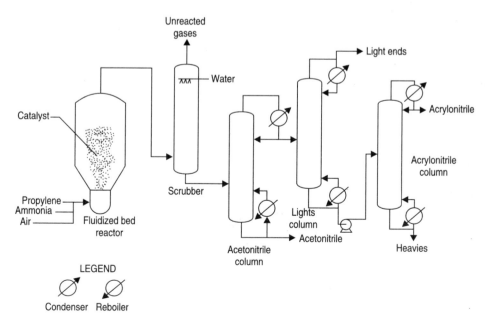

Figure 3.18 Acrylonitrile production.

Regeneration of the catalyst is not required if no sulphur is present in the feed.

The product gases from the reactor are scrubbed with water in the scrubber to produce a solution of acylonitrile and other by-products in water. This is distilled in a series of columns whereby acetonitrile is first separated out in the acetonitrile column, and then, light ends are removed in the lights column. Acrylonitrile is produced as top product in the acrylonitrile column and the heavies are removed at the bottom of this column.

3.4 PETROCHEMICALS

3.4.1 Steam Cracking Products

Petrochemicals are chemicals that are obtained from petroleum excluding the portion that is used as fuel. Here petroleum means crude oil, associated oil and natural gas. The chemicals obtained serve as raw materials for many cost-effective products of daily use in the industrialized world.

Since almost 70% of all synthetic chemicals produced at present will come under petrochemicals, we will first focus on the processing activities of the present-day large petrochemical companies.

Most of the petrochemical products obtained at present are derived by the steam and catalytic cracking of naphtha or natural gas.

Cracking methods and products

Petrochemical companies use basic petroleum products as their raw materials such as natural gas or naphtha. These are first steam cracked with a cracking furnace as in refineries. The difference is that while refineries crack 'heavies like vacuum gas oil', petrochemical companies crack natural gas and naphtha but at a much higher scale. Naphtha cracking is popular only in countries where petrol consumption is controlled. Most new facilities are being built on gas cracking.

The high value products obtained are mainly acetylene, ethylene, propylene, butadiene and butylenes. Products like benzene, toluene, and xylene are obtained by the process of aromatization. Finally, products like linear alkyl benzene, ethylene oxide, linear low density polyethylene (LLDPE), high density polyethylene (HDPE), polypropylene, polyvinyl chloride, monoethylene glycol, etc. are made. Almost all plastics and synthetic fibres are made from petrochemicals.

Steam cracking of naphtha (naphtha cracking)

Steam cracking of naphtha is a widely used method in petrochemical complexes. A typical case of petrochemical production would be the steam cracking of naphtha to produce unsaturated hydrocarbons which are further converted to useful products such as polymers, artificial rubbers, synthetic fibres, etc.

A typical naphtha cracking process is described in Figure 3.19.

Organic Chemicals

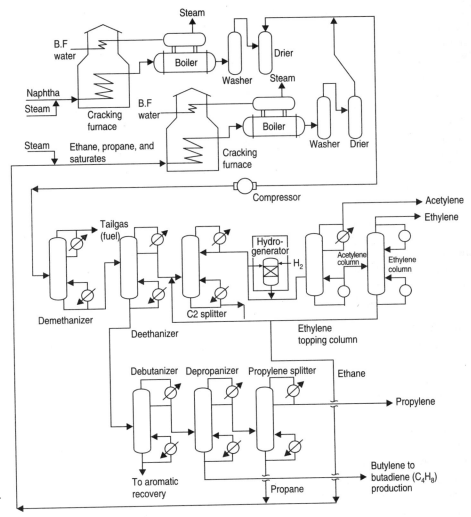

Figure 3.19 Naphtha cracker unit.

A mixture of steam and superheated naphtha is sent to a cracking furnace where the mixture is heated to about 600 to 700°C. Pyrolysis of the naphtha takes place, visualized as per the following approximate equation.

$$C_nH_{2n} + H_2O \rightarrow C_2H_4 + C_2H_6 + C_2H_2 + H_2 + CO + CO_2 + CH_4 + C_3H_6 + C_3H_8 + C_4H_8 + C_4H_{10} + C_4H_6 + \text{Heavies} + C$$

There are two pyrolysis furnaces—one for the initial feed and another for recycled saturated hydrocarbons consisting of mainly ethane, propane and butane. This is because the temperature and residence times required for this mixture are different from those of naphtha and as also for operational convenience. There are boilers at the outlet of the cracking furnaces to absorb the heat from these gases with production of steam.

The cracked gases from both the furnaces are cooled, washed, dried mixed and compressed to a pressure of 35 kg/cm² in the cracked gases compressor. The gases are washed for removing solid impurities and dried in adsorbers or molecular sieves, for removing water.

First, the gases enter a demethanizer column where the lowest boiling component, i.e. methane, is recovered as tail gas and sent as fuel. In the next column, deethanizer, ethane and ethylene are obtained as top product and sent to a 'C_2 splitter column' where ethylene and acetylene are obtained at the top and ethane at the bottom. Ethane is sent to the second pyrolysis reactor for further steam cracking. The acetylene and ethylene obtained at the top are sent to an acetylene splitter column to produce acetylene at the top. The bottom is sent to ethylene column where ethylene is produced at the top and bottoms containing some ethylene is recycled to C_2 splitter. In case there is no demand for acetylene (acetylene is mainly used for gas welding), it can be hydrogenated to ethane and recycled through the bottom of ethylene column to the C_2 splitter.

The bottoms of the deethanizer are fed into a debutanizer where heavier than C_4 components, mainly aromatics, are removed at the bottom. The top product is sent to a depropanizer where 'butylenes to butadiene' cut is obtained at the bottom and the top product consisting of C_3 cut is sent to the propylene splitter column where propylene is produced at the top and propane at the bottom. The propane joins the ethane stream going to the second pyrolysis furnace. After pyrolysis, it joins back at the suction of the cracked gases compressor.

Steam cracking of gas (Gas cracking)

Naphtha is much costlier than natural gas. Even though naphtha cracking has been the dominant source of ethylene, a large number of gas cracking units are being established throughout the world for reasons of economy, and availability of gas. Ethane, propane and butane liquids are obtained during cryogenic processing of natural gas. This is called by the name natural gas liquids (NGL) in the gas industry. The NGL is mixed with steam and sent to a cracking furnace where it is heated to a high temperature. Pyrolysis of the components take place and a mixture of unsaturated hydrocarbons, predominantly ethylene, and lesser quantities of propylene and butylene are formed. Some acetyline will also be formed. The method of separation is similar to that described in naphtha cracking with an array of distillation columns. Different configurations are possible such as heavier components separated first or lighter components separated first or a combination which is most energy efficient.

3.4.2 Vinyl Chloride and Vinyl Acetate

Vinyl chloride

Vinyl chloride is the monomer for making the popular plastic, PVC (Polyvinyl Chloride).

For making vinyl chloride from ethylene, it is first converted into an intermediate, ethylene dichloride, by chlorination. Then the ethylene dichloride is cracked into vinyl chloride and hydrogen chloride (HCl). See Figure 3.20.

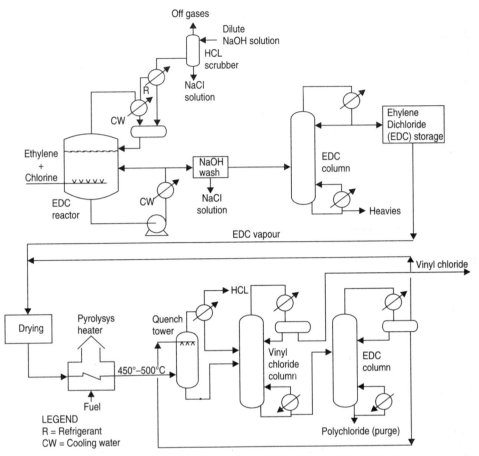

Figure 3.20 Production of ethylene dichloride and vinyl chloride.

Ethylene dichloride (EDC) liquid, being itself the reaction medium, is taken in a reactor and chlorine and ethylene at stoichiometric flow rates are introduced into the liquid. Chlorine reacts with ethylene to produce ethylene dichloride (EDC).

$$C_2H_4 + Cl_2 \rightarrow C_2H_4Cl_2 \text{ (ethylene dichloride)}$$

Trace quantities of ferric chloride and ethylene bromide are used as catalysts. The reaction mixture in the reactor is circulated through a cooler in order to remove the heat of reaction and keep the temperature of the reacting medium at 45 to 50°C.

Top vapours from the reactor are condensed and recycled in two stages: in the first stage by cooling water and in the second stage by refrigerant in a chiller. The off-gases are further scrubbed by a dilute caustic soda solution to absorb free chlorine.

The volume of EDC in the reactor increases and the excess liquid is drawn out on level control and washed with caustic soda to neutralize free chlorine. Then it is sent to EDC column where EDC is obtained as the top product. This is stored in spheres.

Conversion of EDC to vinyl chloride monomer (VCM)

The ethylene dichloride vapour from the top of the storage sphere is dried using silica gel adsorbers and sent to the pyrolysis heater. The heater has stainless steel tubes which are filled with pumice or charcoal which acts as the catalyst for the chemical reaction, splitting the ethylene dichloride molecule into VCM and HCl at temperature of 450 to 500°C.

$$C_2H_4Cl_2 \rightarrow C_2H_3Cl \text{ (vinyl chloride monomer)} + HCl$$

The reacted mass is sent to the bottom of a quench tower where recycled EDC is sprayed from the top. The mixture is then sent to a vinyl chloride column where it is distilled to produce vinyl chloride at the top. The bottoms of this column are sent to an EDC column where EDC is separated at the top. It is drawn out, both as vapour and liquid. The unreacted EDC vapour is recycled to the drying section and the EDC liquid is used for quenching the reacted mass in the quench tower.

Even though the conversion per pass is only 50%, by recycling the vapour and liquid as above, 95% and above conversion can be achieved. The impurities, polychlorides are removed from the bottom of the EDC column.

It may be noted that the HCl produced in the above reaction can be reacted with acetylene (C_2H_2) to produce further quantities of vinyl chloride by the following reaction. This process becomes economic if acetylene, also produced during naphtha cracking, is available.

$$C_2H_2 + HCl \rightarrow C_2H_3Cl$$

The above reaction is conducted in tubular reactors at a temperature of about 200°C, using mercuric chloride impregnated carbon as the catalyst. Depending on availability, the acetylene route by itself can also be used for producing vinyl chloride.

Excess HCl can also be absorbed in water and sold as hydrochloric acid.

Vinyl acetate

Vinyl acetate is used for making poly vinyl acetate, poly vinyl alcohol, poly vinyl formals, etc. It is also used for many copolymers such as vinyl chloride ethylene co-polymer. All these polymers are used in various applications such as coatings, adhesives, textile finishing agents, emulsifiers, lacquers, etc. Among new uses, the copolymer of ethylene and vinyl acetate called Ethylene Vinyl Acetate (EVA) is widely used in manufacture of solar cells. It is used as an encapsulant between the glass cover of a solar module and the glass cover of the silicon solar cells. EVA is used in biomedical engineering because it does not cause any side effects to the body. EVA foam is used as a padding material for sports shoes and equipment.

The older method of manufacture of vinyl acetate is by the chemical reaction of acetylene with acetic acid as per the following equation.

$$C_2H_2 + CH_3COOH \rightarrow CH_3COOCHCH_3$$

The newer method of manufacture of vinyl acetate is from ethylene, which is much less costlier than acetylene. Using ethylene is also safer and considered more environment-friendly. The reaction is conducted in a gas phase at 180°C temperature and at a pressure around 8 kg/cm^2. The catalyst used is palladium chloride. The oxy-esterification reaction is as shown below.

$$C_2H_4 + CH_3COOH + 0.5O_2 \rightarrow CH_3COOCHCH_3 + H_2O$$

The exothermic heat of reaction produced is used to make steam which is used in the subsequent distillation operations for the purification of vinyl acetate.

3.4.3 Phthalic Anhydride

Having discussed vinyl chloride and vinyl acetate we now turn to an important additive that controls the strength and flexibility of polymers. Plasticizers are added to improve the performance of plastic materials as well as that of polyester fibres. The most important plasticizers are phthalates made from phthalic anhydride. The plasticizers embed themselves between chains of polymers. This increases the spacing between the chains, which makes the plastics softer. The hardness of the material decreases and flexibility is increased. Phthalate-based plasticizers are also used where good oil and water resistance are required.

Phthalic anhydride can be produced from two base materials, either from ortho-xylene or from naphthalene. These materials are oxidized by either air or oxygen. The chemical reaction that takes place is given below.

From Naphthalene:
$$C_{10}H_8 + 4.5O_2 \rightarrow C_8H_4O_3 + 2CO_2 + 2H_2O$$

From Ortho-xylene:
$$C_8H_{10} + 3O_2 \rightarrow C_8H_4O_3 + 3H_2O$$

The actual reaction mechanisms taking place during the two routes are shown in Figure 3.21.

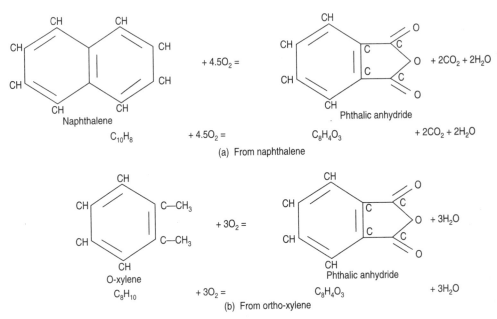

Figure 3.21 Conversion to phthalic anhydride.

The availability of naphthalene has increased because it can be obtained in hydro-alkylation processes. Hence, the naphthalene route is described in Figure 3.22.

The plant consists of a fixed bed reactor with catalyst contained within the tubes. The catalyst consists of a mixture of vanadium pentoxide and titanium dioxide on a carrier. The heat of reaction produced is absorbed by molten salt and this heat is used to produce steam in a boiler.

The products of the chemical reaction are sent to three columns in series. In the first column, water is removed at the top. The bottoms go to the second column where light hydrocarbons are again removed at the top. In the third column, the heavies are removed at the bottom and the product, phthalic anhydride, is obtained at the top.

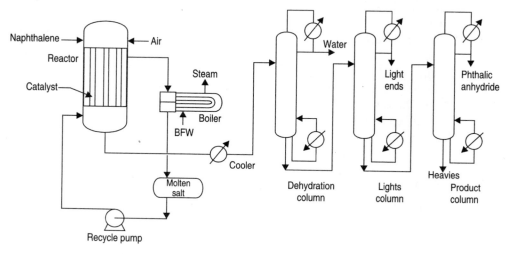

Figure 3.22 Manufacture of phthalic anhydride.

It is to be noted that adequate precautions need to be taken to ensure that no runaway reaction occurs, which may lead to an explosion. The reactor can be designed to be explosion proof and also the reactor can be kept in a suitable concrete explosion proof bunker.

3.5 POLYMERIZATION PRODUCTS

Polymers are made of repeated basic units called monomers, linked in the form of a chain. Hence, polymers usually have a very high molecular weight in the range of 10^3 to 10^7.

A polymer can be built only if the monomer unit is either bifunctional or trifunctional. A bifunctional can bind itself on two sides such as $CH_2 = CH_2$ (ethylene) by breaking the double bond to get a chain '—CH_2—CH_2—'. Similarly, a trifunctional can bind at three sides.

Polymers can be:

1. Linear
2. Cross linked with trifunctional groups
3. Branched chain which will have branches growing from parent chain.

Polymers can be formed by addition reactions or by polycondensation reactions. If two different monomers are involved in the chain it is called copolymerization. The two polymers can be alternating, or alternating in blocks, or it can also be a random alternation.

The polymerization itself may consist of the following activities:

1. Initiation,
2. Propagation
3. Chain transfer
4. Termination.

Energy is usually required for polymerization to take place.

3.5.1 HDPE, LDPE and LLDPE

High Density Poly Ethylene (HDPE)

Polyethylene and polypropylene are polymers which are manufactured by the addition polymerization technique. Ethylene molecules are combined together in the following fashion to form polyethylene.

$$n CH_2 = CH_2 \rightarrow CH_3-(CH_2-CH_2-)_{n-1}-CH_3$$

There are two types of polyethylene, the HDPE which is hard and used for making items like bins, jugs, bottles, pipes, whereas LDPE is softer and used for items like carry bags, packaging, films, etc. Density of LDPE is 0.910–0.940 g/cm^3, whereas that of HDPE is 0.941 g/cm^3 or higher.

There are more types of polyethylenes. One of them, LLDPE, is used for making tough films. It has a straight chain structure but has short branches consisting of butane, hexane and octane made by copolymerization. Films of very low thickness can be made without compromising on strength.

Manufacture of HDPE

There are many methods of manufacturing HDPE, among which the autoclave method is more popular and hence described in Figure 3.23. Ethylene is purified and traces of oxygen, water and CO_2 are removed so that the catalyst is not poisoned. Aluminium triethyl, along with titanium tetrachloride, is used as the catalyst. The catalyst is mixed with purified ethylene and fed into a stirred autoclave where it gets mixed with recycled stream consisting of liquid separated by centrifuge. Polymerization takes place in the reactor at a pressure of about 7 kg/cm^2 and temperature of 70°C.

The effluent from the autoclave is flashed in the flash drum. The vapours are sent to a distillation section where light ends and heavy ends formed are removed and the pure ethylene is recycled back to the catalyst slurry tank. The polymer slurry taken from the bottom of flash drum is sent to a filter centrifuge. The polymer is separated and sent to the drier for drying. The dried HDPE is extruded and converted to powder, flakes or pellets as required for storage and sales.

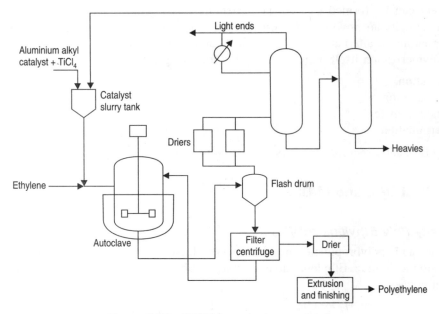

Figure 3.23 HDPE by autoclave method.

Manufacture of Low Density Poly Ethylene (LDPE)[8]

LDPE can be made either in an autoclave or in a pipe reactor which is essentially a double pipe with cooling water circulating in the annular space. There is better cooling and control of the chemical reaction in the pipe reactor. It is also more suitable for high pressures. The pipe reactor system is described here. See Figure 3.24. Purified ethylene vapours are compressed to a

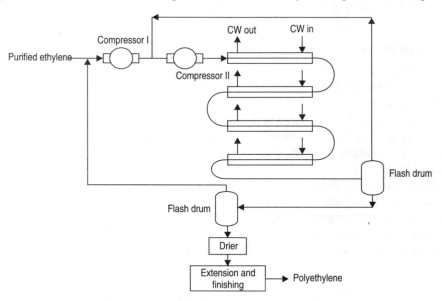

Figure 3.24 LDPE by pipe reactor method.

high pressure of about 300 kg/cm² in two stages and fed into a pipe reactor which consists of a jacketed pipe of 25 mm to 70 mm diameter and 200 to 300 m long, built in sections one on top of the other. Heat transfer fluid circulates in the annular space. The inlet temperature is about 150°C and the outlet temperature is controlled at about 250°C. The pressure at the outlet of the reactor will be about 200 kg/cm².

A reaction initiator (which can be either oxygen or a peroxide which can supply oxygen) is introduced between the first and second compressors to initiate the reaction. A 15 to 20% conversion is achieved which can be increased by providing an additional initiator at other downstream locations or at intermediate points in the reactor.

The high pressure and low pressure separators separate the vapours which are fed to the suction of the second and first compressors respectively.

The polyethylene separated from the second separator gets solidified and this product is extruded, made into powder, flakes or pellets as required for storage and sales.

Linear Low Density Poly Ethylene (LLDPE)

LLDPE has straight chains of CH_2 with many short branches of other comonomers consisting of higher olefins such as 1-butene,1-hexene or 1-octene. Hence, it is not linear in the true sense. It is stronger than LDPE, the short branches being made of comonomer.

A fluidized bed reactor is used for production of LLDPE[9]. See Figure 3.25. Ethylene, mixed with hydrogen and comonomer, is fed into a fluidized bed reactor, where the LLDPE particles

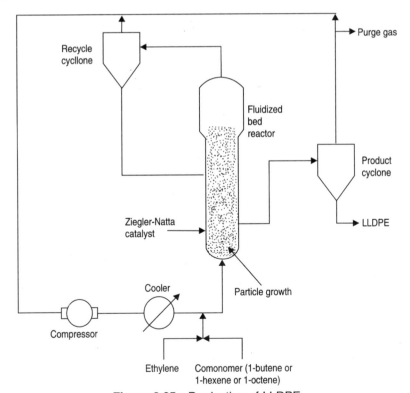

Figure 3.25 Production of LLDPE.

are formed on which the polymer grows with efficient removal of heat. A Ziegler-Natta or chromium-based metallocene catalyst is also fed into the reactor. The reactor is about 25 m high and about 3.5 m in diameter[10]. Polymer particles are separated from the gas in cyclone separators. The recycled gases are compressed and fed back to the bottom of the reactor. The residence time in the reactor is about 5 to 7 hours.

3.5.2 Poly Vinyl Chloride and Poly Vinyl Acetate

Poly vinyl chloride (PVC)

The processes for the manufacture of vinyl chloride and vinyl acetate are already covered in Section 3.5.2

The polymerization of vinyl chloride to PVC takes place as per the following equation.

$$n(CH_2 = CHCl) \rightarrow CH_3-(CH_2-CHCl-)_{n-1}-CH_2Cl$$

Poly vinyl chloride is made by either of two processes, which are 'suspension polymerization' and 'emulsion polymerization'.

Suspension polymerization: In this process, the vinyl chloride is dispersed in water with vigorous agitation. An initiator soluble in water is added to initiate the polymerization and the mixture is heated to 60 to 70°C. When 90% of the monomer is converted the reaction is halted and the slurry is discharged to a degasser vessel. The remaining monomer is returned to the reactor. The resin produced is filtered or centrifuged and dried.

Emulsion polymerization: In this method, equal parts of water and vinyl chloride monomer are taken and persulphate catalyst (1% of monomer) and detergent emulsifier (1.5% of monomer) are added to reactor. The reaction is conducted at 500°C for a long period (approximately 72 hours). The mixture is coagulated to produce PVC and dried. However if vinyl latex is to be produced, the liquid is stabilized by adding emulsifying agents. The PVC latex produced by this method has a fine particle size and is more suitable for use in paints, paper and fabric finishes, and printing inks.

Poly vinyl acetate

Among the many uses of poly vinyl acetate the most common is the glue used for furniture manufacture.

Vinyl acetate is polymerized to polyvinyl acetate by the following polymerization reaction.

$$n(CH_2 = CH \cdot COO\ CH_3) \rightarrow (-CH_2 \cdot COOCH_3-CH-)_n$$

This type of polymerization reaction where a double bond is broken to single bonds is called a free radical polymerization.

The above free radical polymerization reaction can be visualized as shown in Figure 3.26. The manufacture of poly vinyl acetate is done in a jacketed polymerization vessel. Vinyl acetate is dissolved in benzene in the polymerization vessel and heated by steam in the jacket of the vessel, where it is boiled under total reflux for a period of about 5 hours. The product is transferred after the solvent is removed and dried. The molten material is then extruded into rods and cut into flakes and packed to be sold as poly vinyl acetate.

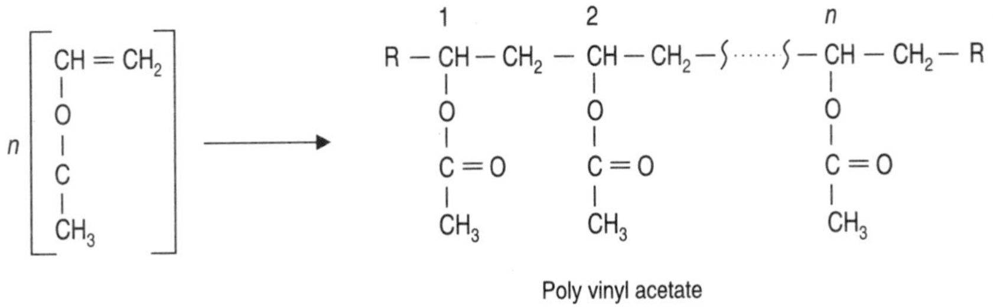

$$\left[\begin{array}{c} CH=CH_2 \\ | \\ O \\ | \\ C \\ | \\ CH_3 \end{array} \right]_n \longrightarrow \overset{1}{R-CH-CH_2} - \overset{2}{CH-CH_2} - \{\cdots\cdots\} - \overset{n}{CH-CH_2-R}$$

(with each repeating unit bearing $-O-C(=O)-CH_3$)

Poly vinyl acetate

NOTES
1. '*n*' is the number of monomers which is about 20,000
2. The two ends of the polymer will be capped by a hydrocarbon radical such as CH_3

Figure 3.26 Polymerization of vinyl acetate.

3.5.3 Oxo-biodegradable Polyethylene

Recent trends indicate that oxo-biodegradable polyethylene is being increasingly introduced into markets in cities, which can partly overcome the menace of the persistence of polyethylene in the land and water. It is claimed that the degradation time is reduced from hundreds of years to just three months. This is a claimed figure, but additional conditions might be required to initiate/propogate the degradation. If properly made, biodegradable polyethylene can prevent the occurrence of following.

1. Choking of drainages.
2. High death rate of cattle due to eating plastics.
3. Threat of extinction of marine wild life due to polyethylene being entangled and mixed up with its food. It has been predicted and also observed that most plastic waste will gradually migrate to sea and finally get trapped in the major ocean currents of Pacific Ocean, the north and south Atlantic Oceans and Indian Ocean.
4. Species reduction of birds for reasons similar to above.

While discussing the biodegradability of detergents we found that the problem was minimized by using only straight chain compounds for making detergents which can be broken down by natural forces in the environment. However, the same is not applicable to plastics since straight chain compounds are not actually made. LLDPE is, in fact, a copolymer and not sufficiently straight chain considering that copolymers form branches in their structure. If littered, they will in time affect water bodies and agricultural land.

However in oxo-biodegradable polyethylene, certain metals (transition metals such as cobalt, iron or manganese) are added while carrying out the polymerization. These metals get oxidized within a certain period such as three to six months. Suppose there are '*n*' iron atoms in a polymer chain, the chain may break into '*n*' pieces, in time, in the presence of oxygen, provided the reaction is initiated by sunlight/ultraviolet rays, etc. If this happens, the plastics, in turn, will lose their undesirable properties and blend with the soil.

Standards for oxo-biodegradable polyethylene

There is a Standard Guide (ASTM D6954) available which specifies procedures to test the degradability of plastics. Even though this standard is not met by oxo-biodegradable polymers in the market, it is likely to be met in the future with better additives. Further, the above standard is only a guide as opposed to a Standard Specification and it does not provide a pass/fail criteria.

It does have some use in deciding the future course of action on biodegradability of polyethylene. It may also be a basis on which expiry dates can be decided.

ASTM D6400 is a Standard Specification, but is appropriate only for compostable plastics, which are of course made from organic matter.

Recycling versus biodegradability

For each plastic we have to decide whether the recyclability path or the biodegradability path is better, mainly based on feasibility of operation. For example, recycling for polythene is very difficult since the cost of collecting the waste will be too high. This is mainly because a kilogram of used and crumpled polythene will occupy a very huge volume equivalent to about 20 buckets and the recycler can never get his effort rewarded. Hence, biodegradability is the answer. For a material like poly vinyl chloride, the opposite may be the case.

Mineralization

Mineralization is the process of complete biodegradation to ultimately form water, minerals, gases like methane and carbon dioxide. This process can occur only after initial cleavage of the long chains by biodegradation or by other processes. Poly lactic acid (PLA) and poly hydroxy butyrate (PHB) are important packing materials which are biodegradable and will get completely degraded or mineralized.

3.6 PESTICIDES

The pesticide industry is both a fast developing and fast changing industry adapting itself to changing environmental and pollution control requirements and at the same time protecting farmlands and agriculture.

Many pesticides are developed to act against specific pests as well as against pests in specific crops. While it is difficult to list all of them, a few are given below.

1. DDT (Dichloro diphenyl trichloroethane) and Dicofol (DDT converted to environment-friendly form) acts against malaria vectors such as mosquitoes.
2. Malathion also acts against mosquitoes but mainly used against agricultural pests.
3. Endosulphan acts against pests in coffee, tea and other plantations.
4. Monocrophos is very effective against pests in sugarcane and tobacco fields and also protects food storage barns.
5. Triazaphos is for protection of coffee, tea and cardamom plantations.
6. Cypermethrin, Acephate, DDVP, Phorate, Mancozeb are other pesticides which protect against pests in horticulture, vegetable, grape fields, etc.

3.6.1 DDT and Dicofol

DDT became a very popular insecticide during Second World War since it was used to increase food production by killing various pests and insects, thus helping in the war effort. However, in the 1970s its use came under heavy pressure from environmentalists since it was found to be non-biodegradable with consequent buildup of DDT in the food chain. However, if DDT is converted into Dicofol, it is biodegradable. Nowadays, Dicofol is used as pesticide instead of DDT. DDT's use in agriculture is banned in almost all countries.

DDT (Dichloro diphenyl trichloroethane)

DDT is a crystalline solid with formula, $(C_6H_4Cl)_2—CH—CCl_3$, and melting point between 108.5 and 109°C. It decomposes when heated further. It is very slightly soluble in water.

DDT is made by the condensation of two intermediates, chloral and monochloro benzene (MCB). These are produced as follows.

The intermediate MCB is manufactured by the chlorination of benzene in glass lined reactor vessels and by undertaking further purification.

Chlorination of ethyl alcohol results in the production of chloral alcoholate. This is distilled in the presence of oleum in a glass lined reactor to form chloral.

Then the condensation reaction is conducted in the DDT Reactor. Concentrated sulphuric acid or Oleum is used as the condensation catalyst.

$$2C_6H_5Cl \text{ (MCB)} + CCl_3CHO \text{ (Chloral)} \rightarrow (C_6H_4Cl)_2—CH\text{-}CCl_3 \text{ (DDT)}$$

The DDT obtained is cast into blocks, and powdered to obtain DDT powder.

Dicofol

Dicofol is a liquid with the formula, $(C_6H_4Cl)_2—CH—O—CCl_3$, and a molecular weight of 370.49. It is made by the replacement of CH group in DDT by the COH group, thereby adding an oxygen atom which is necessary for easy biodegradability. Dicofol is produced mainly in India, Spain and Israel. It is used as an acaricide in tea plantations and vegetable gardens.

Dicofol is manufactured by the selective oxidation of DDT in glass lined reactors as per four steps given below:

1. Dehydrochlorination of DDT to Dichloro Diphenyl Dichloro Ethylene (DDE) with elimination of one molecule of HCl.
2. Chlorination of DDE to Tetramer (Dichloro Diphenyl tetrachloro Ethane).
3. Hydrolysis of Tetramer to Dicofol using Para Toluene Sulphonic acid.
4. Dicofol is separated and vacuum distilled purified and formulated by adding suitable additives.

The product Dicofol, with the formula $C_6H_4Cl_2—COH—CCl_3$, is readily biodegradable since it contains the required oxygen atom that vastly helps in biodegradation.

3.6.2 Malathion and Parathion

Malathion and parathion are organophosphate compounds and are considered biodegradable pesticides, which mean they are superior in terms of biodegradation to other products which are non-biodegradable. Phosphorus, sulphur, chlorine, ethylene and benzene are the raw materials for the production these group of pesticides.

Malathion

Malathion is a yellowish brown liquid with the formula $C_{10}H_{19}O_6PS_2$. Its chemical name is dimethyl dicarbethoxy phosphorodithioate. It has the molecular weight of 330.36. Other properties are melting point of about 3°C and a boiling point of between 156 and 157°C. It is soluble in water to the extent of 130 milligrams per litre. Technical grade malathion is made as 98% malathion. Malathion formulations are available as water dispersable powder, made with 25% malathion and emulsion concentrate liquid. It is also available with 50% concentration.

Malathion is a popular biodegradable insecticide for application to field crops, fruit and vegetable gardens, flower gardens and for removal household parasites. It can even be applied on the hair to remove hair lice.

First, phosphorus pentasulphide is reacted with methanol in a reactor to produce dimethyl dithio phosphoric acid (DMPA). See Figure 3.27. Ethyl alcohol is reacted with maleic anhydride and benzene in the presence of sulphuric acid as the catalyst to obtain diethyl maleate.

Dimethyl maleate and dimethyl dithio phosphoric acid are fed into a reactor with heating/cooling jacket. The vessel is heated. The vapours produced are partly condensed and returned to the column. The reaction produces butanedionic acid diethyl ester which is called malathion. This product can be represented by the formula $(CH_3O)_2 — (P=S) — S — (COOC_2H_5)_2$.

After the reaction the product is taken out from the bottom of the reactor and dried and sent for making solid or liquid formulations.

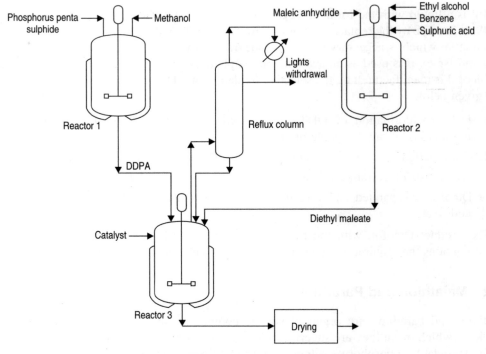

Figure 3.27 Production of malathion.

Parathion

Parathion is a white crystalline solid, but it is mostly sold as a brownish liquid. Its chemical name is diethyl 4-nitrophenyl phosphorothioate. It is synthesized from diethyl dithio phosphoric acid (similar to malathion where methanol substitutes ethanol) with ethanol. It is applied to cotton crops and other cash crops, but rarely used in food crops.

REVIEW QUESTIONS

1. How is methanol made? How is the heat produced in the reformer recovered?
2. Explain how phenol and acetone are coproduced.
3. How is LPG produced from refinery gases?
4. Describe the production of propylene from cracked LPG.
5. Write short notes on: Rayon, Nylon, Polyester, Phthali, and Anhydride.
6. Explain the steps in the manufacture of caprolactam and its conversion to nylon 6.
7. Describe the Sohio process for the production of acrylonitrile.
8. Can Liquefied Natural Gas (LNG) replace LPG for (1) industrial and (2) domestic applications?
9. Explain how viscose rayon is made.
10. What is acrylonitrile and how is it made?
11. Describe steam cracking of naphtha.
12. What is plastisizer? Explain the production of phthalic anhydride.
13. How can we make plastics biodegradable? Explain oxo-biodegradability.
14. Describe the process of making vinyl chloride.
15. Write short notes on: HDPE, LDPE, DDT, Dicofol, and Malathion.

Chapter 4 INORGANIC CHEMICALS

4.1 PRECIPITATED CALCIUM CARBONATE

As in the previous chapters we start with a simple and old but relevant technology along with a view on plant cost. Precipitated calcium carbonate ($CaCO_3$), also called 'chalk', happens to be an important ingredient of the human endeavour of teaching. It has been used for a very long time. Even today, in villages, artisans use collected seashells and make chalk from it. It is also heated to make quick lime (CaO) and then dissolved in water to get slaked lime, $Ca(OH)_2$, and this is used to white wash walls of homes/buildings, which on drying gets converted to chalk ($CaCO_3$) by absorbing carbon dioxide from air by the following reaction.

$$Ca(OH)_2 + CO_2 \rightarrow CaCO_3 + H_2O$$

Calcium carbonate is a white solid and has a molecular weight of 100.9. It is available as a powder of various micron sizes with a wide range of bulk densities up to 900 kg/m^3.

Some of the present-day uses are in the paper industry where precipitated calcium carbonate is added as pigment and filler. It improves the optical properties of paper. It is an additive to PVC and other plastics to increase strength. Material of less than 0.1 micron size is used in the manufacture of automotive and construction sealants. It is also used in toothpaste and as an antacid in medicine.

Limestone is a good source of precipitated calcium carbonate. Limestone is converted into calcium oxide and carbon dioxide by means of calcination at a temperature of 900°C. To ensure a high level of purity, the calcination process is carried out using natural gas. After the calcinated lime has been slaked with water, the resulting milk of lime is purified and carbonized with the carbon dioxide obtained from the calcination process as per the following chemical reactions.

Calcination

$$CaCO_3 \rightleftharpoons CaO + CO_2$$

Slaking

$$CaO + H_2O \rightarrow Ca(OH)_2$$

Carbonation and precipitation

$$Ca(OH)_2 + CO_2 \rightarrow CaCO_3 + H_2O$$

The carbonation and precipitation reaction results in forming a suspension of $CaCO_3$ in water. A cake comprising 40–60% solid matter (depending on particle diameter) is then obtained by filtration. This filter cake is then dried and subsequently disagglomerated in grinders.

4.1.1 Process Description

Limestone and charcoal are mixed and fed to a lime kiln where the mixture is heated by burning natural gas with air. See Figure 4.1. In the kiln, the calcium carbonate is converted into quick lime and carbon dioxide. The quick lime is slaked with water in the slakers and gets converted to pure slaked lime after removing sediments in the slaker itself as well as in the hydrocyclone. Then the slaked lime obtained is sent to a carbonator vessel.

The carbon dioxide produced in the lime kiln may contain dust particles which are removed in a dust catcher. It is also purified by scrubbing with water and fed to the carbonator. Slaked lime reacts with the carbon dioxide in the carbonator to form precipitated calcium carbonate. The product (precipitated $CaCO_3$) is centrifuged and dried in a tray drier and ground to the required size and sent to storage. A portion of the precipitated $CaCO_3$ can be activated using soap solution to produce activated calcium carbonate as a by-product.

The fineness of the grain as well as the crystal form (aragonite, calcite), is controlled by temperature, concentration of reactants and time.

Depending on the composition of the milk of lime (a fine suspension of calcium hydroxide in water) used and on the purifying stages during production, technical as well as foodstuff and pharmaceutical grades can be produced.

Design of the lime kiln

The length of the kiln depends upon the residence time required for first heating to the temperature of 900°C and then for the completion of reaction and then again exchanging heat to the incoming air.

The mechanical grate provided at the bottom is designed so that the product is withdrawn at the correct 'required rate' at any point of time.

Design of the tray drier

The tray drier consists of a cabinet with trays in it. Hot air or gas at optimum temperature must enter the tray drier for drying the calcium carbonate. The gas flows through the cabinet in a zig zag fashion so that the contact time increases and so also the heat given to the product. At the same time the back pressure on the blower is kept low to reduce power consumption. The trays should have a lifting mechanism such as lifting fingers or screw jacks. In the design of the drier, the overall cost has to be minimized, taking into account the labour costs too.

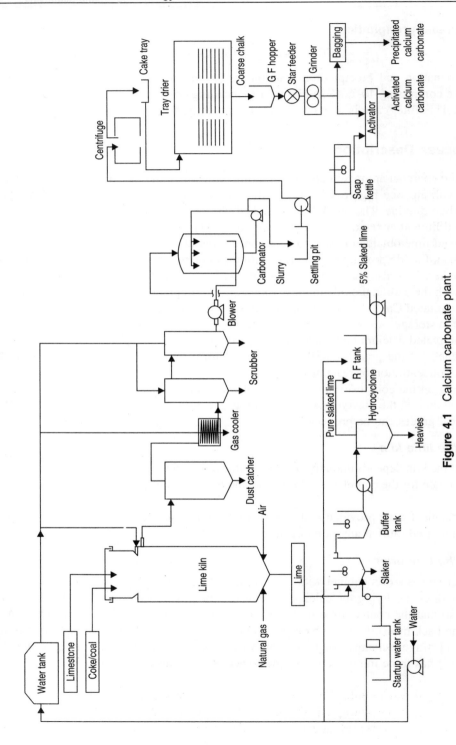

Figure 4.1 Calcium carbonate plant.

4.1.2 Case Study—A Small Capacity Plant

Raw materials and utilities

Quantity of $CaCO_3$ to be produced	= 5 tonnes/day
Limestone required (@ 1.6 tonne/tonne of $CaCO_3$)	= 1.6 × 5 = 8 tonnes/day
Coal for kiln (@ 0.65 tonne/tonne of $CaCO_3$)	= 0.65 × 5 = 3.25 tonnes/day
Water (@ 30,000 litres/tonne of $CaCO_3$)	= 30,000 × 5 = 150,000 litres/day

Capital fixed cost

(All costs are in rupees and given for illustration purpose only)

A. Calcination and slaking

Lime kiln (mixed feed/gas fired, including civil works)	= 2,250,000
Lime crusher	= 375,000
Bucket elevator	= 225,000
Feeder and hopper	= 120,000
Wet slaking system	= 330,000
Slaked lime enrichment and storage	= 240,000
Subtotal	= 3,615,000

B. Carbonation and finishing

Carbonation reactor	= 690,000
CO_2 gas compressors (2 nos.)	= 1,050,000
Settling tanks	= 180,000
Centrifuges (3 nos.)	= 1,440,000
Soap kettle (+ heating arrangements)	= 120,000
Drier (including civil works)	= 675,000
Soft pulverizer and air classifier	= 270,000
Packing, hopper, feeder balance	= 120,000
Subtotal	= 4,545,000

C. Auxiliaries

Pumps and piping	= 930,000
Instruments	= 180,000
Electrical	= 1,395,000
Subtotal	= 2,505,000

D. Building

Production shed	=	2,700,000
Office	=	300,000
Foundations drains and other civil works	=	225,000
Subtotal	=	3,225,000
Total fixed cost (A + B + C + D)	=	13,890,000

E. Optional items

Producer gas plant for lime kiln	=	1,500,000
Grand Total (A + B + C + D + E)	=	15,390,000
	≈	150 lakhs

Direct cost

Raw material	Quantity/tonne of $CaCO_3$	Rate (₹/tonne)	Cost (₹)
Limestone	1.6 tonnes	1500	2400
Coal			
Calcination	0.65 tonne	3000	1950
Drier	0.30 tonne	3000	900
Stearic acid and other chemicals	–	–	750
Packing material (40 bags @ 45 per bag)			1800

Utilities

Electrical energy (600 kWh @ ₹ 3 per unit)		1800
H_2O (30 m³ @ ₹ 15/m³)		450
	Total	10,050/tonne of $CaCO_3$

Indirect cost

	₹ (per month)
Interest on 150 lakhs capital @ 15% per annum	
Interest on 37.5 lakhs working capital @ 15% per annum	234,375
Contract labour and supervision	187,500
Salaries	45,000
Others	135,000
Sales expenses	135,000
Overheads	45,000
Transporting and forwarding	30,000
Total	811,875
Cost per tonne of limestone	5412.50

Profitability

(for 1650 tonnes of $CaCO_3$ per annum, 330 days)

Sales realization @ ₹18,000 per tonne (average)	= ₹29,700,000
Cost of direct + indirect production (10050 + 5412.5) × 1650	= ₹25,513,125
Gross profit	= ₹41,86,875
Return on investment	= 16.4%

Conclusion

The return on investment being reasonable, it is a profitable project. The above calculation does not account for debt equity ratio and interest during construction and some other smaller expenses. The selling price is a decisive factor in the profitability.

4.2 PHOSPHORUS

Phosphorus is an important plant nutrient and mainly used for making phosphoric acid by the process of oxidation and hydration. There are two varieties of phosphorus known as yellow (also called white) phosphorus and red phosphorus. Red phosphorus is obtained by heating yellow phosphorus. Red phosphorus has a higher stability and more resistance to oxidation than the yellow phosphorus.

The important properties of yellow and red phosphorus are given below:

Molecular weight : 123.9
Melting point   : 44.1°C (yellow), 593°C (red)
Boiling point   : 280°C (yellow)
Density of solid : 1.82 (yellow), 2.36 (red)
Density of liquid : 1.74 at 45°C (yellow)
Toxicity   : yellow variety is toxic.

Phosphorus is the second most important chemical for plant growth in agriculture. It is also used for making chemicals such as ammonium phosphate, calcium phosphate, sodium phosphate, phosphor-bronze and organic phosphates.

4.2.1 Electrochemical Process of Manufacture

Phosphorus is made from phosphate rock by the electric arc process, which is described below.

Phosphate rock consists of calcium phosphate and calcium fluoride. The rock is ground into powder and mixed with powdered silica and carbon. This mixture is fed into an electric furnace. See Figure 4.2.

The chemical reaction that takes place can be expressed as follows:

$$2Ca_3(PO_4)_2 + 6SiO_2 + 10C \rightarrow 6CaSiO_3 + 4P + 10CO$$

Phosphorus and carbon monoxide come out of the furnace as gases. These gases pass through a precipitator where dust is taken out, and then through a spray tower where cold water is sprayed. The

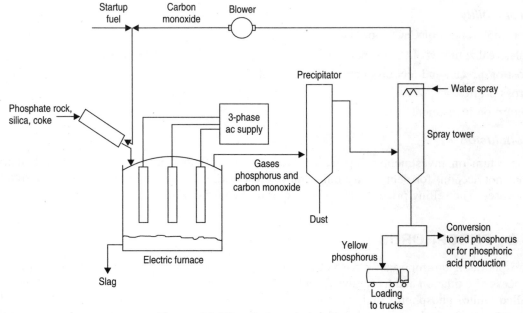

Figure 4.2 Production of phosphorus.

phosphorus condenses, which is collected and taken out. The carbon monoxide produced is used as fuel. Calcium silicate and calcium fluoride are also formed, which come out as slag from the electric furnace. Slag can be processed and used as construction material.

The fluorine present in rock is also converted to silicon fluoride and then to hydrofluoric acid which can be recovered in an effluent treatment plant. Since iron which is present as an impurity in rock combines with phosphorus and forms a liquid ferrophosphorus, this can be tapped out from the furnace as per requirement.

The electric furnace technology described above can also be employed for making phosphorus along with phosphoric acid. In this case, additional equipment is installed to convert part of phosphorus to P_2O_5 and then to phosphoric acid using burners, hydrators, and precipitators to separate other gases from the acid.

4.3 PHOSPHORIC ACID

Phosphoric acid, also called orthophosphoric acid, has a density of 1.885 in the pure form. It is extensively used for the manufacture of phosphatic fertilizers. There are also many other uses for phosphoric acid, such as a food chemical and for the manufacture of soaps and detergents.

Gypsum which is a by-product in the manufacture of phosphoric acid can be used for making building panels which can be used for low-cost housing, plastering of walls, agriculture and pottery, etc.

Phosphoric acid is primarily produced by a "Wet Process", whereby rock phosphate is ground and fed into attack tanks and digested with sulphuric acid. The following chemical reaction occurs:

$$Ca_3(PO_4)_2 + 3H_2SO_4 + 6H_2O \rightarrow 2H_3PO_4 + 3CaSO_4 \cdot 2H_2O$$

The reaction produces phosphoric acid and gypsum. The gypsum precipitates and is filtered. The equipment required includes attack and digestion tanks, flash coolers, filtration units and a phosphoric acid concentration section.

There are, in fact, many processes for the production of wet process phosphoric acid. Some of these processes are given below.

- Dihydrate (meaning two molecules of water per molecule or compound)
- Hemihydrate (meaning two molecules of compound per molecule of water)
- Hemidihydrate process
- Hemihydrate recrystallization (HRC) process
- Dihydrate/hemihydrate (DH/HH) process

Each of the above processes has its own advantages and disadvantages as explained below.

Dihydrate process

The dihydrate process consists of grinding the rock, reacting with concentrated sulphuric acid with agitation in multiple vessels or in a single vessel with compartments. The temperature is controlled, using a flash cooler. Continuous moving vacuum assisted filters are used for separating gypsum. The initial separation is followed by two stages of washing to recover maximum acid. The acid is concentrated by evaporation using heat exchangers, flash chamber, condenser, a vacuum pump and a circulation system. The heat exchangers are mostly made of graphite or stainless steel. The tanks are made of rubber lined steel.

The advantage of dihydrate process is that it is easy to operate and that wet rock can be used. The product acid concentration obtained is between 27 and 32% P_2O_5.

The gypsum obtained can be converted to good quality gypsum boards (or fibre boards by adding fibre) which can be safely used for rapid civil construction purposes.

Hemihydrate process

In the dihydrate process, the gypsum is precipitated as dihydrate whereas in hemihydrate process it is precipitated as hemihydrates. There is saving of energy because somewhat coarse material in rock grinding is sufficient for this process, but the filtration step is more difficult. For certain rock characteristics, this method is more suitable. The phosphoric acid concentration obtained is more than 40%, P_2O_5, which contains two molecules of compound per molecule of water, for example, $2CaSO_4 \cdot H_2O$.

Hemidihydrate process (HDH process)

In this process, calcium sulphate is first precipitated as hemihydrate and the phosphoric acid is separated. Then the hemihydrate is recrystallized with more water as dihydrate.

HRC process

This process, called the hemihydrate re-crystallization process, recrystallizes the hemihydrates to dihydrate even before separating the hemihydrates, which thus saves one filtration step.

DH/HH process

One more method called the dihydrate/hemihydrate method is available in which case the acidulation is conducted under dihydrate conditions and calcium sulphate ($CaSO_4$) is recrystallized to hemihydrates.

4.3.1 Description of the Dihydrate Process

See Figure 4.3. Rock phosphate is ground to the required size of 90% through 100 mesh of Taylor series sieve. The ground rock is fed at a regulated rate, using a weigh feeder, into the first compartment of the attack tank. Here a large quantity of slurry is maintained in circulation. Sulphuric acid (98%) and recycled phosphoric acid are also fed, which results in the formation of fresh phosphoric acid and gypsum. One portion of the hot slurry is circulated through an evaporator cooler called flash cooler. It removes the excess water from the system. It also controls the temperature of slurry to prevent the formation of hemihydrates.

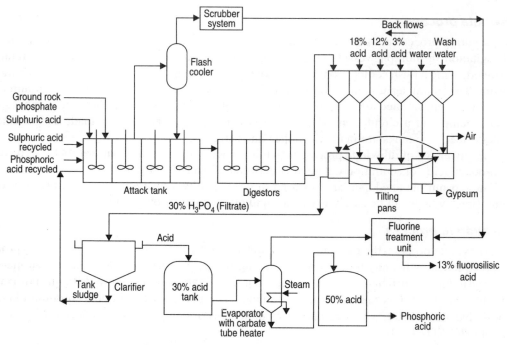

Figure 4.3 Phosphoric acid production.

Another portion of the slurry overflows from the attack tank to digestion vessels that provide residence time to form gypsum crystals. The gypsum crystals are filtered by the aid of vacuum rotating pan filters and then disposed of. The tilting pan operation consists of a series of horizontal, independent trapezoidal pans mounted on a circular structure. This arrangement rotates and is under vacuum during the filtration cycle and inverts under air blow to discharge the cake. The cake is washed countercurrently in three stages with decreasing concentration of phosphoric acid. In the last stage, the filter cloth is dried by evacuating the wash water. Then the pan is brought back to its original position.

The use of 'belt filters', instead of 'tilting pan filters', is also widely prevalent.

The filtrate which is the phosphoric acid goes to the 30% acid settler clarifier. After this the cakes are washed countercurrently, first with water. The water yields 5% acid and this is again used to yield 12% acid and again to yield 18% acid moving countercurrently. This acid goes back to the attack tank along with sulphuric acid as weak recycled acid.

The 30% P_2O_5 acid is concentrated in a concentration section. Forced circulation evaporators are used. These evaporators are provided with Karbate tubed heat exchangers. The water is evaporated under vacuum and the final concentration will be about 50% P_2O_5.

Some of the fluorine content in the rock phosphate is liberated in the attack tank during reaction with sulphuric acid. However, a major portion of the fluorine gets liberated in the evaporator and is recovered by a fluorine recovery system as fluorosilisic acid having 13% H_2SiF_6. The fluorine liberated in the attack section is also recovered in a similar scrubber system. Both streams are sent to storage. More details are given in books[1,2] for further study.

4.3.2 Simulation of the Phosphoric Acid Process

A typical simulation flowsheet is shown in Figure 4.4. The flowsheet represents the addition of phosphate rock, sulphuric acid, and return acids in a multicompartment reactor system. A flash cooler maintains the desired temperature by evaporation of water in the slurry. The reactors are considered adiabatic in nature. The kinetics of the gypsum formation are considered first, followed by the kinetics and growth of crystals. The simulation provides a practical and an evolutionary approach for giving accurate results. The simulation given below is for dihydrate process and the modifications need to be done for other processes, particularly in the precipitation of gypsum with the correct number of water molecules attached to the gypsum.

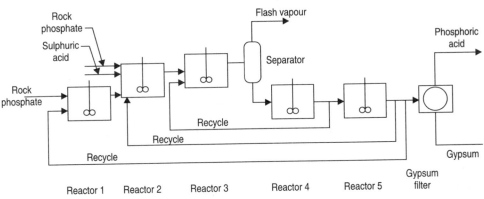

Figure 4.4 Phosphoric acid plant simulation flow diagram.

A simulation package such as 'ASPEN PLUS' and 'DESIGN 2 for Windows' provides electrolyte models for simulation. The requirements for starting a simulation are flowsheet, components list, properties, unit operations, phosphate rock ionized component details.

Apart from the components in the feed and products, the following ionic components should also be fed to the system:

$$Ca^{++}, SO_4^{--}, H_3O^+, HSO_4^-, H_2PO_4^-, OH^-$$

The thermodynamic option to be selected for simulation should be NRTL (Non-Random Two Liquid) Electolyte.

A normal practice in the development of all simulations is that you first model the system with a lesser complication and then gradually increase the complication to the necessary levels. For example, distillation trays are first fed as theoretical trays, by the number of plates given as half of the actual. Once you converge the column with this, then the actual trays are entered. The reaction tanks are modelled as Equilibrium Tanks intitially. Later, they are added as Stirred Tanks to incorporate the kinetics of gypsum crystallization. All the required materials and energy streams are also entered.

Stoichiometry

The principal constituent of phosphate rock is fluorapatite which contains calcium, phosphate, fluoride, carbonate ions held together in a crystal lattice. When the rock is treated with sulphuric acid, the phosphate constituent is solubilized as phosphoric acid. In the simulation, the phosphate compound is treated as a mixture of fluorapatite $Ca_{10}(PO_4)_6F_2$ and calcium phosphate (trical) $Ca_3(PO_4)_2$. In most cases, fluorapatite is the dominant component, but we may also add small amounts of trical to accurately describe the measured rock composition. The carbonate component is treated as calcium carbonate and other fluorides in excess of fluorapatite as calcium fluoride. Impurities such as SiO_2, Al_2O_3, Fe_2O_3, Mg, Na_2O are also included. As already mentioned, all ionic species should be included in the process.

The solubilization of fluorapatite is considered as below:

$$Ca_{10}(PO_4)_6F_2 + 12H_3O^+ \leftrightharpoons 10Ca^{++} + 6H_2PO_4^- + 12H_2O + 2F^-$$

Solubilization of other rock species is done in the same manner.

$$CaCO_3 + H_2O \leftrightharpoons Ca^{++} + CO_2 + 2OH^-$$

$$Al_2O_3 + 3H_2O \leftrightharpoons 2AlOH^{++} + 4OH^-$$

$$Fe_2O_3 + 3H_2O \leftrightharpoons 2FeOH^{++} + 4OH^-$$

$$MgO + H_2O \leftrightharpoons Mg^{++} + 2OH^-$$

$$Na_2O + H_2O \leftrightharpoons 2Na^+ + 2OH^-$$

$$SiO_2 + 6F^- + 2H_2O \leftrightharpoons SiF_6^{--} + 4OH^-$$

Water is dissociated as follows:

$$2H_2O \leftrightharpoons H_3O^+ + OH^-$$

Acids are dissociated as follows:

Sulphuric acid:

$$H_2SO_4 + H_2O \leftrightharpoons H_3O^+ + HSO_4^-$$

$$HSO_4^- + H_2O \leftrightharpoons H_3O^+ + SO_4^{--}$$

Phosphoric acid:

$$H_3PO_4 + H_2O \rightleftharpoons H_3O^+ + H_2PO_4^-$$

Precipitation of gypsum occurs according to the following equation:

$$Ca^{++} + SO_4^{--} + 2H_2O \rightleftharpoons CaSO_4 \cdot 2H_2O$$

The solubility of gypsum in the acids requires a special consideration. The amount of gypsum precipitated depends on the equilibrium constant for solubility of gypsum. (This equation needs changes in hemihydrate and hemidihydrate processes.)

The model is developed starting from the first reactor (a compartment of attack tank) with assumed values for recycled streams. Each reactor should be converged before going to the next reactor. The flash vaporizer is modelled as a separator. Overall convergence should be obtained for the whole flowsheet which will give the final recycle values.

Concentration of phosphoric acid

Weak phosphoric acid obtained as above can be concentrated to the required extent in evaporators using steam as the heating medium. This process can be modelled as heater and flash vaporizer.

Thermodynamics

Electrolyte NRTL option available in ASPEN PLUS or DESIGN2 for Windows is used. Many parameters in the model can be estimated. The temperature variation of equilibrium constants of solubilities may be derived from published values[1]. ASPEN PLUS data bank gives many of the model parameters.

Application

A very valuable application of the above simulation is for mass and energy balances. The model will be able to calculate production rates, acid compositions for specified rocks, sulphuric acid concentrations, true electrolyte compositions and aqueous phases for various conditions. This simulation is useful for design of new systems as well as analysis of existing systems.

4.4 HYDROGEN

Hydrogen is the first element in the periodic table. It is present in water and innumerable other compounds. It is an odourless inorganic gas with a boiling point of $-252.8°C$. It is slightly soluble in water. Its lower and upper flammability limits are 4.1% and 74.2% respectively.

The chemical reaction between steam and fossil fuels (hydrocarbons) produces a mixture of carbon monoxide and hydrogen, two of the very reactive basic building chemicals of the modern chemical industries. This reaction as well as the water gas shift reaction, serves as the front end of production of many chemicals. The reactions are used in the production of four major basic chemicals which are hydrogen, ammonia, methanol as well as for the production of a group of other chemicals known as oxo-alcohols. In refineries, hydrogen is used for hydrogenation, hydrocracking and hydrodesulphurization processes. Hydrogen is also used in the hydrogenation of vegetable oils.

Apart from many general uses of hydrogen, it may also become the fuel of the future for automobiles. The main reason is that hydrogen engine gives out only pure water as an effluent. However, safe storage of hydrogen for automobiles is a subject under intense research.

4.4.1 Process Description

A hydrogen plant consists of the following sections:
1. Reforming of hydrocarbons
2. CO conversion to produce further hydrogen by shift reaction
3. Hydrogen purification by pressure swing adsorption (PSA)

Reforming of hydrocarbons

The reforming of natural gas uses two simple reversible reactions:

$$CH_4 + H_2O \rightarrow CO + 3H_2 \qquad \Delta H = +206 \text{ kJ/mol}$$

$$CO + H_2O \leftrightharpoons CO_2 + H_2 \qquad \Delta H = -41 \text{ kJ/mol}$$

The forward reforming reaction is endothermic, favoured by high temperature and low pressure. Reforming requires large amounts of energy. About 3500 kcal of heat is required to reform 1 kg of natural gas. An energy intensive operation of this type also opens up better opportunities for optimization of energy.

Refer to Figure 4.5. A stochiometric mixture of hydrocarbon and steam is admitted into an array of reformer tubes, loaded with nickel catalyst. In the reformer tubes, conversion of hydrocarbons and steam into hydrogen and oxides of carbon takes place. The endothermic heat of reaction requirement is supplied by burning fuel by burners situated outside the tubes.

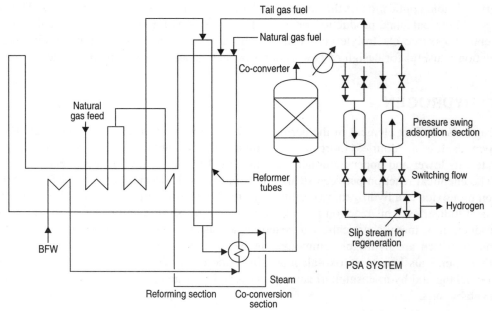

Figure 4.5 Hydrogen production plant.

CO conversion

The gas from the secondary reformer is sent to shift reactors to produce further hydrogen from the CO conversion reaction. The shift reaction takes place in the CO shift reactors to form CO_2 and H_2 from CO and H_2O:

$$CO + H_2O \rightleftharpoons CO_2 + H_2 \quad \text{(forward reaction exothermic)}$$

The equilibrium is favoured by lower temperatures and more steam, while the reaction rate will be higher at higher temperatures. To get optimum results, the CO conversion is performed in two steps in the high and low temperature shift reactors.

Hydrogen purification by pressure swing adsorption

The PSA unit (Pressure Swing Adsorption unit) essentially consists of adsorbent vessels containing the molecular sieves, surge tanks and a microprocessor-based control unit. The control unit provides a staggered sequence of cyclic operations which gives a constant hydrogen product stream as well as a constant tail gas stream. The tail gas stream is recycled to the reformer and used as fuel. The operation of the PSA unit involves the following steps:

1. Adsorption step where the impurities in the feed hydrogen stream are adsorbed in the bed.
2. Depressurization.
3. Purge step.
4. Repressurization step.

The actual arrangement of vessels is the proprietary design of the PSA system vendors.

Typically, there will be many PSA vessels in multiples of two in the PSA unit. Just for illustration, a simplified two-bed operation option will be as given in Figure 4.6.

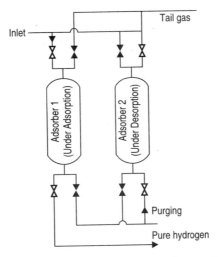

Figure 4.6 Two-bed adsorber unit operation.

One adsorber is shown in the adsorption (Adsorber 1) mode which produces high grade hydrogen by adsorbing the other components. The second adsorber (Adsorber 2) is shown in regeneration mode wherein a slip stream regenerates from the adsorber as it gets heated, stripping off the heavies, a portion of which goes back to the plant as recycle.

The adsorption is carried out in a vertical fixed bed of adsorbent, with feed gas flowing down through the bed. The process is not instantaneous, which leads to the formation of a mass transfer zone within the bed.

At any particulat time, there are three zones in an adsorbent bed. It consists of the equilibrium zone at the top where no additional adsorption occurs due to saturation, the mass transfer zone where the mass transfer takes place and the active zone at the bottom, where adsorption is yet to take place, as shown in Figure 4.7. The switchover is done when the mass transfer zone reaches the bottom.

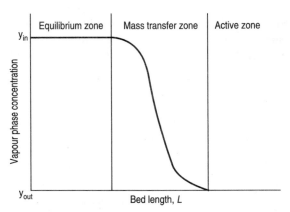

Figure 4.7 Vapour phase concentration profile of an adsorbate.

The PSA unit is capable of obtaining 99.999% purity hydrogen, highest among all methods, and fit for use in fuel cells.

It may be noted that hydrogen is considered as a fuel of the future since vehicles running on hydrogen fuel cells produce zero pollution.

The product hydrogen is obtained at about 18 kg/cm^2 and 45°C. It is pressurized by using a compressor to the desired pressure for transportation/sales.

Design considerations for reformers (this portion applies to hydrogen, ammonia and methanol plants)

The reforming reaction takes place inside tubes filled with nickel catalyst, the endothermic heat of reaction being supplied by burning fuel on the outside of tubes. Basically, three types of reformers are usually built—top fired, side fired and terrace walled (see Figure 4.8).

In top-fired reformers, the tubes are arranged in many rows and the burners are also placed in rows between the tube rows and also on either side. The advantage of this type of arrangement is its compactness.

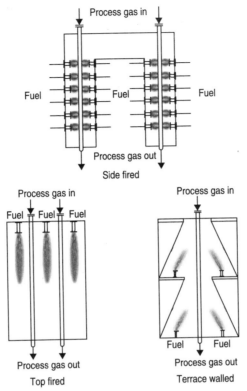

Figure 4.8 Reforming types.

The side-fired reformer consists of either one or two rows of tubes with a number of burners placed horizontally on both sides of each row of tubes. The number of burners required for side-fire reformers is many times more than that required for top-fired reformers. However, the burners are of smaller size. The space required for a side-fired reformer is also much higher than that for a top-fired reformer.

In the terrace-walled reformer, firing is done in an upward direction from two levels: one at the bottom and another at the middle of the reformer.

Reformers for production of hydrogen and methanol work at higher temperatures, typical outlet temperatures being 850–875°C. Ammonia plants have secondary reformers which operate by the addition of air and hence the primary reformers operate at lower temperatures, typical outlet temperature being about 800°C.

Laying out the tubes for a reformer is an exercise in optimization. For a typical new reformer, the pitch of tubes ranges from 2.0 to 2.5 tube diameters for top-fired reformers, whereas it is 1.5 to 2 diameters for side-fired reformers. The new trend is to reduce the pitch. High pitches are not only wasteful of space but it also becomes difficult to keep the flow of flue gases uniform across the cross section, which is important to achieve uniform skin temperatures. However, keeping the pitch very less will increase the circumferential temperature difference factor and thereby reduce the life of tubes. Distance between the tube rows and burner rows are usually fixed by the flame diameter.

Two feedstocks

Either of two feedstocks are generally used for steam reforming. These are natural gas and naphtha. The catalyst activity will be more in the case of natural gas than that for naphtha. Stoichiometrically, naphtha is a better raw material for production of methanol (not for hydrogen and ammonia), but considering the international prices of naphtha and methanol it is difficult to competitively make methanol from naphtha, in a greenfield plant.

Nickel catalyst is used for reforming natural gas, whereas nickel catalyst with some potash is used for reforming naphtha. Potash is incorporated to prevent carbon laydown on catalyst by cracking of hydrocarbons in naphtha. The potash has a tendency to get carried over along with the process gas and hence a guard catalyst is required to catch the potash. Hence, almost all naphtha reformers use two catalyst beds in the same tubes — one is the top layer containing potash and the other is the bottom layer without potash.

Tubes, conversion and temperatures

The progress of the chemical reaction is shown in Figure 4.9 and temperature changes in Figure 4.10.

Ideally, the heat added to a reformer should be absorbed by the feed at an equal rate in such a way that a uniform temperature is obtained across the length of the tube, so that the tube thickness required is minimum.

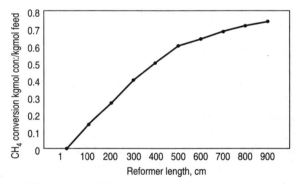

Figure 4.9 CH_4 conversion across reformer.

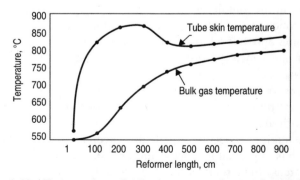

Figure 4.10 Temperature distribution across the top-fired reformer.

Practically, however, the temperature changes as the reaction proceeds. This makes it important for us to find out the variation in temperature with length. A mathematical model can be built using reaction kinetics, thermodynamic equilibrium, radiation and convection heat transfer equations to find out the heat flux and the tube skin temperatures along the length of the tube. A computer program can be developed based on the mathematical models, which gives the temperatures, heat fluxes and composition at every cm length of the tube. Using this calculation the reformer design can be optimized and both top-, side- and terrace-walled reformers can be designed which can result in near ideal temperature profiles.

The trends of methane conversion, tube skin temperature, process gas temperature across the length of the reformer for a typical ammonia plant and typical hydrogen/methanol plant are shown in Figures 4.9 and 4.10.

Reformer tubes are the costliest part of a reformer and the cost of reformer tubes amounts to about 30% of the total cost of the reformer. Reformer tubes are made by various alloys, based on the percentages of Cr, Ni, etc. Most new reformers are made from niobium stabilized chrome-nickel alloy in the ratio 25:35. The reformer tube material should possess:

1. Good creep rupture strength
2. High creep rupture ductility
3. Good resistance to oxidation
4. Good weldability

The design of tube thickness is normally done by trial and error procedures. First, a particular thickness is assumed. Then, the maximum temperature that the tube of this thickness can withstand for 100,000 hours of operation, using proven creep stress values for the particular material selected, is found out. The design temperature is fixed based on the maximum temperature obtained by computer simulation of the tube skin temperatures. The two temperatures are compared and a new tube thickness is selected and the procedure is repeated until the maximum temperature that the tube can withstand is just above the design temperature. This procedure ensures that we do not end up with a high value of tube thickness which is wasteful and inefficient.

The design of burners for reformers should take into consideration not only the heat released per burner but also the diameter and the length of flames. The proper maintenance and control of burners need to be stressed, on which depends the life of a reformer. For top-fired furnaces, the flow of flue gases should be essentially parallel to the tubes till the end of the tubes. Any unevenness in the construction of reformer boxes can also lead to unevenness of flue gas flow, which can then be a factor for causing hot spots on the tubes. The flue gas tunnels (for top-fired reformers) at the bridge walls should be adequately sized and made perfectly even.

While all reformers have inlet pigtails, outlet pigtails are optional. In reformers without outlet pigtails, the tubes are directly connected to the outlet headers. This saves both money and space. A disadvantage of this arrangement is that the individual tubes cannot be pinched out of use, in case of tube failure. However, it is found that in well-run modern reformers such individual tube failures are not encountered in the 100,000 hours (designed) duration of life of tubes.

Reformed gas boiler

From the reformers (primary and secondary in the case of ammonia plants) the reformed gas passes into the reformed gas boiler (sometimes called process gas boiler) through a transfer line.

Care needs to be taken for the design and construction of transfer line since it handles gas at very high temperatures. The line can be water jacketed to reduce the heat loss.

The reformed gas boiler is a simple horizontal water tube boiler made of carbon steel tubes. The gases are cooled down to about 370°C in the boiler. The boiler has a central bypass for temperature control in case of hydrogen and ammonia plants since the gases should enter the high temperature shift converter at a fixed temperature. (A methanol plant does not have this bypass since there is no shift converter in a methanol plant.) The steam drum is mounted on top of this boiler. There may be other boilers in the flue gas side (also called convection side) of the reformer and also after shift converter (except in methanol plants where shift is not performed). Hence, the elevations at which the steam drum is placed depends upon the position and number of boilers it serves. The inlet and outlet circulation lines of boilers need to be designed carefully. The amount of circulation obtained through the boilers depends upon the balancing of the pressure drop with the static head differential of the inlet and outlet pipes.

Convection section design

In any 'top-down' design approach the entire plant has to be visualized first, before taking up the design of the parts.

The typical arrangements of convection section for hydrogen, methanol and ammonia plants are shown in Figure 4.11. The overall arrangement is finalized using pinch technology[2]. However, it should be noted that no single arrangement can be said to be the best since it depends upon on the utility systems of the plant complex, which again depends upon the climate of the place and availability of water and other dependent factors.

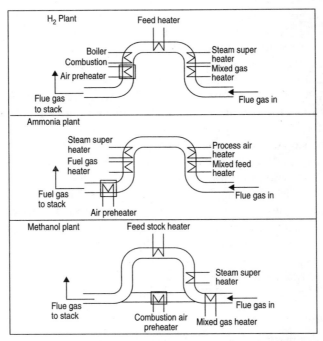

Figure 4.11 Typical convection section of different reformers.

The calculations of individual convection section coils are done serially. It is inevitable that finned tubes are used in convection section since the outside film coefficient (h_o) tends to be much below the inside film coefficient (h_i), and this is offset by increasing the outside area of tubes by using fins. There are four types of fins in normal use. They are stud fins, solid fins, serrated (or cut) fins and longitudinal fins. Sketches of each type are shown in Figure 4.12.

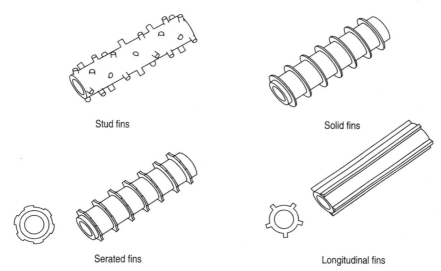

Figure 4.12 Extended surface tubes.

It is a good practice to use only bare tubes (without fins) in coils with convection section temperatures above 750°C. Carbon steel fins are used only if the fin-tip temperature is below 450°C. Various alloy steel fins can be used for intermediate temperatures and stainless steel fins up to fin temperatures of about 720°C.

The prediction of heat transfer at finned tube bundles is best based on empirical correlations derived from experimental data which are given in handbooks such as HTFS[6] handbook. For bare tubes, the average Nusselt number in cross flow over a bundle of smooth tubes is obtained by the equation suggested by Gnielinski[3] for turbulent flow.

The determination of duct side pressure drop of coils is important. For finned tubes the pressure drop is best computed by correlations given by Robinson and Briggs[4].

The material of construction for convection section depends on the temperatures encountered. SS 321 H is used for services where the design temperatures are above 700°C. For lower design temperatures, SS 304 H, P22, P11, A 209-Ti and carbon steel tubes are used.

The combustion air preheater is the last exchanger in the convection train. Corrosion is encountered in the air preheater due to the presence of sulphur in the feed which gets converted to sulphur dioxide and sulphur trioxide in the furnace, and condenses on the tubes if the temperature of the tube is below the flue gas dew point. The dew point can be found from the composition of the flue gases.

Proprietary models of air heaters are available with materials such as carbon steel, glass lined steel, cast iron aluminium, etc. In selecting air heaters, compactness is a major criterion. Large air preheaters tend to get damaged during transportation. Once damaged it is almost impossible to find out where it has occurred and a good portion of air will bypass from FD fan to ID fan through the damaged portion, thereby increasing power consumption and also limiting the capacity of the reformer.

The penthouse temperature should be kept in mind while laying out a reformer. The penthouse should have good cross ventilation. Further, cold air can be pumped into the penthouse to keep the temperature under control especially for people attending on the burners. This will act as a morale booster for the maintenance crew attending on burner maintenance since maintaining the burners in excellent conditions is one of the keys to good reformer performance.

Improvements

The conventional reforming may be replaced by heat exchange reforming when the next generation of plants gets constructed. In heat exchange reforming the heat generated in the secondary reformer heats the primary reformer. The advantages are compactness, heat efficiency, lesser compression costs, etc. The only disadvantage (which does not apply to ammonia plants) is the necessity of incorporating an air separation plant for making oxygen for feeding into the secondary reformer.

4.5 AMMONIA

Haber and Bosch, two scientists from Germany, were responsible for developing a process for making ammonia by combining nitrogen and hydrogen at a very high pressure in the presence of catalysts. However, pure hydrogen and pure nitrogen are costly to manufacture. Undeterred by this, they continued experimenting, turning to hydrocarbons and steam as raw materials, which finally resulted in the 'high pressure ammonia synthesis' process in the year 1913. The only other remarkable development since then has been that ammonia is now produced by 'low pressure ammonia' process, thereby saving on compression costs.

Ammonia, NH_3, is a pungent smelling gas at room temperature. It is soluble in water. The other important properties of ammonia are given below.

Molecular weight = 17.03

Boiling point = $-33.4°C$

Melting point (at atmospheric pressure) = $-77.7°C$

Ammonia is an important fertilizer intermediate. All major fertilizers like ammonium sulphate, ammonium phosphate, ammonium nitrate and urea require ammonia. Reacted with sulphuric acid and phosphoric acid it forms ammonium sulphate phosphate mixture which is an ideal fertilizer for a variety of agricultural crops.

4.5.1 Process Description

An ammonia plant consists of a reforming section and a synthesis section. The reforming section is similar to the one already described under Section 4.4. An additional 'secondary reformer' is incorporated in the case of ammonia plants. In the secondary reformer, air is added at the outlet of primary reformer, as a source of nitrogen. The oxygen in the air is used to burn part of hydrocarbons producing carbon dioxide which provides the heat to the conversion reaction.

Ultimately after absorption and desorption for removal of carbon dioxide and purification (for which there are many alternative methods not described here), a mixture of hydrogen and nitrogen in the molar ratio of 3:1 (synthesis gas) is obtained.

'Ammonia synthesis' is the heart of an ammonia plant. The process flow diagram of ammonia synthesis with explanation is given in Figure 4.13.

Synthesis loop

The feed to the ammonia systhesis section consists of synthesis gas which is a mixture containing three volumes of hydrogen and one volume of nitrogen. The synthesis gas (makeup) is compressed in two stages to a pressure of about 80 kg/cm^2 and joins the circulating loop just before the second ammonia chiller. This mixture enters the synthesis loop to join with the circulating unreacted gas recycled to the synthesis converter. In the ammonia synthesis converter, the nitrogen and hydrogen present in the gas react to form ammonia which is separated in the refrigeration circuit. The chemical reaction taking place is:

$$3H_2 + N_2 \rightleftharpoons 2NH_3 \quad \text{(forward reaction is exothermic)}$$

The above reaction is an equilibrium reaction and therefore only part of the hydrogen and nitrogen will be converted into ammonia while passing through the catalyst beds. The composition of gas at chemical equilibrium is influenced by pressure and temperature. Thermodynamically, high pressure and low temperature tend to favour ammonia formation. But the rate of reaction will be low at low temperatures and more contact time will be required and hence the quantity of the catalyst required will be more. An optimum temperature should be adopted to obtain maximum conversion efficiency.

Historically, the normal operating pressure in the synthesis loop has ranged from 80 to 240 kg/cm^2 and the normal temperature in the catalyst beds has ranged from 250 to 500°C.

Modern plants operate at low pressures, thereby saving substantial compression energy.

The gases coming out of the synthesis gas converter will be at a high temperature as explained above. It has to be cooled to a low temperature for liquifying ammonia. The cooling is done by transferring heat to other fluids in the cooling train. The cooling train consists of a loop boiler, boiler feed water heater, hot hat exchanger, water cooler, first cold exchanger, first ammonia chiller, second cold exchanger and second ammonia chiller.

Inert gas purging

In order to avoid build-up of inert gases such as argon in the synthesis gas, a part of the synthesis gas is purged from the synthsis loop. The quantity of the purge is kept so as to match the inert gas content in the incoming gas.

The purge gas contains ammonia vapour, unconverted nitrogen, hydrogen and inerts such as argon and methane. The ammonia from the purge gas is recovered by cooling the gas in a purge gas cooler and purge gas chiller and then sent through a purge gas absorber. Part of the hydrogen recovered from the purge gas recovery unit is used in the pre-desulphurizer unit for hydrogenation and the rest is mixed with the makeup synthesis gas in the synthesis compressor. The offgases from the purge gas recovery unit containing argon, methane, hydrogen and very small amount of ammonia are used as secondary fuel in the primary reformer.

Refrigeration circuit

The ammonia produced in the synthesis converter is condensed in a water cooler, two cold exchangers and two ammonia chillers which are part of the refrigeration circuit.

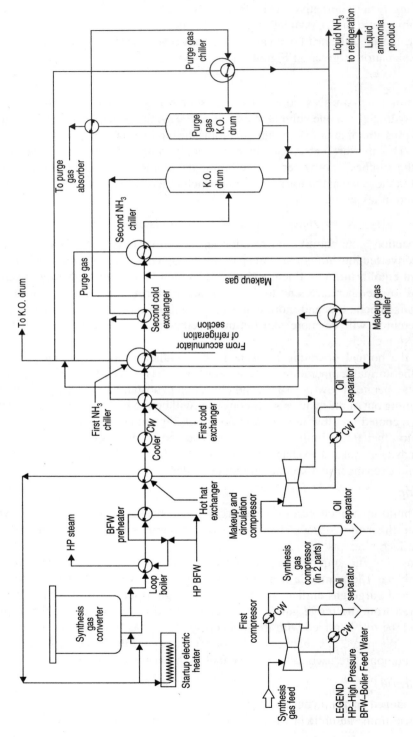

Figure 4.13 Ammonia synthesis.

In the refrigeration circuit, there are many chillers in which cooling of ammonia is effected by evaporation of ammonia at different temperatures.

The uncondensed ammonia vapour is recycled to the synthesis converter along with makeup gas from methanator.

4.5.2 Ammonia Plant—A Case Study

The investment cost, the cost of production and profitability given below is for an ammonia plant of capacity 1350 tonnes per day to be set up in a rice cultivating region.

A sulphuric acid plant and a phosphoric acid plant along with a complex fertilizer plant needs to be set up, resulting in the manufacture of ammonium sulphate phosphate mixed fertilizer. The ammonia will be sold to these units and the sales price is taken as ₹21,000 per tonne. The economics is shown only for the ammonia plant.

First, the capital cost of equipment and machinery forming the plant is estimated from in-house cost data after obtaining cost from vendors of major equipment. To this, add erection, civil works, land development, engineering expenses, commissioning costs, duties and taxes, etc. to get the total investment required.

Second, the variable cost of production of ammonia is arrived at by taking the appropriate consumption ratios of raw materials and utilities and multiplying by the unit costs of the same. Cost of labour, maintenance, interest on working capital, insurance, etc. are also accounted for. Thus, the total cost of production of ammonia is calculated.

Capital investment, cost of production and profitability

Considering the realization price (selling price) of ammonia as ₹21,000 per tonne, the profit per annum is estimated and the payback period is evaluated. See Tables 4.1 and 4.2.

Table 4.1 Investment summary

S.No.	Identification	Investment (₹ in lakhs)
1.	Equipment and Materials	
	1.1 Main Plant	50,012
	1.2 Offsites	17,684
	1.3 Catalysts and Chemicals	5276
	Subtotal (1.1 to 1.3)	72,972
2.	Erection and Subcontracts	13,550
3.	Civil Works	6892
4.	Excise Duty, Sales Tax	5228
5.	Freight and Insurance	4076
6.	Customs Duty	10,682
7.	Land and Land Development	1384
8.	License and Basic Engineering	2880
9.	Detailed Engineering	6516
10.	Preliminary and Preoperatives	2394
11.	Working Capital	3626
	Subtotal (1 to 11)	130,204
12.	Preproduction Interest	6988
13.	Contigencies	
	13.1 Civil Works (3% of Serial No 3)	206
	13.2 Others (4% of Subtotal of 1 to 11)	5208
	Total Investment (1 to 13.2)	142,606

Table 4.2 Cost of production and profitability [1350 TPD (tonnes per day), 330 days/year]

S. No.	Identification	Quantity Q	Cost per unit or Rate (₹) R	Cost of production (₹ in lakhs)
A	Variable Cost (* indicates multiplication)			
	1.1 Natural Gas (Q*R*1350*330)	0.68 tonne	14,600	44,230
	1.2 Water (Q*R*1350*330)	12 m^3	0.1	5.34
	1.3 Power (Q*R*1350*330)	5 kW	5	111.38
	1.4 Consumables (Q*R*1350*330)	1	424	1889
	Total Variable Cost			46,235.72
B	Fixed Cost			
	1. Labour and Overheads			1804.8
	2. Maintenance			6106.24
	3. Interest on Working Capital			718.16
	4. Insurance			930.6
C	Total Fixed Cost (excluding depreciation and interest)			9560
D	Fixed Cost per tonne (excluding depreciation and interest)			2146
E	Cost of Production (excluding depreciation and interest)			55,794.68
F	Cost of Production per tonne (excluding depreciation and interest)			12,524.06
	Depreciation		5%	7130.3011
	F1 Depreciation per tonne			1600.52
	F2 Interest on Loan per tonne			4164.82
	[50% of Investment/days (330)/TPD (1350)]			1921.00
G	Total Cost of Production (F + F1 + F2) per tonne			15,045.52
	Total Cost of Production per Annum		(G*330*1350 /10^5)	67,625.00
H	Sales (1350TPD*330 days)	445,500	21,000	93,555.00
I	Profit per Annum (H – G)			25,930
J	Payback Period in Years (Total Investment/Profit per Annum) = 5.5 years			

The project is attractive since the investment will be paid back in about 5 to 6 years. The IRR (Internal Rate of Return) can also be evaluated using standard software which will come to a value of around 15%. However, selling ammonia at ₹ 21,000/tonne is difficult without government subsidy.

4.6 SODA ASH

Soda ash, chemical name 'sodium carbonate' with chemical formula, Na_2CO_3, is a stable solid material with molecular weight of 106 and melting point of 852°C. It decomposes on further heating. Maximum density of solid material in anhydrous form is 2.54 g/cc, but it is much lower for normally available hydrated forms. There are two commercial varieties supplied in powder form: light soda ash with bulk density of about 450 kg/m^3 and dense soda ash with bulk density about 950 kg/m^3.

Soda ash is used in the manufacture of glass, sodium silicate, pulp and paper, chemical detergents, dyes and textiles.

There are three methods for the production of soda ash.

1. Solvay process, using limestone and common salt as raw material.
2. Dual process, also using limestone and common salt as raw material.
3. Exploitation of underground deposits of soda ash.

The Solvay process, which is most popular is described below in detail.

4.6.1 Solvay Process

In the Solvay process, limestone (calcium carbonate, $CaCO_3$) is reacted with common salt in the presence of ammonia which acts as an intermediate to produce soda ash and calcium chloride.

The overall chemical reaction can be picturized as given below.

$$CaCO_3 + 2NaCl \rightleftharpoons Na_2CO_3 + CaCl_2$$

Common salt, limestone and coal along with makeup ammonia are the materials required for the process.

Process description

See Figure 4.14. At the centre of the Solvay process plant is the lime kiln into which limestone is fed along with coke and heated to between 950°C and 1050°C. The following chemical reaction takes place.

$$CaCO_3 \rightarrow CaO + CO_2 \quad (1)$$

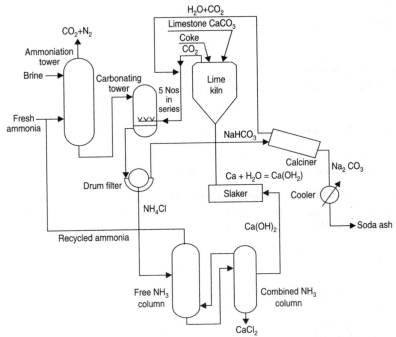

Figure 4.14 Manufacture of soda ash.

The calcium oxide (also called quick lime) which comes out at the bottom of the kiln falls into a lime slaker. The carbon dioxide produced is led to a vertical carbonating tower.

The other raw materials are sodium chloride and ammonia. The sodium chloride is dissolved in water to produce brine which is fed to an ammoniation tower. Ammonia vapour is passed into the bottom of this tower and the ammonia gets dissolved in the brine solution. This brine solution is fed to the top of the carbonating tower. To the bottom of this tower, the carbon dioxide produced by reaction (1) is introduced. The following chemical reaction takes place.

$$NaCl + H_2O + NH_3 + CO_2 \rightarrow NaHCO_3 + NH_4Cl + H_2O \qquad (2)$$

The product obtained is a solution of ammonium chloride in water and the insoluble sodium bicarbonate which remains as suspended matter in this solution. This slurry is sent to a drum filter where the sodium bicarbonate is separated and sent to a calciner and the ammonium chloride solution is sent to the twin ammonia columns consisting of combined ammonia column and free ammonia column.

The quicklime that was produced as per reaction (1) is dropped into a slaker where it is slaked with water and sent to the twin ammonia columns.

The reaction between the ammonium chloride and the calcium hydroxide slurry takes place as per the following reaction, all in water solution.

$$2NH_4Cl + Ca(OH)_2 \rightarrow 2NH_3 + CaCl_2 + 2H_2O \qquad (3)$$

Free ammonia separated in the free ammonia still of twin columns is recycled to the ammoniation tower bottom. Calcium chloride comes out of the combined ammonia still of the twin columns and is sold or disposed.

The insoluble sodium bicarbonate obtained in reaction (2) is sent to the calciner. The following reaction takes place in the calciner, forming soda ash (Na_2CO_3).

$$2NaHCO_3 \rightarrow Na_2CO_3 + H_2O + CO_2 \qquad (4)$$

The water and carbon dioxide produced in the calciner are recycled back to the carbonating tower. The soda ash (sodium carbonate) product obtained is cooled and sent to storage for bagging and sale.

Other processes and sources of soda ash

Dual process: In the dual process, ammonium chloride is taken as a product instead of recycling ammonia. It is precipitated and crystallized, centrifuged dried and sold as a by-product.

Natural soda ash deposits: There are huge deposits of natural soda ash mineral in many places. One of them is in Wyoming in the USA. Either it can be mined by the conventional method or solution mining can also be undertaken wherein it is dissolved in hot water and taken out.

4.7 GLASS

Glass is a mixture of silica, soda and lime. Obsidian is a naturally occurring glass. It is learnt that ancient Egyptians, and Tamils to some extent, knew the technique of making glass. In the early days, it was also used for cheaper quality ornaments. It is used extensively for making window panes, the practice of which started during the middle ages.

According to physics, glass can be called an amorphous solid or a solidified liquid. When a liquid mixture of silica, soda and lime is cooled sufficiently but rapidly, the orderly crystallization is prevented and instead the disordered non-crystallized configuration is frozen into a solid state. Its molecules do not bind as in solids and allow photons to pass through, making it transparent. Glass is similar to a supercooled liquid, but has all the other physicomechanical properties of a solid. Most glasses break easily.

The usefulness of glass arises from its transparency, resistance to chemical attack, electrical insulation property and its ability to contain vacuum. Hundreds of combinations of raw materials can provide different types of glasses. Magnesia (magnesium oxide) is for 'sheet glass' or alumina (aluminium oxide) for 'bottle glass'. Fused silica is an excellent glass but expensive because of pure silica's very high melting point and consequent cost of equipment. Lead oxide is used for fine sparkling tableware. Other specialized glasses are optical, photosensitive, metallic and fibre-optic glass for making fibre-optic cables.

4.7.1 Glass-making Furnaces

Glass is made by melting the correct mixture of raw materials in a furnace, drawing it out, cooling and cutting to shape. The usually used furnaces are of two types—pot furnace and tank furnace.

Pot furnace

Pot furnace is used in batch processes. The furnace is of pot shape, made of special clay or platinum. When platinum is used, melting of the pot itself is completely avoided. The pot furnace is mainly used for optical glass and art glass. The melted material is cast to different shapes as required.

Tank furnace

The tank furnace is fed batchwise but the final products are drawn out continuously. At the centre of the furnace a pool of molten glass is formed due to the heat of firing of fuel from two sides alternatively in cycles. The sensible heat of flue gases is largely recovered by giving heat to incoming air and fuel. The outcoming molten glass is formed or shaped by several methods.

4.7.2 Glass Blowing and Automation

Earlier, glass used to be blown by glass workers into moulds for making glass. Later, a blast of compressed air was used for this job. In the suction feed method, the molten glass is drawn into the moulds, where it waits for a period of time and is then dropped down automatically. The operation is conducted at a speed of about 1 bottle per second.

In another automatic method, there are two rotating tables called Parison-mould table and blow table. The molten glass flows through a small orifice where it is cut to required length and fed through a funnel into the Parison-mould. Then the mould switches to the blow table where two blows of air take place. The settle blow settles the glass in the mould and the counter blow makes it into the desired bottle form. It is reheated and a final blow is made to make the exact shape. The Parison-mould opens and the bottle comes out.

4.7.3 Glass Rolling (for Plate and Float Glasses)

Molten (melted) glass from melting furnaces at a very high temperature is passed between two rollers and comes out as a continuous ribbon, and then it is rolled on smaller water-cooled rollers. The stretching and shrinking of glass occur due to cooling aids used in flattening of glass. It is then annealed and cut into required lengths. These glasses can be used for automobiles and buildings or for any other application.

Float glass is made on a pool of molten tin in an inert atmosphere. The melted glass passes through a pool of molten tin at a desirable temperature. The glass is cooled while still on the pool of tin. This produces a float glass without any irregularities, both sides being flat and parallel.

4.7.4 Annealing and Finishing

Annealing is done to reduce the strain developed by the glass during the manufacturing stage. It is done by first holding the glass at a particular critical temperature and slowly cooling it to the room temperature. The rate of cooling is carefully controlled.

If one side of the glass sheet is cooled rapidly and at the same time the other side very slowly, then the side which is cooled quickly will have stress fault lines but the glass will be very strong due to the other side being cooled very slowly. Such glass will not break easily and when it breaks there will not be any sharp edges, since the other side has fault lines along which it breaks into rectangular cuboid pieces. This type of glass can be used on windshields of vehicles.

Finishing may consist of one or many of the operations such as cleaning, grinding, polishing, sandblasting, etc.

4.7.5 Other Types of Glasses

Very high purity fused silica glasses are made, which have negligible thermal expansion. Hence, they are suitable for making items of high accuracy such as space telescopes. Photochromatic (also called photochromic) glasses are those glasses that darken in sunlight. These are useful for making cooling glasses. There are others that fade in the dark and are used for night vision glasses.

Fibreglass is an important glass-based material used for strengthening plastics for making sports goods and other applications. It serves as an additive to composite materials. The fibres produced can be very thin, about ten microns in diameter.

'Glass ceramics' are produced by a process called 'controlled crystallization'. The most important property of glass ceramics is their ability to sustain repeated quick temperature changes up to 800–1000°C. They can also be used for making stoves and cooktops apart from scientific applications.

'Chalcogenide glass' contains elements such as sulphur, selenium and tellurium. The physical properties of this material make them ideal for mouldable improved opticals such as optical fibres. They are useful for rewritable optical discs and nonvolatile memory devices such as PRAM[5].

LCD computer screens

There are many types of LCD display screens which are made of layers of glass, typically two of them coated with polymers, one having a liquid crystal layer rubbed on the side, again followed by glass filter, polarizing film and cover glass. The types of glass used are sodalime glass, borosilicate glass or boroaluminosilicate glass.

4.8 CAUSTIC-CHLORINE PRODUCTS AND METALS

Electrical energy is used in chemical industries not only to drive electrical machinery and devices, but also to cause chemical reactions.

Caustic soda and chlorine, metals such as aluminium, sodium, potassium, magnesium, cadmium, chromium, copper, nickel, etc. are also manufactured by electrolytic methods. Many of them are described in this chapter. The cost of electrical energy plays a major part in the decision-making process of setting up plants and generally all these industries are established in places where electrical energy is available in abundance.

4.8.1 Caustic Soda and Chlorine

Caustic soda and chlorine are manufactured by the electrolysis of brine which is essentially salt water.

The properties of caustic soda and chlorine are as given underneath:

Property	Caustic soda	Chlorine
Molecular weight	40	70.9
Melting point	318°C	–101.6°C
Boiling point	1390°C	–34.2°C
Toxicity	Toxic	Toxic

Caustic soda is used for making pulp and paper, polyvinyl chloride, paraffinwax, pesticides, and rayon grade pulp, etc. Chlorine is also used for the treatment of water. Caustic soda was earlier produced by the lime soda process, but now it is completely produced by electrolysis of brine.

The industry grew very rapidly in the last century but its growth slowed down in the later part due to lack of demand for chlorinated plastics and solvents. One reason could be evolution of chlorine containing gases if chlorinated plastics are burned as litter.

There are three types of electrolysis cells which can be used for the production of caustic soda and chlorine. These are:

1. Mercury cell
2. Diaphragm cell
3. Membrane cell.

The basic chemistry for production of caustic soda is given by the following equation.

$$NaCl + H_2O \rightarrow NaOH + 0.5H_2 + 0.5Cl_2$$

The above electrochemical reaction happens in different ways at the anode and the cathode of the cell as shown below separately. However, in all the three cases half a mole each of chlorine and hydrogen are produced per mole of sodium hydroxide produced.

Reaction at anode

A. For mercury, diaphragm and membrane cells

$$Cl^- - e \rightarrow 0.5Cl_2$$

where 'e' represents an electron.

Reactions at cathode

A. For mercury cell

$$Na^+ + e + Hg \rightarrow NaHg \text{ amalgam}$$
$$NaHg + H_2O \rightarrow NaOH + Hg + 0.5H_2$$

B. For diaphragm and membrane cells

$$Na^+ + H_2O + e \rightarrow NaOH + 0.5H_2$$

Description of cells

Mercury cell: The mercury cell is mainly made of rubber coated steel. It is a rectangular box (see Figure 4.15(a)) with the anode plate supported from the top. The cathode consists of a stream of mercury flowing at the bottom from one side to another. Saturated brine is fed from one side. The sodium ion generated gets amalgamated (meaning united) with mercury and the mixture goes to the denuding tower to separate the two and recycle the mercury. The sodium combines with water to form sodium hydroxide, and hydrogen is liberated. Chlorine gas produced at the top goes to the chlorine drier.

Diaphragm cell: The diaphragm cell is a rectangular box with a copper base plate on which the anode is supported by copper plates from the bottom. It has a diaphragm across the middle of the rectangular box (see Figure 4.15(b)) made by asbestos fibre deposition. The brine enters the anode chamber. The sodium ions generated will pass through the diaphragm and taken out as sodium hydroxide solution. The chlorine is liberated in the anode chamber and goes to the chlorine drier.

Membrane cell: The membrane cell consists of a rectangular box separated by a membrane (see Figure 4.15(c)) consisting of two layers: per-fluoro carboxylic acid and poly-fluoro sulphonic acid. The anode and cathode plates are situated in two compartments. The saturated sodium chloride solution is fed to the anode compartment. The sodium ions migrate to the cathode. At the start of the operation this compartment is filled with 32% NaOH. This is converted to 35% by NaOH absorption. A part of it is taken out as product and the other part is diluted with water and sent back to the compartment. In this way, continuous production of 35% NaOH is maintained.

Description of overall process

The overall process for diaphragm cell plant is described below. See Figure 4.16.

Common salt stored in bins is dissolved in water to saturation level by directing water jets on the salt, and the brine formed is drained to a pit and pumped to a brine storage tank. The brine is purified by removing magnesium, calcium and barium salts by adding caustic soda, soda ash, barium carbonate, etc. The sludge is removed from the bottom of the purifier tank. The brine is again filtered and heated and sent to the diaphragm cells.

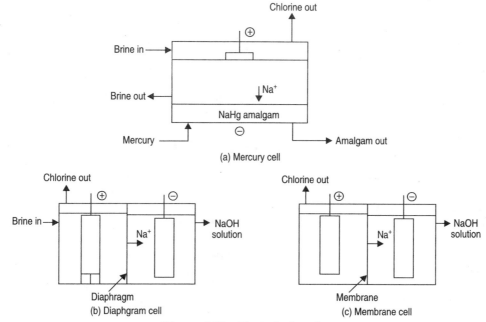

Figure 4.15 Electrolysis cells.

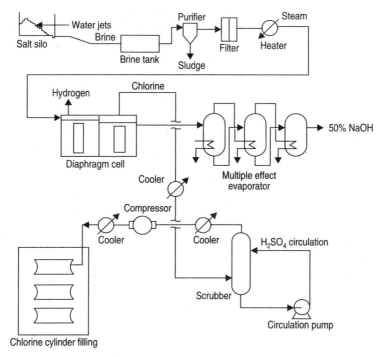

Figure 4.16 Production of caustic soda and chlorine.

As per electrolysis already explained, a 35% NaOH solution obtained is concentrated to 50% NaOH by means of multiple effect evaporators and then sent to storage for sale. It can also be further concentrated and made into flakes, granules or blocks or sold as liquid to many user industries such as textile, soap, paper, refineries and a host of other users.

The chlorine gas obtained from the cells is first cooled and then dried using sulphuric acid circulating through an absorber. Then it is compressed, cooled again and filled in cylinders for sale.

4.8.2 Hydrogen Chloride and Hydrochloric Acid

Hydrogen chloride is a gas at normal temperatures. An aqueous solution of hydrogen chloride is called hydrochloric acid. Concentrated hydrochloric acid is a 38% solution of HCl in water. Higher concentrations may result in the evaporation of HCl.

Hydrochloric acid is an important acid used for the processing of metals. It is widely used in the pickling of steel. Pickling is done so as to remove the iron oxide scale formed on the surface of iron and prepare the steel for coating and painting operations. Calcium chloride, nickel chloride and zinc chloride are made by treating the particular metal with hydrochloric acid. These chlorides are used in electroplating and galvanizing industries.

Hydrogen chloride is manufactured by directly burning chlorine with hydrogen. The reaction involved is:

$$H_2 + Cl_2 \rightarrow 2HCl$$

The reaction is highly exothermic and spontaneous. The reaction is carried out in a reaction chamber with a chlorine burner and the product is taken out as gas from the top at about 200°C. The HCl gas is then absorbed in water using graphite absorbers to produce hydrochloric acid of strength more than 30% HCl for delivery to consumers in road tankers.

Hydrochloric acid from organic chemicals production plants

Hydrogen chloride and hydrochloric acid are also obtained as by-products in many plants producing various organic compounds. For example, while chlorinating benzene to monochlorobenzene, hydrogen chloride is obtained as a by-product. The main reaction occurring is shown below:

$$C_6H_6 + Cl_2 \rightarrow C_6H_5Cl + HCl$$

In such operations, the by-product HCl is absorbed in water, purified and sent for sale. HCl is also produced while fluorinating chlorides.

4.8.3 Metals

Metals are chemical substances with high electrical and thermal conductivity. Characteristically, they are also shiny, malleable, fusible and ductile. Electrolytic processes are common in metal production.

Metals are generally alloyed before they are put into use. This is dealt within the subject of metallurgy. The properties of these materials depend on the proportion of components present while alloying them. However, chemical engineers have to put in their knowledge and get involved in the various production processes of metals.

Metals have a wide variety of uses including making of cans and utensils, cables and wires, pots and crockery, bells and jewellery, pipes, equipment and machinery, vehicles, ships, aeroplanes, bridges, guns, etc.

Groups of metals are generally known by certain names, based on the periodic table of elements, which are useful for identifying their properties. A partial list is given below.

1. Alkali metals, for example, sodium (Na) and potassium (K). Alkali metals are soft, with low density and low boiling points, with white or silvery colour and are highly reactive.
2. Alkaline earth metals, for example, calcium (Ca) and magnesium (Mg). These are metals having two electrons in the outermost shell, with resultant characteristics.
3. Light metals, for example, aluminium (Al) and scandium (Sc). Light metals have low densities.
4. Transition metals, for example, iron (Fe), copper (Cu) and zinc (Zn). They react readily and form numerous compounds with various colours. They also form stable complex ions. Some metals such as lead (Pb) and tin (Sn) are further sub-classified as post transition metals in the periodic table.
5. Metalloids, for example, boron (B) and silicon (Si). They are nearest to nonmetals in the periodic table and have properties between those of metals and nonmetals. One of them, silicon (Si), is covered in Chapter 7.

The characteristics and production processes of the above mentioned example metals (Na, K, Ca, Mg, Al, Sc, Fe, Cu, Zn, Pb, Sn) are given below very briefly.

Sodium

Sodium is a silvery white metal with a specific gravity of 0.971. It reacts with water violently. It is stored under a blanket of inert gas. It is used in the metallurgical industries as a reducing agent. It is also used in the manufacture of sodium vapour lamps and sodium alloys.

Sodium is manufactured by splitting sodium chloride (common salt) into sodium metal and chlorine gas by the below given chemical reaction. The sodium chloride is electrolyzed in a special electrolytic cell called Down's cell. See Figure 4.17.

$$2NaCl \rightarrow 2Na + Cl_2$$

The voltage required is small, only about 4 volts, but the current required is very large, about 40,000 amperes.

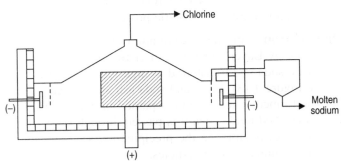

Figure 4.17 Production of sodium.

As in the case of aluminium, another salt, calcium chloride is introduced. Thus, the melting point of sodium chloride which is about 800°C, is brought down to 600°C. The cell does not produce calcium metal because the electro-winning of sodium occurs at a less negative cathode potential than does the electro-winning of calcium.

It is interesting to note that calcium is also produced in a manner similar to sodium by using calcium chloride as raw material.

Potassium

Potassium is a silvery white metal with a specific gravity of 0.862. Potassium is an important ingredient for the growth of plants and vegetation. It is an important ingredient in the development of leaves and barks. When leaves are heated to a high temperature the residue left behind is mainly potassium. Because of this observation the ash obtained was called pot ash or potash. This is the origin of the word potassium. Different essential salts are made from potassium. Potassium is also used as a heat transfer alloy along with sodium (Na-K).

Potassium is made by the reaction of high temperature sodium at atmospheric pressure with molten potassium chloride.

$$Na + KCl \rightleftharpoons K + NaCl$$

A continuous process is used. Molten potassium chloride is introduced into a packed column and brought in contact with ascending sodium vapours. There will be a reaction zone and a distillation zone. In the reaction zone, equilibrium vapour of sodium and potassium is produced. A distillation zone, above the reaction zone separates the lighter boiling potassium to pure potassium. The sodium chloride formed is continuously withdrawn from the reaction zone below.

Aluminium

Aluminium is a silvery metal with a specific gravity of 2.6. It is one of the most useful metals used by humans as it combines a high strength with a low density. Produced in large quantities the metal has become cheap and useable because of its low weight, high strength, high electrical conductivity, high ductility, high temperature and corrosion resistance. It has thus become an essential part of our industrialized world. Aluminium has been slowly replacing steel for many applications. It has already replaced iron in many day-to-day applications like cooking vessels, and in high-end applications like aircraft bodies.

The production of aluminium from bauxite consists of two steps:

1. Refining of bauxite to produce commercial alumina
2. Production of aluminium by electrolytic process.

Production of alumina from bauxite: Aluminium oxide (Al_2O_3) is called alumina. Alumina is produced from bauxite by Bayer's process (Josef Bayer, 1888). The ground bauxite is made into a slurry with caustic soda and transferred to digesters which are kept under high pressures and temperatures (see Figure 4.18). Lime and soda ash are also added to the digesters. Sodium aluminate ($NaAlO_2$) is formed. The solution is passed through clarifiers where flocculants are added to remove sand and red mud. Then it is passed through filters to remove the balance red mud, and then fed into precipitators. In the precipitators, seeds of alumina hydrate are added which form into bigger alumina hydrate particles.

The spent caustic is returned to the digesters. The alumina hydrate crystals are sent to calcination kilns. In the kilns, the hot gases at a temperature of 960°C reduce the alumina hydrate to alumina. The high grade alumina obtained is sent to a cooler and then stored in bins and sent to consumers. On an average about 10% of alumina produced goes into the making of refractory bricks and the remainder 90% for making aluminium.

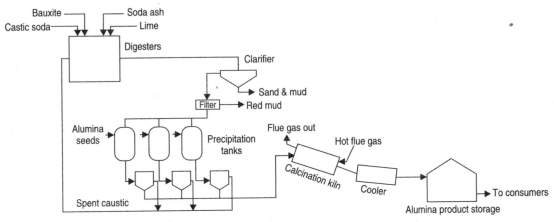

Figure 4.18 Refining of bauxite to produce pure alumina.

Production of aluminium from alumina: The raw material for the manufacture of aluminium is bauxite which contains alumina or Al_2O_3. The cryolite, a sodium aluminium fluoride salt ($3NaF \cdot AlF_3$) required for dissolving alumina is fed to the reduction pots along with finely powdered bauxite. To reduce the melting point of cryolite, an excess of aluminium fluoride is also added. The resulting mixture melts at about 900°C and dissolves the incoming bauxite.

A large number of aluminium reduction pots are lined up in rows depending on the capacity of the plant. The pot consists of an insulating container made of steel lined with graphite with cathode collector embedded at the bottom.

An electric bus bar that runs over the rows of reduction pots supplies the current (see Figure 4.19). The current is passed into the electrolyte by means of a set of three or more electrodes dipping into the molten electrolyte. The following overall chemical reaction takes place, carbon being consumed from the anode electrodes.

$$2Al_2O_3 + 3C \rightarrow 4Al + 3CO_2$$

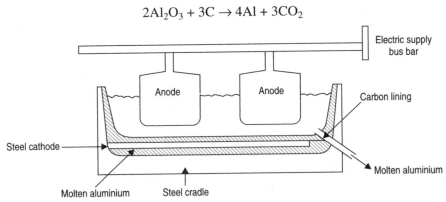

Figure 4.19 Aluminium reduction pot.

Molten alumina, being heavier, gets separated at the bottom of the pot and is drained off. The voltage required is only about 7 volts, but the current required is very large, say, 100,000 amperes.

There are two types of electrolytic cells in general use. One is called the Hall Heroult, which has multiple, three or more anodes dipping into the pot. In the other type, a single large anode is used with more facilities for adjustments.

The aluminium produced contains 99.7% aluminium. If super pure aluminium is required, the aluminium can be further electrolysed to increase its purity.

Aluminium formation from alumina in the pots takes place at about 900°C. However, the melting point of aluminium is only 660°C. In many production units, this high temperature heat is used to melt the recycled metal, which is then blended with the new metal. Recycling of aluminium is widely practised and the quality of aluminium is not affected if the recycled material is used.

The aluminium pots are kept in operation all round the year. If a pot cools down and the melt solidifies, it is extremely difficult to restart it.

Scandium

Scandium is mainly used for alloying during the production of aluminium. The scandium oxide is first converted to scandium fluoride and then it is reduced with calcium to form scandium. Only a few kilograms are required to be produced in a year as pure scandium.

Magnesium

Magnesium is a silvery white metal which is light in weight (specific gravity 1.738). Its main industrial use is in the production of aluminium magnesium alloys. Magnesium chloride which is present in brine is separated. The same is converted into molten magnesium in an electrolytic cell at a temperature of about 670 to 730°C. During the electrolysis process, magnesium oxide is added to bring down the melting temperature. Chlorine is released which reacts with water to form hydrogen chloride vapours.

Calcium

Calcium is a silvery white metal with a specific gravity of 1.55. As previously mentioned, calcium is made by splitting the calcium chloride salt into calcium metal and chlorine by the following chemical reaction. The calcium chloride is electrolysed in a special electrolytic cell called Down's cell as in the case of sodium.

$$CaCl_2 \rightarrow Ca + Cl_2$$

Similar to aluminium, the voltage required is small, but the current required in a cell is high. The electro-winning of calcium occurs at a higher negative cathode potential than that in the case of sodium.

Iron

Iron is a heavy lustrous, grayish metal with a specific gravity of about 7.85. Steel is an alloy of iron made by the addition of other elements, especially carbon. Steel is the most usable form of iron. Everything that has to be hard and strong, from bolts and nuts to rockets and ships, is manufactured of steel.

Beneficiated iron ore is the starting material for making iron. Blast furnaces can be made with internal or external firing. Iron ore, coke and limestone are fed in sequence such that they form thin layers at the top of the furnace. These layers will get thoroughly mixed as their materials go down, countercurrent to the flow of gas from the bottom. See Figure 4.19(a).

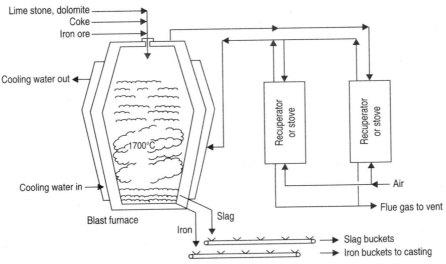

Figure 4.19(a) Typical blast furnace for iron production.

The material in the blast furnace gets heated up to about 1700°C by the time it reaches the reaction portion. The oxygen in the ore will get converted to carbon dioxide by reacting with coke and is discharged through the flue gas. Fluxes such as limestone and dolomite will be added to assist in the production of slag (the liquid that is formed) which can easily flow out of the furnace.

The reactions take place in the furnace are as follows: First, the carbon reacts with oxygen in the presence of limited oxygen to form carbon monoxide.

$$2C + O_2 \rightleftharpoons 2CO$$

The CO reacts with the molten ferric oxide by gas-liquid contact to produce iron and carbon dioxide

$$Fe_2O_3 + 3CO \rightleftharpoons 2Fe + 3CO_2$$

A part of the ore is also directly reduced to iron in the liquid state by coke as per the following reaction.

$$2Fe_2O_3 + 3C \rightarrow 4Fe + 3CO_2$$

The furnace is lined from the inside as well as cooled from the outside with water so that it can withstand the high temperatures in the furnace which reach up to about 1700°C at the bottom.

The flue gases are used to heat the air fed to the blast furnace using recuperative heat exchangers known as stoves. They work alternatively, collecting heat from the flue gases and releasing it to the feed gases.

The molten iron formed is drained to buckets travelling on rollers, which are then cast into ingots and sold or processed further. The balance liquid which forms a slag is also removed from the furnace in travelling buckets. The slag is recovered and used in making slow setting, high performance cement-mix for heavy concrete constructions like bridges.

Direct reduction using reducing gas: In newer processes, iron ore is reduced by using natural gas which has already become a very common fuel in most places of the world. This reduction is carried out by forming a reducing gas. This is a mixture of carbon monoxide and hydrogen, obtained by partial oxidation of natural gas by the following reaction.

$$CH_4 + H_2O \rightleftharpoons CO + 3H_2$$

The 'reducing gas' reacts with molten iron ore (ferric oxide) by a sequence of reactions. Fe_3O_4 is first converted to F_2O_3 and then to FeO and finally to iron by the following reaction.

$$FeO + CO \rightleftharpoons Fe + CO_2$$

The precipitated sponge iron is removed as product. It can be powdered or pelletized or made into blocks as per market requirements.

Manufacture of steel from iron: Pure iron is soft and not readily usable. It has to be made hard and tough to become useful. This is done by heating it in the presence of a certain amount of carbon. The amount of carbon in the steel decides its hardness, but toughness or strength is also important. Hence, most steels contain between 0.3% and 1.5% of carbon. This results in the production of steel or more specifically carbon steel which has become the most useful product of the industrialized world.

Steel

The production of steel is done in open hearth furnaces or converters such as Bessemer converter, where the iron is melted to a liquid. The main process is to mix the various components into the molten iron, and by blowing oxygen or air by lances resulting in the evolution of gases such as carbon monoxide. Steel is made over a period of many hours since carburizing and other reactions take a long time. Steel can also be made directly from iron ore by an integrated process.

Copper

Copper is a reddish-coloured heavy metal having a specific gravity of 8.93. It is produced from copper ores, which consists of copper sulphide, copper oxide or copper iron sulphide (formula $CuFeS_2$).

First, the ore is crushed and then ground to fine powder. The impurities are then removed by means of froth flotation. This is done by increasing the hydrophobic nature of copper minerals by adding suitable flotation agents such as xanthates. The impurities are carried away by water, leaving the ore behind.

Once the beneficiated ore is obtained, there are two routes for making pure copper. In one route, leaching and precipitation are done whereby the ore is leached with weak acid and the copper is precipitated to get pure copper.

The other newer method of copper production is electrowinning, which is an electrochemical process in which the copper from the solution migrates to the cathode made of ultra-thin sheets of copper, which then grows to a substantial thickness by the migration of pure copper to it. Further purification of copper can also be done, using electrowinning. The impure copper is made into the anode and very thin sheets of pure copper are made into the cathode. The copper migrates from anode to cathode, resulting in bars of 99.99% pure copper. See Figure 4.19(b).

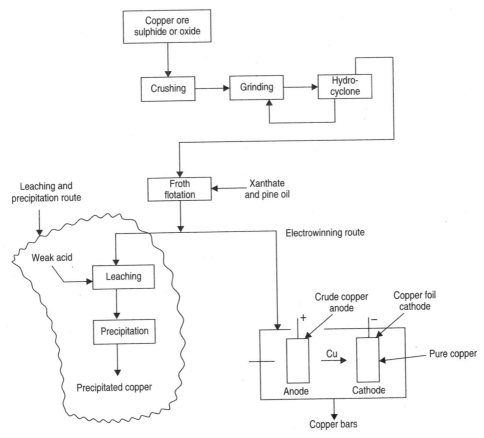

Figure 4.19(b) Production of copper.

Zinc

Zinc is a bluish, pale grey heavy metal with a specific gravity of 7.135. One of the major present-day uses of zinc is in electrogalvanizing, which is same as electroplating steel materials with zinc. Zinc is also used in steel castings. Both uses are mainly pertaining to automobile and construction industries. See Figure 4.19(c).

Similar to copper, zinc ore, consisting of zinc sulphide, is crushed and ground to a fine powder. Then, the zinc sulphide is roasted in the presence of oxygen whereby it is converted to zinc oxide.

$$2ZnS + 2O_2 \rightarrow 2ZnO + SO_2$$

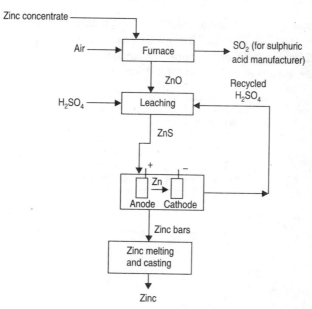

Figure 4.19(c) Production of zinc.

The sulphur dioxide produced is converted into sulphuric acid by the DCDA process (see Section 4.6 on sulphuric acid).

The zinc oxide is leached with weak sulphuric acid to produce zinc sulphate by the following reaction.

$$ZnO + H_2SO_4 \rightarrow ZnSO_4 + H_2O$$

The zinc sulphate solution is electrolysed by electrowinning process in an electrolytic cell, whereby the pure zinc migrates and settles on top of the aluminium cathode. The overall reaction is given below.

$$2ZnSO_4 + 2H_2O \rightarrow 2Zn + 2H_2SO_4 + O_2$$

The regenerated sulphuric acid can be recycled or sold. From the aluminium cathode deposited with pure zinc, the zinc is melted into liquid and cast into ingots.

Lead

Lead is a grey coloured, heavy and soft metal with a specific gravity of 11.34. The major use of lead is in the manufacture of lead acid batteries, which has indeed revolutionized the automobile sector. The demand for lead will continue to increase further as the demand for vehicles and batteries increases. Lead is also used to make paints. Lead carbonate is used as a paint colour medium. White, yellow, orange and red paints are made from lead compounds. Tetraethyl lead was used as an anti-knocking agent in petrol but the practice has been phased out due to pollution problems and substituted by MTBE (methyl tert butyl ester) and the other newer materials. See Figure 4.19(d).

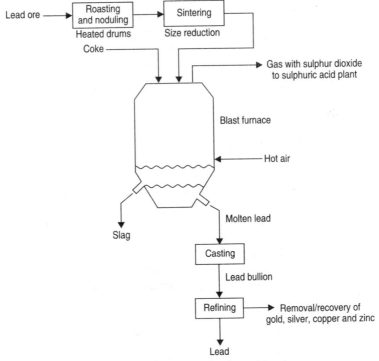

Figure 4.19(d) Production of lead.

Lead is mainly produced from the ore, galena with lead sulphide as its main constituent. The ore also contains small quantities of other metals such as copper, zinc, silver and gold. The ore is first roasted in a roaster where it is converted to lead oxide. Then the oxide is sent to a blast furnace as in the case of other metals like iron, where it gets reduced by coke to elemental lead. The impurities are removed as slag. The product still contains traces of gold, silver and bismuth as impurities, which are removed by liquid–liquid extraction to obtain pure lead.

Tin

Tin is a silvery-white metalloid with a specific gravity of 7.28. The major industrial use of tin is in making tins and cans as well as for coating other metals because of the excellent corrosion resistant properties of tin. Bronze is produced by combining copper with tin. Bronze is a very strong alloy, the use of which by man gave the name 'bronze age' to a period of history, spanning almost two thousand years, until it was overtaken by 'iron age'. In the bronze age, major tools, shields and weapons were made of bronze.

The production process of tin consists of concentrating the mined product, which is then heated in a furnace up to a temperature of 1400°C along with coke, limestone and sand. The carbon monoxide produced will reduce the oxide of tin to metallic tin. The process is again repeated to get substantially pure tin, which is again mixed with coke and heated to get pure tin as product, and the other impurities are taken out as slag.

Further refining of tin is done by taking advantage of the lower melting point of tin compared to other metals such as iron. The product is melted and liquid tin is tapped out of the furnace as soon as it melts. The electrolytic process can be used for further refining. This is done by electrowinning process in an electrolytic cell, whereby pure tin is preferentially transferred to the cathode which initially consists of a thin film of tin. The built-up cathodes are melted and made into ingots or bars of tin, for further processing or sale.

4.9 SULPHUR

Historically, sulphur is obtained from sulphur mines. In some places prone to volcanic activity, sulphur is pushed to the earth's surface which can be mined more easily.

Nowadays, almost all sulphur is being made by converting hydrogen sulphide to sulphur, the hydrogen sulphide being invariably available as a constituent of natural gas which is being abundantly produced from onshore and offshore gas wells. Similarly, all modern refineries produce sulphur from H_2S and SO_2 evolved from refinery operations. These are clean technologies compared to mining. Hence, there is no specific requirement for mined sulphur.

Elemental sulphur has a melting point of 112.8°C and boiling point of 444.6°C. The most important use of sulphur is in the production of sulphuric acid. It is also used for making carbon disulphide, bleachimg agents, vulcanized rubber, detergents, pharmaceuticals (one of its oldest uses was for fumigation), insecticides and fungicides.

4.9.1 Sulphur Production from Mines by Frasch Process

This method of sulphur extraction was followed mainly in the USA and Mexico and other places where sulphur deposits were found. The drilling was done using ordinary petroleum drilling equipment. The extraction procedure is given below.

A nest of three concentric pipes is inserted into the sulphur bearing rock deposit (see Figure 4.20). Superheated water is let in through the annular space between the outermost pipe and the middle pipe. Hot compressed air is let in through the central pipe. The sulphur melts and forms a pool of liquid at the bottom of the well. Because it gets aerated by mixing with compressed air, its density decreases and the mixture rises to the surface. The sulphur is separated from the mixture and made into blocks for sale. The rate of water and air injection has to be controlled such that a balance is maintained throughout the operation. The sulphur is also shipped in liquid form.

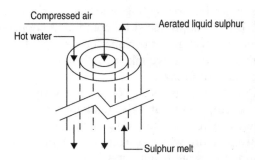

Figure 4.20 Sulphur production from underground deposits Frasch process.

4.9.2 Sulphur Production from Hydrogen Sulphide by Modified Claus Process

Hydrogen sulphide is normally removed from natural gas during natural gas processing and also obtained from refinery hydrogenation operations. This becomes an ideal source for sulphur. This process was originally invented and patented by Carl Friedrich Claus in the year 1883, and hence referred to as Claus process. The original Claus process was a single step process and the yield was poor. The modified Claus process uses two or three steps to give upto 97% yield.

H_2S burning and thermal step

In the first step, a part of the hydrogen sulphide is converted to SO_2 by burning sulphur in controlled substoichiometric quantity of oxygen at temperatures around 850°C.

$$2H_2S + 3O_2 \rightarrow 2SO_2 + 2H_2O \quad \text{(irreversible, exothermic)}$$

One part of hydrogen sulphide gases is burned with lances in the H_2S furnace to produce SO_2 and the heat evolved is used to produce steam in a waste heat boiler, and also for cooling down the gases. A part of the converted SO_2 gas reacts with balance unreacted H_2S in the first reactor and forms liquid sulphur. See Figure 4.21.

$$2H_2S + SO_2 \rightarrow 3S + 2H_2O$$

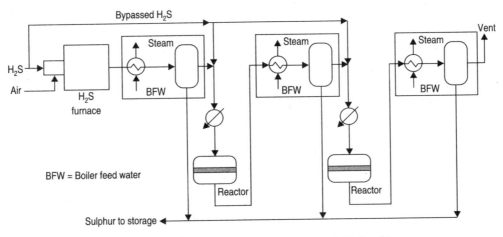

Figure 4.21 Sulphur production from hydrogen sulphide by Claus process.

The liquid sulphur is collected and drained to storage. The above operation is called "thermal step".

First catalytic step

In this step, the balance gas is mixed with a portion of the bypassed H_2S, heated by steam and fed to the first catalytic reactor. The chemical reaction forming sulphur continues in this reactor at a temperature of 315–330°C and the heat is used to produce steam. The catalytic conversion is actually maximized at lower temperatures but care must be taken not to condense any sulphur on the catalyst. The liquid sulphur is collected and drained to storage.

Second catalytic step

Again the operation as in the first step is repeated with another portion of bypassed H_2S. Here the chemical reaction takes place at about 240°C. The process gas is cooled to 130–150°C and the heat is used to produce steam. The liquid sulphur is collected and drained to storage. This step is optional.

The tail gas produced is treated in tail gas scrubbers. It is also possible to make more sulphur catalytic steps but then two converters will have to be operated in parallel since some sulphur condenses on the catalyst.

An overall yield of 97% sulphur is achieved with two catalytic steps. About 2.6 tonnes of steam is generated per tonne of sulphur produced.

4.10 SULPHURIC ACID

So diverse were the uses of sulphuric acid that its quantity of production used to be an index of industrial development of a country. A very large number of chemicals are made using sulphuric acid which include salts, fertilizers, insecticides, pesticides, and pharmaceuticals, etc.

The molecular weight of sulphuric acid (formula H_2SO_4) is 98.08. The melting point of 100% sulphuric acid is 10.5°C. It boils with decomposition at 340°C. Sulphuric acid being miscible in water, many grades, depending on the amount of water contained in it are being marketed. Oleum is produced by dissolving sulphur trioxide in pure or 100% sulphuric acid.

In the early days, sulphuric acid was produced by the Chamber process and thereafter by the Contact process. The Contact process itself consists of two varieties: Single Contact process and Double Contact Double Absorption process, commonly known as DCDA process. The DCDA process is less polluting and it easily meets the pollution control norms of different countries. Hence, the DCDA process is detailed below. Sulphuric acid is also produced as a by-product while processing zinc sulphide and pyrite ores.

4.10.1 DCDA Process

The sulphur having a purity of 99.5% is melted in a melting pit and then burned in a sulphur furnace using air from an air drying tower. The drying is done in an air drying tower by spraying concentrated sulphuric acid. See Figure 4.22.

In the furnace, the sulphur is converted to sulphur dioxide, using a controlled quantity of air. The furnace is provided with baffles for complete burning and mixing.

$$S + O_2 \rightarrow SO_2 \quad \text{(exothermic, irreversible)}$$

The gas from the funace chamber containg 10–12% sulphur dioxide is cooled from between 1000–1200°C to 430°C in a waste heat boiler where high pressure steam is produced. The temperature is maintained by adjusting the boiler bypass valve.

The cooled gas at 430°C enters the reactor and passes through the the first bed of vanadium pentoxide catalyst. The sulphur dioxide is converted to sulphur trioxide by the following equation.

$$2SO_2 + O_2 \rightleftharpoons 2SO_3 \quad \text{(exothermic, reversible)}$$

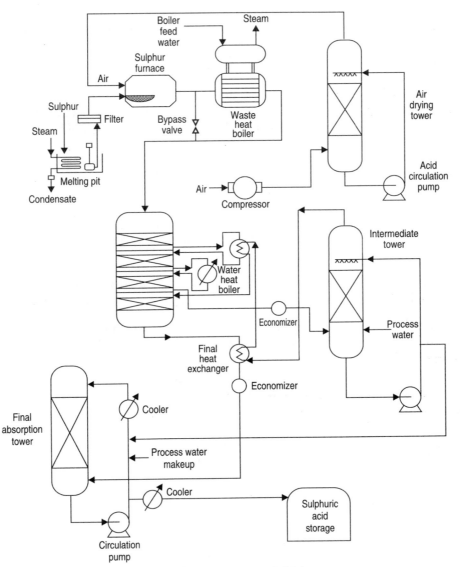

Figure 4.22 Sulphuric acid by DCDA process.

After each bed the outlet gas is cooled by heat exchange to get optimum conversion.

The conversion of sulphur dioxide in the first bed is 57% and the outlet temperature is 590°C. It is cooled to 460°C in the intermediate heat exchanger where it gives heat to the SO_2 gas coming from the intermediate absorption tower after passing through the final heat exchanger.

The total conversion becomes 85% after the second bed at a temperature of 538°C. This gas is cooled in the waste heat boiler to 450°C and enters the third bed.

The total conversion at the outlet of the third bed will be 99.5% and hence requires a reduction in this value for further reaction. The gas is cooled in an economizer to 200°C and then fed to the intermediate absorption tower. The liquid in the tower is circulated by a circulation pump. Here the SO_3 in the gas is absorbed and a portion of circulating liquid is diverted to the final absorption tower.

The gas from the top of the intermediate absorption tower is at a temperature of 65°C and is heated by the final heat exchanger as well as the intermediate heat exchanger to a temperature of 420°C required by the fourth bed of catalyst. In this bed, too, the conversion rises to 99.5% and the temperature rises to 439°C. The gas is cooled to 200°C in the final heat exchanger and then in the economizer, and then goes to the final absorption tower where the SO_3 is absorbed in the circulating acid to produce the product sulphuric acid.

The above arrangement and parameters for heat exchangers and heat recovery are only typical values and can vary depending on a specific plant design.

Oleum

Oleum can be produced in the sulphuric acid plant by circulating the acid through the oleum tower where SO_3 from the reactor is injected at the bottom of tower. The SO_3 is absorbed in the acid to produce oleum.

Procedure for process design of new DCDA sulphuric acid plant with better catalyst

Fix the capacity of the plant in tonnes per day and convert it to kgmol/h.

1. Determine the kgmoles per hour of sulphur required which is equal to the kgmoles per hour of SO_3 formed assuming that 100% of SO_3 is absorbed as H_2SO_4, correct to the required efficiency.
2. Determine the amount of air required. First, determine the moles of oxygen theoretically required and excess air requirement (say 40%) to get the dry oxygen required. Convert it to air and include water vapour based on humidity and fix the air requirement in kg per hour.
3. Design the air drying tower. Calculate the circulation rate of 98% H_2SO_4 based on the absorption of 1% moisture per circulation (the H_2SO_4 at the bottom is considered to be 97%).
4. Specify the combustion burner.
5. Design the parameters of gas leaving each bed with conversion achieved which depends on equilibrium constants at different temperatures.
6. Typical temperature change and conversion are as given below[6,7].

   First bed
   Inlet temp 410°C, outlet temp 601.8°C, conversion to SO_3 74%
   Second bed
   Inlet temp 438°C, outlet temp 485.3°C, conversion to SO_3 18.4%
   Third bed
   Inlet temp 432°C, outlet temp 443°C, conversion to SO_3 4.3%
   Fourth bed
   Inlet temp 427°C, outlet temp 430.3°C, conversion to SO_3 1.3%

The percentage conversion is based on the equilibrium constant which is based on vapour pressures of the constituents, the equilibrium constant is itself based on temperature. The recommendations of catalyst vendors may also be taken into account.

7. Design the intermediate absorption tower based on equilibrium constants between the third and fourth beds.
8. Design the final absorption tower.

4.11 NITRIC ACID

In the past, nitric acid was obtained by treating saltpeter (potassium nitrate) with sulphuric acid. Nitric acid is used for making ammonium nitrate fertilizer and many other organic and inorganic nitrates like cellulose nitrate, sodium nitrate, silver nitrate, etc. Nitric acid is also used for the parting of gold and silver which means separating gold from silver and silver from gold, since, as naturally occurring they are contaminated with each other in small amounts.

Nitric acid is miscible in water and consequently many grades are made and used in industry. The pure or 100% nitric acid has a molecular weight of 63.03, specific gravity of 1.502, melting point of $-42°C$ and boiling point of $83°C$.

There are two main processes for making nitric acid with ammonia as raw material. These are 'single pressure' process and 'dual pressure' process. The difference is that in the dual pressure process, the catalyst loss is less. This is because when the reactor works at low pressure the attrition is low. However, the reactor and other upstream equipment need to be bigger in size and a gas compressor is required to feed the absorber which works at higher pressure. Costwise, it is found that both processes are almost equally profitable.

4.11.1 Process Description of Single Pressure Process

See Figure 4.23. Air is compressed and mixed with vaporized and heated ammonia and sent to the reactor at a temperature of $230°C$ and a pressure of 8 to 9.5 kg/cm^2 (there are some medium pressure processes where this pressure is 7 kg/cm^2). The reactor has a catalyst bed which consists of stacked layers of platinum-rhodium wire mesh. The ammonia gets oxidized to nitric oxide and then to other nitrogen oxides. The chemical reaction taking place is as follows.

Oxidation of ammonia is carried out by the following reaction.

$$4NH_3 + 5O_2 \leftrightharpoons 4NO + 6H_2O$$

The nitric oxide is further oxidized to nitrogen peroxide and dinitrogen trioxide as per the following reactions.

$$2NO + O_2 \leftrightharpoons 2NO_2$$
$$NO + NO_2 \leftrightharpoons N_2O_3$$

The absorption of nitric acid takes place by the following reaction.

$$3NO_2 + H_2O \leftrightharpoons 2HNO_3 + NO$$

The NO produced is partly converted back to NO_2 because of the absence of oxygen due to precise quantity of secondary air that is fed.

$$2NO + O_2 \rightleftharpoons 2NO_2$$

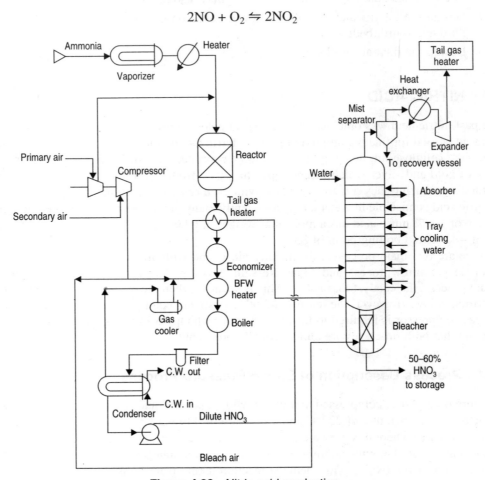

Figure 4.23 Nitric acid production.

The converted gas passes through the tail gas heater, the economizer, the boiler and the boiler water heater and other exchangers (as per design), which cools the gas to a temperature of 200°C, and is then sent to the nitric acid condenser after passing through a filter which catches any carryover of catalyst particles. The condensed dilute nitric acid is sent to the fifth tray from the bottom of the absorber, whereas the uncondensed gases containing nitric oxide are sent to the bottom of the absorber.

The absorber has about 25 or more actual trays with cooling coils in each tray. Cold water is fed at the top, the weak nitric acid at the fifth tray and the uncondensed gases from the condenser at the bottom of the column.

The bottoms of the absorber section are 55–60% nitric acid and go to the bleacher which is attached to the bottom of the column vessel. The bleacher can also be a separate vessel in some designs. The nitric acid may contain small quantities of nitrogen peroxide which gives a red colour to the product. In the bleacher, compressed air is fed at the bottom which carries away the impurities. The pure nitric acid is sent to storage.

The tail gas coming out of the top of the absorber goes to a mist eliminator to separate liquid particles, and is then expanded in an expander which supplies part of the power to the feed compressor. The tail gas is treated to remove nitrogen oxides so that the effluent gas conforms to the pollution control norms.

4.11.2 Dual Pressure Process

In the dual pressure process, the air compressor is a low pressure compressor, and the reaction is carried out at about 2.5 kg/cm^2 pressure instead of 9.5 kg/cm^2. An intermediate compressor compresses the gases from the condenser to the absorber column to about 10 kg/cm^2 for the absorption and bleaching operations. The greatest advantage of this change is that the loss of very expensive platinum-rhodium catalyst is considerably reduced. This advantage is partly offset because capital investment is higher mainly owing to the higher dimensions required for equipment in the low pressure section because the actual volume of gas handled is higher.

4.11.3 Concentration of Nitric Acid

There are two main processes for the concentration of nitric acid from 57% to more than 90% acid. The earlier method was to distill the acid with the addition of 93% sulphuric acid in a stoneware tower. The water nitric acid azeotrope is broken by the sulphuric acid so that concentrated nitric acid is obtained at the top and dilute sulphuric acid is obtained at the bottom of the tower. The concentrated nitric acid may be bleached with air for making it colourless.

The newer method was developed by Hercules Powder Company in 1958. It uses extractive distillation technique with the help of a 72% magnesium nitrate solution which is fed into the column along with nitric acid. Enhancement of relative volatility in the nitric acid–water system by the presence of magnesium nitrate as dissolved salt component makes it possible for making high strength nitric acid. The process has a continuous extractive distillation stage, producing 90% HNO_3 vapour which is further rectified to nearly 99.5% concentration in another tower.

Diluted magnesium nitrate solution from the still base is reconcentrated to the feed strength of 72% magnesium nitrate by vacuum evaporation and recycled. Steam provides process heat supply to the reboiler and evaporator.

4.11.4 Design and Simulation

Softwares such as Aspen plus, Design 2 for Windows or Prosim can be used for simulation. Thermodynamic models to be used are Electrolyte NRTL or equivalent.

The process can be simulated by configuring the reactor as a stoichiometric reactor followed by separators. However, the absorber trays are externally cooled and reaction takes place in the trays. Hence, the 5-theoretical trays (say, 25 actual) may be configured as coolers followed by reactors with stoichiometric equations giving the outputs from the plates. The dissolution of NO_2

in the liquid phase by which NO is released is to be considered, the other reaction being the conversion of nitric oxide (NO) to nitrogen peroxide (NO_2).

4.12 UREA

Urea is a major nitrogenous compound of formula, $NH_2-CO-NH_2$. It is a solid with a specific gravity of 1.335 and a melting point of 132.7°C. It is a major nitrogenous fertilizer sold in the form of granules or prills.

Urea is produced from ammonia and carbon dioxide. First, ammonia and carbon dioxide are combined to form ammonium carbamate by the following reaction.

$$2NH_3 + CO_2 \rightleftharpoons NH_4COONH_3$$

Ammonium carbamate is decomposed to form urea and water by the following equation.

$$NH_4COONH_3 \rightleftharpoons NH_2CONH_2 + H_2O$$

There are many processes developed by reputed companies, such as Urea Casale, Montecatini, Stamicarbon, Toyo Engineering Company, etc. for making urea as per the reactions shown above. There are several variations but the aim is to recycle all unconverted ammonia and carbon dioxide back to the reactor to get maximum efficiency of conversion.

A simplified flowsheet of a traditional Montecatini total recycle process is described in Figure 4.24.

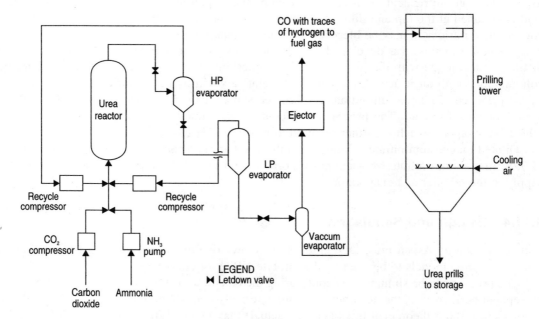

Figure 4.24 Simplified urea production plant using total recycle process.

4.12.1 Process Description

Liquid ammonia and carbon dioxide are mixed in a molar ratio of 3:5 and at a high pressure of about 200 kg/cm^2 and a temperature of 190°C and sent to a urea reactor. Ammonium carbamate will be formed in the reactor which will be further converted into urea and water as per the second reaction.

The product from the reactor is reduced in three stages to 34 kg/cm^2, 2 kg/cm^2 and 0.1 kg/cm^2, respectively. The gases produced in the first stage and second stage pressure letdown are compressed and sent back to the urea reactor as a combined recycle ammonium carbamate. The urea solution from the last separator is evaporated under a slight vacuum and the product is pumped to the top of the prilling tower. The liquid is sprayed from the top against a countercurrent flow of air. The urea prills get solidified, cooled and hardened. They are collected in belt conveyors and sent to storage for bagging and sale.

Urea hydrolyzer

The urea hydrolyzer is installed to treat and recover urea containing waste water streams including the ejector gas absorbed in water from the third stage pressure letdown operation shown previously in Figure 4.24 (in urea plant) as going to fuel gas.

The waste water from the urea plant will contain about 2% urea and 3% ammonia. However, by using the hydrolyzer this is reduced to 50 ppm of urea and ammonia each. The waste water collected is pumped to a distillation column where most of the ammonia is distilled off. The solution taken out from the middle of this column is heated to 160°C and sent to hydrolyzer which is maintained at 13 kg/cm^2 pressure and temperature of 195°C for a period of one hour. The following reaction takes place.

$$NH_2CONH_2 + H_2O \rightleftharpoons 2NH_3 + CO_2$$

A distillation column is provided along with the hydrolyzer. The vapours from the hydrolyzer are sent to top section of the column which are recycled to the urea plant. The solution containing less than 50 ppm ammonia is sent to the bottom section and then to an effluent treatment plant.

4.13 HYDRAZINE

Hydrazine, also called by the name diazane, is also a chemical obtained from ammonia with the formula NH_2-NH_2. It is a liquid with a density of 1.02 g/cc and a boiling point of 113.5°C. Hydrazine is a good corrosion inhibitor. Anybody who visited a boiler feed water preparation plant, has seen hydrazine hydrate being dosed into the boiler feed water. But little do we know that hydrazine has many other uses such as adsorbent for acid gases, stabilizer for plant growth promoters, photographic chemicals, chemicals for making blowing agents, rocket fuel, propellant for space vehicles, fuel cells, raw material for pesticides and pharmaceuticals, etc. It is used as a hydrate for common purposes since hydrazine is highly explosive.

At one stage, especially during the cold war period, hydrazine use for rocket purposes was very large, but at present its use as rocket fuel is much less compared to agricultural and other uses.

Hydrazine can be made by many processes, three such processes are given below.

1. *Olin Raschig process.* In this process ammonium chloride is reacted with ammonia to produce hydrazine as per the following equation:

$$NH_2Cl + NH_3 \rightarrow NH_2-NH_2 + HCl$$

2. *Pechiney-Ugine-Kuhlman process.* In this process ammonia is reacted with hydrogen peroxide as per the following equation:

$$H_2O_2 + 2NH_3 \rightarrow NH_2-NH_2 + 2H_2O$$

3. *Ketazine process.* It is a variation of the Raschig process. Ammonia is oxidized by chlorine or chloramine in the presence of an aliphatic ketone, usually acetone. The resulting ketazine is then hydrolysed to hydrazine.

4.13.1 Olin Raschig Process

In the Olin Raschig process, sodium hypochlorite (NaOCl) is mixed with excess ammonia at 5°C to form ammonium chloride (chloramine) and sodium hydroxide. The following reaction takes place.

$$NaOCl + NH_3 \rightarrow NH_2Cl + NaOH$$

This mixture is rapidly added to excess ammonia at a pressure of about 250 kg/cm^2 and heated to 130°C. The following reaction takes place forming hydrazine. Ammonia is used about 30 times excess.

$$NH_2Cl + NH_3 \rightarrow NH_2-NH_2 + HCl$$

Excess ammonia is removed and the hydrazine water solution obtained is distilled to azeotropic composition. Extractive distillation of the product by aniline or 50% caustic solution yields pure hydrazine. If ultra pure hydrazine is required, freeze crystallization is used.

It may be noted that the NaOH produced in the first reaction reacts with the HCl produced by the second reaction and produces sodium chloride which dissolves in the water and is continuously removed by evaporative crystallization.

4.14 COMPLEX FERTILIZER

Nutrients are required for plant growth. In the olden days, natural fertilizers were used in agricultural fields to increase the yield of crops. Barks, leaves, bone powder, ash, etc. supplemented the decreased nutrients in the soil. Complex fertilizers are those which supply many important nutrients to the soil.

The main nutrients are:

1. Nitrogen—mainly for the growth of stems and leaves.
2. Phosphorus—for stems, leaves and seed production.
3. Potassium—for the growth of starch.

Apart from the above, plants also need secondary nutrients.

Some organic manure is also required which the soil should be able to supply.

Production of ammonia and phosphoric acid is already described in the respective sections. In this section, we examine how to produce complex fertilizer which contains the major nutrients, that is, nitrogen, phosphorus and potassium (in short, NPK).

It is preferable that a complex fertilizer plant should be able to produce various products containing different ratios of N, P and K so that the different farmers have the choice of selecting the product as per the soil conditions and requirements of the produce. Mixing of different fertilizers is an option for the benefit of the farmer. Complex fertilizers can be produced in varying NPK ratios such as 17:17:17, 28:28:0, 14:28:14, 20:20:0, 18:46:0. Ease of manufacture by the producer is also considered in making these ratios. For example, 18:46 is di-ammonium phosphate.

4.14.1 NPK by Pipe Reactor Process

See Figure 4.25. Sulphuric acid, phosphoric acid and ammonia are introduced into a pipe reactor and allowed to react with each other according to the following chemical reactions.

$$NH_3 + H_3PO_4 \rightarrow NH_3H_2PO_4 \text{ (Mono-ammonium phosphate)}$$

$$NH_3 + NH_3H_2PO_4 \rightarrow (NH_4)_2HPO_4 \text{ (Di-ammonium phosphate)}$$

$$2NH_3 + H_2SO_4 \rightarrow (NH_4)_2SO_4 \text{ (Ammonium sulphate)}$$

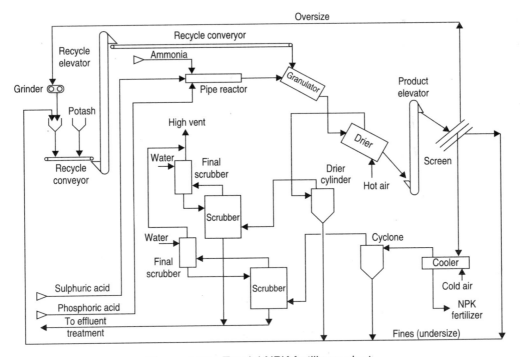

Figure 4.25 Typcial NPK fertilizers plant.

The feed to the 'pipe reactor' is given at the correct stoichiometric ratio. After leaving the pipe reactor, the reaction product mixture, in the form of a slurry enters a granulator.

Recirculating fines are also fed into the granulator by a recycle conveyor and elevator system. Muriate of potash (potassium chloride, KCl, mostly mined from deposits or produced from sea water) is also fed into this system. This forms a rotating bed of particles in the granulator. Ammonia vapour in quantities as required by the chemical reactions is also introduced into the granulator. The granules formed move into a drier where they are dried by a countercurrent flow of dry air. Then it goes to an elevator which feeds a screening system. The screens divide the product into:

1. Oversize particles
2. Correct size particles and
3. Undersize particles or fines.

The oversize material is ground and mixed with the fines and both are fed to granulator by the recycle conveyor/elevator system. The correct size material is cooled with cold air and sent to storage for packing and sale.

A dust and fume recovery system is also provided for the vapours evolved in the product drier and the product cooler. Dilute phosphoric acid is also used for scrubbing the fumes from the drier and cooler. Then effluent gases are further scrubbed with water in the respective water scrubbers so that any residual acid is removed. The effluent water is sent to an effluent treatment plant.

Secondary nutrients

To obtain optimum results, crops have to be supplied with secondary nutrients in addition to primary nutrients. The secondary nutrients are calcium, magnesium and sulphur. They are all important for plant growth, but in relatively smaller quantities than those of primary nutrients. Any deficiency of the secondary nutrients and other essential elements can reduce the efficiency of primary nutrients. Sulphur is usually available as ammonium sulphate if care is taken to involve ammonia and sulphuric acid in the production processes and retain the ammonium sulphate in the product.

Calcium and magnesium are the elements that are largely responsible for the acidity or alkalinity of soil. In areas of low rainfall, and where the soil is derived from limestone, the soil is high in calcium and magnesium and is usually alkaline. In regions of high rainfall they are washed from the soil and carried away. Soils in these regions are usually acidic. Lime (calcium carbonate) is added to make the soil less acidic.

4.15 CEMENT

Modern civilization is totally dependent on cement for building houses, roads, tunnels, dams, etc. Our cities are almost entirely built on cement and concrete. However, cement was known in some form or the other to older civilizations as well.

Two materials are most important for the production of cement. These are calcium-containing materials such as limestone and silica-containing materials such as clay. Limestone varies widely in its composition based on the place where it is mined. Beneficiation of limestone will be required to make it suitable for cement manufacture.

Portland cement is the most common form of cement. As per ASTM specification, Portland cement is defined as the product obtained by pulverizing clinker, essentially of 'hydraulic' calcium silicates, usually containing one or more forms of calcium sulphates as an interground addition. Here 'hydraulic' means ability to harden with water alone (i.e. without drying or reacting with carbon dioxide).

4.15.1 Designation of Components

Cement is mainly a mixture of lime, silica, alumina and other clinker compounds. In the cement industry practice, these components are designated by letters as given below.

$C - CaO$

$S - SiO_2$

$A - Al_2O_3$

$F - Fe_2O_3$

$M - MgO$

$S - SO_2$

$N - Na_2O$

$K - K_2O$

$C - CO_2$

$H - H_2O$

As an example of the designation the compound $3CaO \cdot SiO_2$ will be designated as C_3S.

There are five types of Portland cement. These five types of Portland cement and their composition and features and uses are given in Table 4.3.

Table 4.3 Types of Portland cement

Type	Type I	Type II	Type III	Type IV	Type V
Composition	C_2S, C_3S, C_3A	C_2S, C_3S	C_2S(high), C_3A	C_3S(low), C_3A, C_4AF	C_3A (less than 4%)
Features	Full strength in 28 days			Low heat of evolution	
Uses	General construction use	General and sulphate resisting use	Roads for early use	Massive structure use	Good for sea water contact

The compressive strengths of these cements after 1, 3 and 28 days are given in Table 4.4.

Table 4.4 Compressive strength of cements during setting

Type days	Type I	Type II	Type III	Type IV	Type V
1 day	37	28	103	20	28
3 days	120	83	240	49	88
28 days	340	260	440	177	214

4.15.2 Compounds in Cement

Portland cement contains systems of $CaO—SiO_2$ or more closely systems of $CaO—SiO_2—Al_2O_3$ and $CaO—SiO_2—Al_2O_3—Fe_2O_3$ and then $CaO—SiO_2—Al_2O_3—Fe_2O_3—MgO$. The first system $CaO—SiO_2—Al_2O_3$ can be expressed as a ternary diagram. Some of the chemistry of these systems can be found in books[8,9].

Many theories exist for the mechanism of setting of the cement. It is known that hydration and hydrolysis are involved.

Portland cement is now manufactured by igniting a mixture of raw materials mainly composed of limestone ($CaCO_3$ with the impurities) and clay ($AlSiO_3$ with the impurities).

A typical raw meal (feed-mix) composition is given below in percentages:

SiO_2 (14.3%), Al_2O_3 (3.03%), Fe_2O_3 (1.11%), CaO (44.38%), MgO (0.59%), SO_3 (0.07%), LOI (35.86%), K_2O (0.52%), Na_2O (0.13%).

In the above composition of raw meal, LOI indicates 'loss on ignition' which takes care of gases (CO_2, etc.) escaping from calcium carbonate.

The source of calcium carbonate can also be (apart from limestone) chalk, shale, marl, slag, and substitutes of clay can be alkali waste, sand, waste bauxite and iron oxide. The limiting percentages of various chemical species expressed as oxides for making acceptable cement are given below.

SiO_2 (16–26%), CaO (58–67%), TiO_2 (0–2.5%), Al_2O_3 (4–8%), MgO (1–4%), K_2O/Na_2O (0–2.5%), Fe_3O_2 (2–5%), Mn_2O_3 (0–3%), P_2O_5 (0–1.5%).

In addition to the above, various module values are practised in the industry. As an example, the module value for CaO is given by:

$$\text{Module} = \text{mass of CaO/mass of } (SiO_2 + Al_2O_3 + Fe_2O_3)$$

The module value for CaO ranges between 1.7 and 2.3. The ideal value is 2.

4.15.3 Beneficiation of Limestone

Process description

See Figure 4.26. While calcium carbonate is the main component of limestone, sometimes it contains excess quantities of iron. Such excess quantities are first removed by physical separation methods. Then the rock is wet ground and fed to a hydro separator from where the overflow goes to the final thickener after removing a portion of silica, mica and talc as required. The underflow is sent to a classifier and then to the froth flotation equipment. Here the fines are floated and sent to the intermediate thickener and then to the final thickener. Those particles which settle in the process of flotation are sent to the second stage of flotation and then the unrecovered solids are rejected.

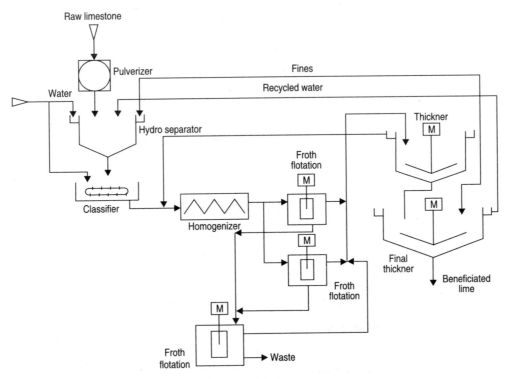

Figure 4.26 Beneficiation of limestone.

Cement manufacture

See Figure 4.27. Clay and limestone are passed through a pulverizer and fed into a tube mill grinder. There are both wet and dry grinding processes but dry grinding is more popular. Close circuit grinding, is usually followed, in which fines are directly taken out and the coarse material is recycled.

The dry powder is directly fed to the kilns. The kilns can be very long when wet grinding is used. The internal diameters are in the range of 2.5 to 6 metres and they are rotated at a speed of between 0.5 and 2 rpm. Lengths can be in the range of 45 to 150 metres. The kilns are slightly inclined.

The material reaches the tail end in 1 to 3 hours. They are fired from the downstream end and the residual heat at the upstream end is used for making steam in waste heat boilers. The kilns are computer-controlled for making adjustments based on rock composition and operating conditions so that quality is maintained.

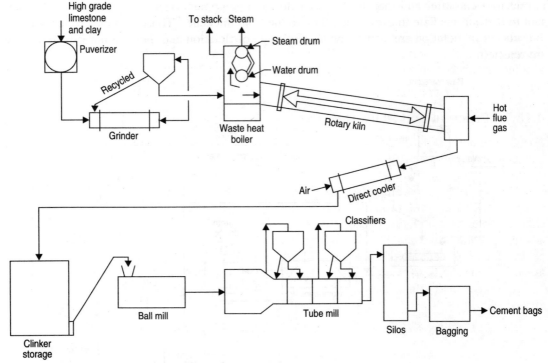

Figure 4.27 Production of cement.

The following chemical reactions take place at temperatures above 1000°C.

$$CaCO_3 \rightarrow CaO + CO_2$$
$$CaO + Al_2O_3 + SiO_2 \rightarrow C_2S + C_3S + C_3A$$

The C_2S, C_3S, C_3A are clinker compounds defined earlier.

The clinkers are cooled by air circulation and stored. From there they are sent to the ball mill and the tube mill. In the tube mill, there are different sections and recycling of higher-sized particles is done. Then the cement is stored in silos and bagged and sent for sales.

Details about nano scale bonding and other aspects of cement chemistry are given in books[12,13] for further reading.

4.16 LIME

The most common marketed lime is calcium oxide, also called quicklime, and contains about 90% CaO, about 5% MgO and small quantities of $CaCO_3$, SiO_2, Al_2O_3 and Fe_2O_3. It can also be converted to hydrated lime (also called slaked lime) in the same plant depending upon the demand. The customers have the choice to buy quicklime and convert it to slaked lime in his own plants, in which case the transportation cost will be less because of smaller weight of quicklime compared to hydrated lime. On the other hand, customers need to spend on additional equipment if they want to convert quicklime to hydrated lime in their own locations.

Due to the abundant availability of limestone in most locations throughout the world, its use is also widespread. Lime has been used for medicinal purposes, as insecticide and fertilizer. It is also used in gas absorption, manufacture of cement, steel, soap, rubber, varnish, refractories, bricks, paper, water softening, etc. Quicklime can be used to dehydrate rectified spirit to obtain absolute alcohol. It is also a good preservative for vegetables. Lime is also a constituent of 'chewing pan' used in many Asian countries along with betel leaves and areca nuts.

4.16.1 Process and Plant Description

For good quality lime the starting material should be preferably low in impurities like silica, clay, and iron. In some cases, beneficiation will be required which is already described under cement.

The kiln is the most important part of the process. There are basically two types of kilns. They are shaft kilns and rotary kilns. Within shaft kilns, there are different types such as vertical shaft, countercurrent shaft, regenerative shaft, rotary shaft, etc. Each type of kiln has its own advantages and disadvantages. In the North and South American continents, the production centres are more centralized since material transportation is cheaper and hence rotary kilns are extensively used.

A vertical shaft kiln is shown in Figure 4.28. The chemical reaction taking place in the lime kiln is:

$$CaCO_3 \rightleftharpoons CaO + CO_2 \quad \text{(reversible, endothermic)}$$

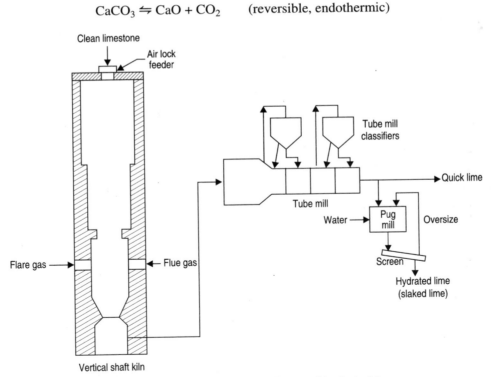

Figure 4.28 Production of quicklime and hydrated lime.

The above decomposition reaction starts taking place at around 900°C at the surface of the limestone lump and further increases as limestone flows down the shaft. Higher temperatures may arise at the middle of the lumps. This is because the CO_2 formed at the centre of the lumps will exert high pressure and consequently increase the temperature. The heat supplied is not only for heating the limestone but also for supplying the endothermic reaction heat. Of the total heat supplied, 40% is for sensible heat and the remainder for reaction heat.

The lime moves downwards towards the cooling section where it is cooled in contact with air. After further cooling it can be stored as lumps or else sent to the tube mills to grind it to powder, stored and/or packaged in bags and sent for sale.

Hydrated lime

Presently more consumers are opting for hydrated lime. In the process of hydration the lime obtained is crushed to about 20 mm size or less and sent to a pug mill containing horizontal rotating arms. Water is fed into the pug mills where the following chemical reaction takes place.

$$CaO + H_2O \rightarrow Ca(OH)_2 \quad \text{(irreversible, exothermic)}$$

The product need not be dried since there is no excess water. But the product is classified to remove any unreacted quicklime. It is stored and/or packaged in bags and sent for sale.

4.17 PAINTS

Paints and varnishes are expected to have the following qualities:
1. Covering power to form thin films of about 1 micron in thickness.
2. Colour. This is not required for varnishes, transparency being its main quality.
3. Weather resistance.
4. Washability.
5. Consistency and ease of application.
6. Non-toxicity.

Paints differ from varnishes mainly because they contain a pigment. Hence, pigments are the most important constituent of paints.

Earlier white lead ($2PbCO_3$—$Pb(OH)_2$), zinc oxide (ZnO) and lithopone (ZnO—$BaSO_4$) were used as white pigments. At present, titanium dioxide is the most important white pigment. Carbon black is the most important black pigment. Chromium oxides and hydrates supply yellow, orange and green colour to paints. Metal thiocyanine is an important blue pigment. Other blue pigments include Prussian blue and ultramarine blue.

4.17.1 Manufacture of Titanium Dioxide from Ilmenite

There are two processes for the manufacture of titanium dioxide. They are the sulphate process using the ilmenite ore (FeO-TiO_2) and the chloride process using the rutile ore (TiO_2). The sulphate process is more polluting compared to the chloride process. The rutile grade produces better quality paint. Of late, almost all paint manufacturers use the rutile quality pigment. Hence, modern plants convert ilmenite ore into rutile grade (called synthetic rutile) and produce titanium dioxide from it by the chloride process. A plant based on the chloride process is shown in Figure 4.29. In this process, the titanium dioxide pigment production plant consists of three sections:

1. Ilmenite beneficiation
2. Acid regeneration
3. Pigment production

Ilmenite beneficiation

Sand containing ilmenite ($FeO\text{-}TiO_2$), obtained from ilmenite rich washed beach sands from seashore, is separated from other constituents using first milling and drying and then using electrostatic, magnetic and gravitational separation techniques.

HCl digestion and recovery

Sand is roasted and leached, by which most of the ferrous iron is removed. Then it is digested with HCl and leached to obtain beneficiated ilmenite. The HCl reacts with the ferric iron (Fe_2O_3) in the digestion process and forms $FeCl_3$.

$$Fe_2O_3 + 6HCl \rightarrow 2FeCl_3 + 3H_2O$$

The HCl is reclaimed by a process of spray roasting and heating, converting high purity ferric chloride to ferric oxide and HCl.

$$2FeCl_3 + 3H_2O \rightarrow Fe_2O_3 + 6HCl$$

Titanium dioxide pigment production

See Figure 4.29. The beneficiated ilmenite and coke are fed to a fluidized bed reactor using chlorine gas as the fluidizing agent fed from the bottom.

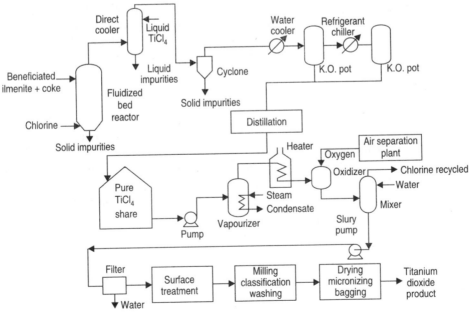

Figure 4.29 Production of titanium dioxide pigment.

The following chemical reaction takes place, forming titanium tetrachloride ($TiCl_4$)

$$C + TiO_2 + 2Cl_2 \rightarrow TiCl_4 + CO_2 + CO + \text{metal chlorides}$$

The other chlorides are removed from $TiCl_4$ by direct cooling using liquid $TiCl_4$. Solid impurities are removed by cyclonic separation. Again $TiCl_4$ is cooled with cooling water and then with chilled refrigerant. It is further concentrated by distillation to obtain pure $TiCl_4$ in liquid form which is sent to storage.

Oxidation, surface treatment and finishing

The pure $TiCl_4$ is vaporized and preheated and sent to an oxidizer where it is oxidized by an oxygen stream (99.98% purity) produced by an air separation plant. The following chemical reaction takes place.

$$TiCl_4 + O_2 \rightarrow TiO_2 \text{ (Rutile)} + 2Cl_2$$

This reaction is exothermic. The chlorine gas evolved is recycled back to the fluidized bed reactor.

The product is quenched in water which forms a slurry of titanium dioxide in water. The slurry is stored and then sent to filtration, milling, polishing and suface coating. Micronizing is done using superheated steam as the medium. Finally, the product is packaged in bags and sent to storage for sale.

4.18 VARNISHES

A varnish is an unpigmented colloidal dispersion or solution of resins in an oil/thinner which forms a transparent layer when dried. There are three varieties, oleo resinous, alkid and urethane varnishes.

Oleo resinous varnishes are being replaced by alkid and urethane varnishes because of their greater durability, less yellowing, and ease of application. Similarly, water thinned varnishes are replacing solvent thinned varnishes to avoid pollution of air by the evaporation of solvents. Shellac is a varnish dissolved in spirit, either ethanol or methanol.

The production process of varnishes consists of:

1. Mixing the constituents, and
2. Rapid stirring and heating.

The composition of varnishes varies widely and it is difficult to standardize it. The product should be capable of being applied by spraying or brushing.

Varnishes are made of a resin and a solvent. The solvent evaporates after the varnish is applied leaving behind a smooth and tough finish.

Varnishes can be classified as follows on the basis of the solvent used and the resin used.

1. Using water as solvent.
2. Using organic solvents such as methanol, alcohol or acetone as solvent.

According to the resin used, varnishes are divided into three categories:

1. Oleoresins.
2. Alkyds.
3. Urethane and epoxies.

Oleoresin varnishes can also be called turpene varnishes or turpentine varnishes. The oleoresin that is produced or oozes out or exudes from pine trees is collected and distilled to get turpentine varnish. The distillation is conducted in specially designed batch reactors. The product contains alpha and beta pinenes. The pinene polymerizes to form turpentine resin. Once it is applied on the surfaces, polymerization continues for a long time (up to a month). Exposure to air, heat, ultraviolet rays and moisture help polymerization. Oleoresin varnishes are also used for varnishing stringed musical instruments.

Alkid resins are synthetic resins, more used for outdoor applications and for coating on vehicles. Alkid resins are formed by condensing dicarboxylic acids with polyhydric alcohols. Since they find diverse applications, they are replacing natural oleoresin varnishes for many applications. Other similar resins are acrylic, silicones and vinyls which are being increasingly used for furniture.

Epoxy resins are also synthetic varnishes used for furniture and other applications. Urethane is used as finish coatings. These finishes can consist of two types as follows.

1. Toluene di-isocyanate in combination with glycerine, methyl glycicide and linseed oil.
2. Toluene di-isocyanate in combination with a hydroxyl bearing substance.

Coating of boats, ships and sea-going vessels has been of historical importance. Cardanol obtained from cashewnut shell liquid was discussed Chapter 1. Many other oleoresins are also used for this purpose. For steel ships, use of copper oxide with vinyl resin binder is common.

Touch the front of your teeth with your tongue. You will feel both, a smoothness of surface as well as the toughness of the teeth which can cut food materials. The surface is, in fact, a protein-based coating called enamel.

Enamel paints have hardness which has been exploited by painters like Picasso for the preservation of famous paintings. Enamel varnishes are also similar and give toughness to the varnished surface.

4.19 WATER

4.19.1 Water for Municipal and Industrial Use

The quality of water obtained at any particular location depends upon the geology and geography of the land. In most cases, treatment will be required before water is supplied to the municipality or industry. The type of treatment depends upon the impurities in the water. These include temporary hardness due to calcium and magnesium bicarbonates, and permanent hardness due to calcium and magnesium sulphates and chlorides. Soluble gases, turbidity, colour and microorganisms are other impurities which have to be removed.

A typical water treatment system is shown in Figure 4.30.

Water from natural sources like ponds, tanks, rivers, etc. is taken to a mixing tank where it is mixed with soda ash (sodium carbonate), alum (aluminium sulphate, which is usually obtained by adding sulphuric acid to bauxite) and lime (calcium hydroxide) for removing hardness and turbidity. After mixing, it is sent to a clarifier. In the clarifier a slow moving agitation is given so that all the coagulated impurities settle down, which are then removed from the bottom.

194 Chemical Process Technology and Simulation

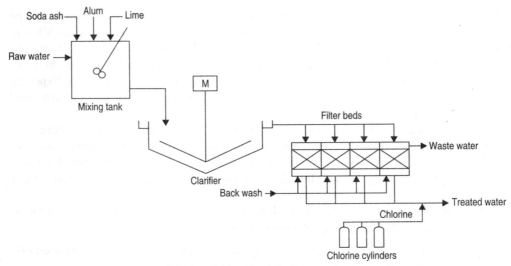

Figure 4.30 Municipal water treatement.

A high alkalinity is also useful to kill the bacteria. After settling out the impurities the water is recarbonated by injecting carbon dioxide produced by burning any local fuel.

The clarified and carbonated water is passed through filter beds where solid impurities are removed. The beds are regenerated by 'back washing' with water. Chlorine is added in controlled quantities to prevent bacterial growth. The water is stored and pumped to the consumer network.

Ion exchange treatment

Ion exchange was first noticed in the middle of the nineteenth century during an experiment when ammonia was passed through soil. It was found that ammonium ions were displacing calcium ions in the soil. Synthetic organic resins came into existence in the early twentieth century. Now, most demineralized water is produced by ion exchange processes.

The materials used for producing ion exchange resins were zeolites such as aluminium silicates which have low exchange capacity (kilograms exchanged per kilogram of resin). In the next stage, organic ion exchangers were introduced which have high exchange capacity. The next improvement was resins made from sulphonated natural products such as coal and lignite. But the highest exchange capacity resins produced are based on polystyrene-divinyl benzene (PSDB) which is now extensively used in resin production.

Softening of hard water

The calcium and magnesium ions in water (hard water) do not allow soap to lather while sodium ions in water (soft water) have no such effect while using soap for washing with water.

Softening of hard water is a process of exchanging the calcium and magnesium ions in water with sodium ions.

$$\text{Ca or Mg salt} + 2\text{NaR} \rightarrow \text{Ca or MgR}_2 + \text{Na}_2 \text{ salt}$$

In the above given equation, 'salt' indicates bicarbonates, sulphates and chlorides and 'R' denotes the exchanger radical.

When the ability to exchange the cation is lost by the depletion of sodium ions, the bed is regenerated to sodium by passing sodium chloride solution through it.

$$\text{Ca or MgR}_2 + 2\text{NaCl} \rightarrow 2\text{NaR} + \text{Ca or MgCl}_2$$

In the next stage of development, hydrogen cation replaces the sodium cation. Now the exchange reaction and the regeneration reaction are respectively as follows.

$$\text{Ca or Mg salt} + 2\text{HR} \rightarrow \text{Ca or MgR}_2 + \text{H}_2 \text{ acid (exchange)}$$

$$\text{Ca or MgR}_2 + 2\text{HCl} \rightarrow 2\text{HR} + \text{Ca or MgCl}_2 \text{ (regeneration)}$$

Regeneration is done by sulphuric acid or hydrochloric acid. Acidic water effluent can either be neutralized or it can be sent through anion exchangers.

4.19.2 Potable Water from Sea Water

Fresh water is in great shortage in many countries and in several regions within countries and this shortage will become more acute in the future due to pollution and climate change. Hence, sea water to potable water conversion is gaining importance. There are many ways in which this can be done. Evaporation and condensation techniques, freezing, reverse osmosis, electro-dialysis are the more important processes. Within evaporation-condensation methods, multiple effect evaporation is a well-known method; however, the capital investment is high. The vapour recompression method is described first.

Vapour recompression process

See Figure 4.31. Filtered sea water is heated by outgoing potable water and then again heated by outgoing concentrated brine which is being discharged back to the sea. The hot brine enters the evaporator. In the evaporator, it is heated by steam condensing outside the tubes of a calandria. The evaporator is maintained in partial vacuum by means of a vapour recompressor which compresses the vaporized water to higher pressure steam. This steam condenses in the calandria

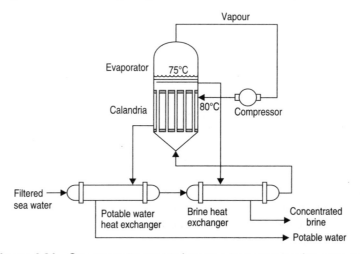

Figure 4.31 Sea water treatment by vapour recompression process.

at a temperature of 80°C and comes out as pure potable water. The produced water is cooled by heat exchange to incoming sea water. The water is stored and supplied to consumers. The above process is energy efficient and has low capital cost in comparison to multiple effect evaporators.

Reverse osmosis process

Osmosis is the process of diffusion of molecules across a semi-permeable membrane. The natural tendency of salt is to diffuse from sea water side to the pure water side when separated by a membrane. While this is called osmosis, flow of water will occur instead of salt if the pressure in the chamber is more than the osmotic pressure of sea water at that temperature. This is called reverse osmosis (RO) and is equivalent to pushing the water selectively through the specially developed membranes. A pressure of 50 to 100 kg/cm^2 is used. The RO process gives very good efficiencies of separation without high investment.

A typical RO process scheme is shown in Figure 4.32. The saline water is filtered and pumped through a set of RO modules kept in parallel. The water flows into a serpentile tube which is made of semi-permeable membrane. Pure water flows out of the tubes leaving behind the salinity which is taken out as waste. Part of pure water is taken out as product and the other part recycled. This is a single-stage process. A two-stage process can also be devised in which case the product goes through another set of similar modules to further enhance the purity.

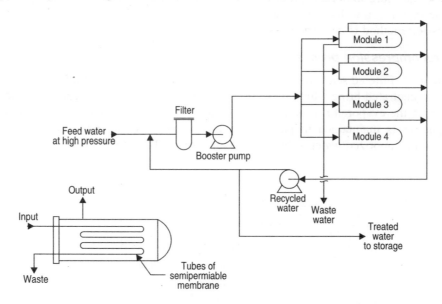

Figure 4.32 Reverse osmosis process for sea water purification.

A packaged unit of size 3 m length, 3 m width and 3 m height, working at 68 kg/cm^2 pressure, can produce typically 6 m^3/h of treated water reducing the total dissolved solids from 40000 ppm in the inlet to about 500 ppm in the treated water. A number of such units can be assembled to get the required capacity.

Electrodialysis process (membrane process)

This process technically called 'electrodialysis process' is also called 'membrane process'. It is an interesting process since it is the salt that is separated from the brine rather than the water. See Figure 4.33.

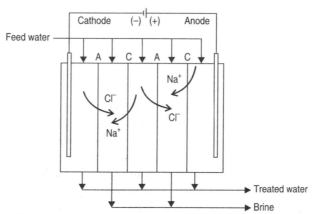

Figure 4.33 Electrodialysis process for sea water.

Cation membranes contain negatively charged subgroups which repel anions. In the same manner, the anion membranes contain positively charged subgroups which repel cations. Sodium chloride in brine consists of anions and cations and by keeping alternate layers of cation and anion membranes and imposing a current across the pack of membranes, separation occurs.

The cations pass through cation exchange membranes, the anions being repulsed and the anions through anion exchange membranes where the cations are repulsed, and so on. This results in salinity increase and decrease in alternate spaces. We get a brine solution on alternate compartments and pure water in the other compartments. Each unit consists of 200 to 1000 compartments and the voltage applied is about 110 volts.

The power requirement is directly proportional to the concentration of salt. Hence, energy requirement will be huge to process sea water. But it will be reasonable for brackish waters which contain less salt. Brackish water is a mix of sea water and natural water, obtained in places where a river meets the sea, or in backwaters or ponds and lakes formed near sea shores.

4.20 URANIUM AND HEAVY WATER

Nuclear energy has become a significant part of the world's energy production. Participation of chemical engineers in this field, especially in the manufacture of fuel grade uranium and heavy water is significant. Chemical engineers are involved in both the design and operation of such plants.

Nuclear energy is produced by a typical chain reaction such as the one shown below.

$$_0n^1 + {}_{92}U^{235} \rightleftharpoons X + Y + 2 \text{ or } 3 \text{ }_0n^1$$

This means that, when a neutron, with atomic number 1 and an atomic mass 0, strikes a uranium atom (with atomic number 92 and atomic mass 235), the latter breaks into two atoms of lesser atomic number (fragments X and Y) radiating a huge amount of energy which is produced, and at the same time releasing two or three fresh neutrons. These neutrons in turn will set up a chain reaction and an explosion occurs due to the multiplication effect.

For controlling the excessive energy release we make use of moderators which can absorb neutrons without releasing energy. Graphite is a very good natural moderator. Heavy water is another very important moderator. The production of uranium and heavy water is described below.

4.20.1 Uranium

Uranium is a heavy metal with specific gravity of 18.7, melting point of 1133°C and boiling point of 3900°C. Uranium is generally utilized as a source of energy, mostly constructive, except for a lone case which unfortunately happened during the Second World War.

Uranium is produced from uranium containing ores. Uranium ore is extracted by either acid or alkaline leaching methods and then purified or upgraded. Acid leaching is more popular but not used if there is a high lime content in the ore and then further purification is done by solvent extraction. Alkaline leaching is suitable for high lime content ores and further purification in this case is done by ion exchange methods.

See Figure 4.34. In the acid leaching process, a complexation reaction produces either uranium sulphate or nitrate. The uranium salt is then upgraded by the ion exchange process, resulting in pure uranium sulphate or uranium nitrate. The pure uranium nitrate ($UO_2(NO_3)_2 \cdot 6H_2O$) is denitrated by roasting in the presence of air to give uranium trioxide (UO_3) by the following approximate chemical reaction.

$$UO_2(NO_3)_2 \cdot 6H_2O \rightleftharpoons UO_3 + NO + NO_2 + 6H_2O + O_2$$

The uranium trioxide produced is reduced by hydrogen using a fluidized bed reactor to give pure uranium dioxide, UO_2, by the following chemical reaction.

$$UO_3 + H_2 \rightarrow UO_2 + H_2O$$

Part of the UO_2 produced is sent for making uranium carbides which are constituents of fabricated fuel elements. The reaction takes place as shown below.

$$UO_2 + 4C \rightleftharpoons UC_2 + 2CO$$

The balance UO_2 is fluorinated to uranium tetrafluoride by using hydrogen fluoride as follows. The fluorination is done using the fluidized bed reactor.

$$UO_2 + 4HF \rightarrow UF_4 + 2H_2O$$

The primary reason for the use of fluidized beds is that there are no moving parts, which minimize chances of leakage of dusts or fumes, especially because all uranium salts are toxic when inhaled ingested or brought in contact with the skin.

The uranium tetrafluoride is reduced by calcium by the following reaction.

$$UF_4 + 2Ca \rightleftharpoons U + 2CaF_2$$

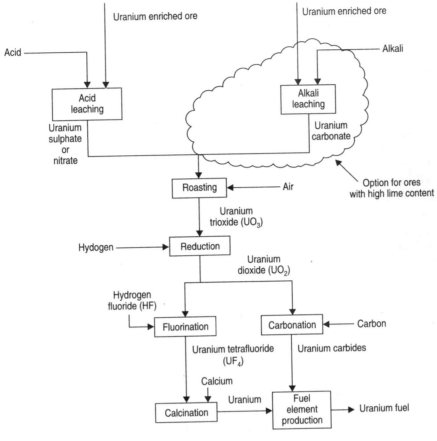

Figure 4.34 Uranium production from ores.

This uranium along with uranium carbide already obtained from carbonation of uranium dioxide is fabricated to produce nuclear fuel elements, which can release energy in a moderated manner.

Uranium from other sources

Uranium can also be produced in phosphoric acid plants since the freshly prepared phosphoric acid contains about 100 to 120 ppm of uranium as U_3O_8. This can be converted to uranium by the solvent extraction process. Copper tailings from copper refineries also contain U_3O_8 to the extent of 100 to 120 ppm. The same can also be used to recover uranium.

4.20.2 Heavy Water

Heavy water is the most important moderator used for nuclear reactors. Some of its important properties are given below.

    Density   – 1.1056 kg/m^3
    Viscosity   – 1.25 cP at 20°C

Freezing point — 3.82°C
Boiling point — 101.4°C
Heat of vaporization — 10864 kcal/kg

All the property figures given above are higher than those of normal water. But simple physical separation will be very costly since the normal water contains only 0.0143% of heavy water (D_2O). This means we have to enrich it by other means before using plane distillation.

One method of manufacture is by making enriched deuterium (D_2) adjacent to ammonia plants where a large quantity of hydrogen is available to enable separation of D_2. This D_2 is reacted with oxygen to produce heavy water.

Heavy water can be produced from normal water by many combinations of methods. The most popular process consists of dual temperature deuterium exchange using hydrogen sulphide followed by a separation process. The separation process can be either a distillation process or an electrolytic process or a combination of both. The gas liquid contact process utilizes the difference in volatility of water and heavy water whereas the electrolytic process utilizes the preferential release of hydrogen molecules instead of deuterium molecules during electrolysis.

A combined process for the production of 99.75% D_2O is shown in Figure 4.35.

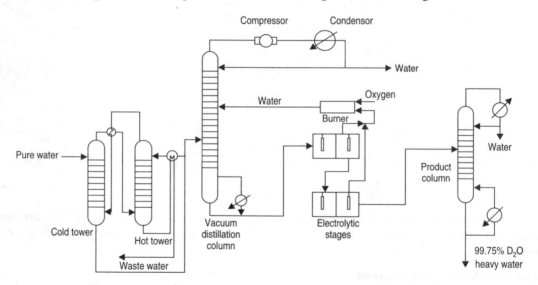

Figure 4.35 Production of heavy water by the combined method.

In the cold tower, an isotopic exchange between hydrogen sulphide and water takes place. The deuterium from the H_2S molecule is transferred to the water molecule by the following reaction.

$$H_2O + HDS \rightleftharpoons HDO + H_2S$$

In the hot tower, the second deuterium atom is also transferred to water by the following reaction.

$$H_2O + D_2S \rightleftharpoons D_2O + H_2S$$

The water vapour from the top of the hot tower goes back to cold tower. The enriched water from the cold tower bottom goes to the vacuum distillation column. In this column, water rich in deuterium (15% D) is obtained at the bottom which goes to the first stage of the electrolytic section. The hydrogen is preferentially produced as hydrogen gas in comparison to deuterium, during the electrolysis. This hydrogen is burnt with the pure oxygen, also produced during electrolysis. The water obtained is fed back to the vacuum distillation column. The heavy water produced goes to the second stage where it is concentrated further (to about 90% D) and sent to the product column.

In the product column, 99.75% D_2O is produced as the bottom product. Water (H_2O) is obtained at the top which is rejected or recycled.

Another method of heavy water production is by ammonia hydrogen exchange. This will be conducted at high pressures and is to be preferentially done near the ammonia production plants.

REVIEW QUESTIONS

1. Explain the production of precipitated calcium carbonate.
2. How is milk of lime converted to chalk in the atmosphere?
3. Why is a mechanical grate provided at the bottom of a lime kiln?
4. Compare the three processes for the production of phosphoric acid.
5. Describe the method of making NPK fertilizer.
6. What catalyst is used in a hydrogen reformer?
7. Explain the manufacture of sulphuric acid by the DCDA process.
8. Describe the production process of titanium dioxide pigment by the chloride route.
9. How is ammonia produced from natural gas?
10. How does vapour recompression work for purifying water?
11. Explain three types of reformers, considering the tube arrangement inside them.
12. What are the new developments in glass industry? (for further reading)
13. Describe the electrolytic process for the manufacture of caustic soda.
14. Explain the features of single and dual pressure processes for the production of nitric acid.
15. How is cement manufactured? What are the steps involved?
16. What are the properties of heavy water? What are the methods for making it? Explain one process with the help of a suitable diagram.
17. What are the moderators used in nuclear technology other than heavy water? (for further study)

Chapter 5: REFINERY OPERATIONS

5.1 CRUDE DISTILLATION

Petroleum contains more than 5000 components mainly consisting of carbon, hydrogen, oxygen, nitrogen and sulphur plus traces of metals. The amount of carbon in the crude ranges from 83 to 87% by weight, and hydrogen ranges from 10 to 14% by weight. It is quite impossible to analyze and work on the actual components for making calculations. Refinery operations are complex and special methods are adopted during operation as well as design.

Oil is measured in barrels. A barrel is about 159 litres (exact figure is 158.987 litres). The gravity (or liquid density) of petroleum is measured in degrees API. It is a measure of how heavy or light the petroleum liquid is, compared to water. It is done for practical reasons such as ease of measurement and communication. Note that if the specific gravity of oil is ρ, then its 'degrees API' will be $((141.5/\rho) - 131.5)$.

Any mixture will not boil at any one temperature. It boils over a temperature range based on the composition and that is why distillation curves are used to depict an oil. There are many types of curves. The True Boiling Point (TBP) curve is one of them. Then there are other ASTM distillation curves, the most popular being the ASTM D86 and ASTM D1160. D86 is used for light to medium fluids and D1160 is used for heavier fluids.

Petroleum has constituents with a boiling range of about $-160°C$ (methane) to more than $+1000°C$. The mixture of gas, liquid and solid, is converted into a number of cuts of small boiling range. These cuts are later processed and tailored to suit the requirements of consumers. Simulation and manual techniques have been developed to recover as many fractions as possible from crude, discarding almost nothing including tar.

The fractions in demand are a feature of a country or region (although most of the fractions are commonly in good demand, but not so to the same extent) and hence a refinery should have the facilities for producing such fractions as required in the markets. This in turn makes each refinery unique in its design. The most important petroleum refinery products and their uses are as follows:

1. **LPG (Liquefied Petroleum Gas)**—It is also known as cooking gas. It consists of propane, propylene, butane and butylenes and is used as domestic/industrial fuel.

2. **MS (Motor Spirit)**—It is generally known as petrol and used in automobiles.
3. **Naphtha**—It has almost similar boiling range as petrol but without any additives, and is used as petrochemical feedstock.
4. **SKO (Superior Kerosene Oil)**—It is mainly used as domestic fuel or converted to jet fuel.
5. **ATF (Aviation Turbine Fuel)**—It is used as turbine fuel in aircraft.
6. **HSD (High Speed Diesel)**—It is generally known as diesel, a transportation fuel.
7. **Fuel Oil**—It is mainly used as boiler fuel.
8. **Bitumen**—It is used for tarring of roads.
9. **Petroleum Coke**—It is used in the manufacture of carbon electrodes.

The present and future demands, future expansion plans, etc. are some factors to be considered while planning a refinery; hence a designer has to consider all of the anticipated situations. In some places, the petrol is the chief motive fuel whereas in other places it is the diesel. Many such differences exist and many more changes are also expected in the future.

Primary distillation of crude can be considered in three stages.

1. Topping Column Operations.
2. Atmospheric Distillation Unit (ADU).
3. Vacuum Distillation Unit (VDU).

The topping column operations consist of removing the soluble gases in the crude. This will also reduce the gas load on the atmospheric column.

5.1.1 Atmospheric Distillation Column

The atmospheric column is the column which separates the petroleum fractions at pressures typically of about 2 kg/cm^2. In a typical design, the pressure in an atmospheric column is about two atmospheres at the bottom, and at the top of the column the pressure is only 100 millimetres of mercury above atmospheric pressure. Such a design requires the use of a compressor to handle the gas separating at the top of the tower.

In some designs, the atmospheric column is operated at about 3 kg/cm^2 pressure at the top to avoid the use of a compressor to handle the overhead gas.

The reboiler has been done away with by heating the feed to the maximum permissible temperature only once and allowing it to flash in towers. The maximum temperature allowed in the feed heater is about 375°C. Higher temperatures are not permitted to avoid degradation of crude by thermal cracking.

When the crude contains a good amount of soluble gases, to avoid load on ADU, a preflashing or lights column is employed. Preflashing is also done when crude has to be transported to a long distance. Light-ends-free crude gives no problem in transportation. A crude containing less than 6% light ends (gases), such that the vapour pressure of crude is about 0.5 kg/cm^2, can be considered suitable for transportation. Preflashing is conducted at 100°C under a pressure of 3–5 atmospheres to remove these light ends.

Arrangement of ADU towers

The feed to the atmospheric distillations column consisting of heated crude oil (part liquid, part gas) is fed to the bottom sections of the column (around third tray from the bottom). See

Figure 5.1. The top vapours from the column are condensed. A part of the condensate is refluxed to the column and part of it is drawn out as naphtha which is the top product. Liquids are drawn out from different intermediate plates and stripped in side stripping columns and the bottom products of the strippers consisting of heavy naphtha, light distillate, kerosene and diesel are sent to tankages where they are further treated or blended. The bottom product of the atmospheric distillations column becomes the feed to the vacuum distillation column.

The reflux arrangements in ADU towers are classified into three distinguishable types.

1. Top tray reflux.
2. Pump back reflux.
3. Pump around reflux.

Top tray reflux: This reflux is at the top tray and the reflux is cooled and sent into the tower. In some towers, no reflux is given to any other plate. Operating the tower only with top reflux has some disadvantages. Heat input to column is through heated crude at the bottom and removal is at the top. This creates a large traffic of vapour, necessitating larger tower diameters. Recovery of heat is also not efficient. However, the unit is simple in design and operation.

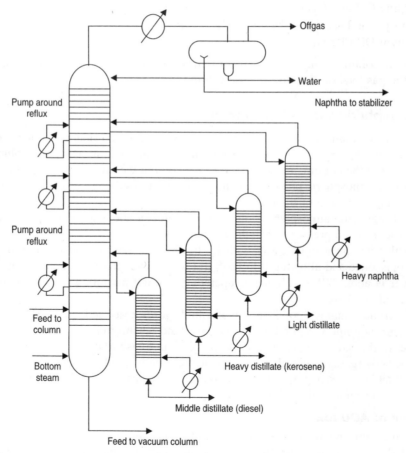

Figure 5.1 Typical atmospheric distillation.

Pump back reflux: In this arrangement of Figure 5.2(a), reflux is provided at regular intervals. This helps every plate to act as a true fractionator. The vapour load in the tower is fairly uniform and hence a uniform and smaller diameter tower will do. The rejected heat at the reflux locations can be effectively utilized. Because the tower is at progressively higher temperatures as we go downwards, the reflux locations can be where the temperature is adequate for transferring heat usefully to another stream. Even though the operation of such towers is more complex, these towers provide excellent service. Many refineries use this reflux arrangement.

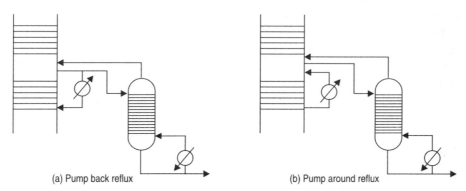

(a) Pump back reflux (b) Pump around reflux

Figure 5.2 Pump back reflux and pump around reflux.

Pump around reflux: In this arrangement of Figure 5.2(b), reflux from a lower plate is taken, cooled and fed into the column at a higher level by 2 to 3 plates. This creates a local problem of mixing uneven compositions of reflux and liquids present on the tray. To overcome these disadvantages, designers treat all the plates in this zone as one single plate, the result of which is reflected by an increase in the height of column and the number of plates in the column. Usually, this pump around is not placed at more than two sections in a column (see Figure 5.1).

The number of side draws in an atmospheric column may go up to seven or eight. Side stream strippers are provided to all crude fractionation units, to ensure quality and close control of products to specification limits.

Process simulation

Simulation software is used for process calculations. Manual calculations which were done earlier can also be used to make an overall check of the simulation results. Hence both methods are shown below.

Design and simulation: The design of atmospheric column is based upon experience and needs. Because of the unpredictable nature of crude, special features of design have been developed. Crude, even though it contains innumerable components, falls into small close boiling cuts and so individual separation is not possible; hence, the design is significantly based upon boiling curves such as True Boiling Point (TBP) and EFV (Equilibrium Flash Vaporization) curves.

The atmospheric towers can be simulated using any of the simulation softwares including ASPEN PLUS, DESIGN 2 FOR WINDOWS, HYSYS or UNISIM. Some of the required guidelines are given below.

A refining column simulation is done according to the following steps:

1. Create a unit set using FPS (Foot Pound Second) or MKS (Metre Kilogram Second) units along with common refining industry-specific units.
2. Choose a property package, say, PENG ROBINSON.
3. Select and add water, carbon dioxide, hydrogen, methane, ethane, propane, two butanes, two pentanes, hydrogen sulphide and any other components present which can be considered as free components.

   A laboratory Crude Assay Data is taken which appears as explained below:

   The True Boiling Point (TBP) distillation curve will contain liquid volume percentages starting from Initial Boiling Point (IBP) up to about 85 volume percentage with corresponding True Boiling Points. Along with this a light ends analysis will also be given. Another important input is the API gravity of the crude given in degrees API.

   API (American Petroleum Institute) gravity is defined as given below.

   Degrees API = $(141.5/\rho) - 131.5$ where ρ is specific gravity at 15.6°C

   The above details (TBP, light ends and gravity) are entered in the simulation through appropriate tables.

4. The next step is the creation of pseudo components or hypo components (both mean the same). This process is also called "Cut and Blend". Usually the crude is cut into about 30 components. But this can be changed as per our discretion. Then a stream called by the name such as "crude oil" or "Feed" is produced in the program. Move on to the simulation environment and check all properties of this stream.
5. Model preflash drum as a Separator. Model the crude oil furnace as a Heater. Model the crude column as a Refluxed Absorber equipped with Pump Arounds and Side Stream operations. Fix the number of trays for the main column and the strippers. Configure each of the strippers as yielding a product such as naphtha, kerosene, etc.
6. Install other unit operations, specify the feed and utility conditions.
7. Make a list of products and results you expect from your simulation. Run the simulation. See if you can match your expected figures after ensuring item by item that there are no bugs in your data fed.
8. Check quality and quantity of products obtained, and change the design parameters as required to obtain the required quality and quantity.

Overall check of simulation using the older manual method[1,2,3]: It is a good practice to check the overall design by means of manual methods since it may throw up some good suggestions.

The design is based upon three terms, namely, "degree of separation", "degree of difficulty of separation" and "ASTM gap"

  (i) Degree of difficulty of separation is expressed as the difference of ASTM 50% boiling points of two successive cuts. Obviously, a larger difference renders separation easier.
  (ii) Degree of separation is encountered in separation of two close boiling cuts, irrespective of the attainable purity. It otherwise closely resembles relative volatility.

(iii) ASTM gap is defined as the difference between 5% boiling point of heavy fraction and 95% boiling point of preceding cut.
(iv) When the ASTM gap is not available, TBP overlap may be taken into account. TBP overlap is the simple difference between FBP (final boiling point) and IBP (initial boiling point) of successive fractions. These are explained in Figures 5.3 and 5.4.

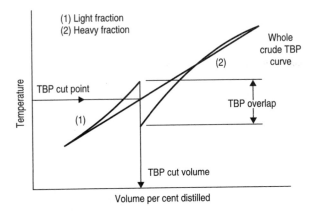

Figure 5.3 TBP overlap and cut point.

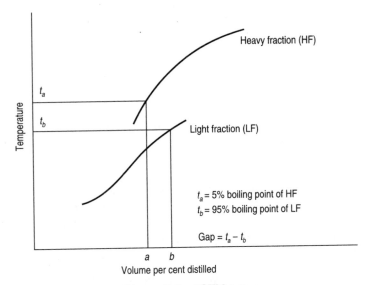

Figure 5.4 ASTM gap.

Separation capability is denoted by F factor given as:

F = reflux ratio (or reboil ratio) multiplied by the number of plates in that section.

This F is related to ASTM 50% difference of fractions (successive) and the gap of these fractions, as shown in Figure 5.5.

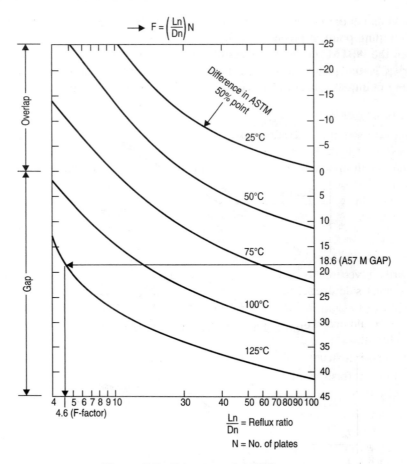

Figure 5.5 F-factor and ASTM gaps.

Example To find out the relation between the number of plates in side stream stripper and the product flow rate from it[1]. (A middle distillate (diesel) side stream stripper is shown in Fig. 5.1). In a refinery side stream operation, the fraction to be collected is diesel. The diesel entering the side stripper is 3500 bbls (barrels) per hour, the 50% point of the cut is 275°C and contaminated with kerosene whose mid-boiling point is 145°C. If the stripper is having 8 plates, find the actual amount of diesel coming out of the stripper, if the ASTM gap is 18.6.

Solution

1. Difference of 50% points of the streams is 275 – 145 = 130°C.
2. Draw a curve parallel to 125 degree curve for 130 degrees in Figure 5.5.
3. Find the F-factor value point on this curve corresponding to an ASTM gap of 18.6. We get the F-factor value equal to 4.6.
4. The F-factor is also equal to reflux ratio multiplied by the number of plates. Dividing F-factor by the number of plates, we get 4.6/8 = 0.575, which is the reflux ratio required.

Assuming equimolal counterflow this can also be defined as the ratio of reboiled liquid to input liquid to stripper column. The fraction which is likely to come out of the stripper is 1/1.575 of the total feed entering the stripper column.

5. Product draw from stripper for 8 trays = 3500/1.575 = 2222 bbls (barrels) per hour. This is only a rough value. Correct value can be obtained from simulation programs.

Other aspects of design

A large quantity of steam is utilized during distillation. Steam is inert and also causes reduction in partial pressure of hydrocarbons in the tower. Reduction of partial pressure of hydrocarbons may contribute to an apparent boiling point drop of 10–20°C in distillation columns and side stream strippers. This saves heating fuel.

The area in which the hot crude is flashed in a column is called a flash zone and the significance of flash zone is immense due to the following facts:

(i) An increase in flash zone pressure increases the draw temperatures.
(ii) An increase in overflash decreases the side draw temperatures from the second draw onwards. Overflash is that portion of total vapour leaving the flash zone boiling above the nearest side draw fraction, but never included in that fraction. Overflash allowance is kept to an extent of 2% of total crude processed in the column. This maintains a good pool of liquid and reflux on plates. This overflash is mostly from the bottom product of the column.
(iii) An increase in steam in flash zone decreases the product temperature.
(iv) Pressure in flash zone is reflected throughout the column in the form of plate temperatures.

An increase in flash zone pressure demands more quantity of steam to maintain the designed product pattern. It may be stated that while distilling heavy crudes, a reduction in flash zone pressure furnishes a good separation between products, and the yields of products will be as shown by the equilibrium flash vaporization (EFV) curve at the flash zone pressure.

Generally, the ASTM gaps followed in practice are as follows:

Light naphtha to heavy naphtha	12 to 18°C
Heavy naphtha to light distillate	15 to 30°C
Light distillate to middle distillate	2 to 6°C
Middle distillate to first draw	2 to 6°C

5.1.2 Vacuum Distillation Column

In refineries, the vacuum distillation unit (VDU) follows the atmospheric distillation unit (ADU).

The bottom of the atmospheric distillation column is called reduced crude. It contains large quantities of distillable components which can be recovered by vacuum processing. For a 30 degrees API crude, the products obtained by vacuum distillation can be to the extent of 30% of the whole crude. There are two types of vacuum distillation units based on product-mix. If the reduced crude is used for producing lubricating oils, the tower is said to be of Lubes type. If the vacuum distilled products are fed to cracking units, then the vacuum distillation unit is called to be of Fuels type. Vacuum distillation towers are usually designed for a limited number of products. A tower designed for one set of operations may not be suitable for other products.

There are also two types of vacuum distillation columns based on operation. These are 'wet type' and 'dry type'. The wet type unit uses steam to reduce the partial pressure of oil in the column. The dry type unit depends solely on the effectiveness of the vacuum inside the column to vaporize the heavy oil. Usually, ejectors are used in the tower overhead to withdraw vapours and create vacuum.

The wet type unit requires a huge quantity of steam which is again based on the amount of vacuum. The plates should have a low pressure drop. A lubes oil type of vacuum column is shown in Figure 5.6.

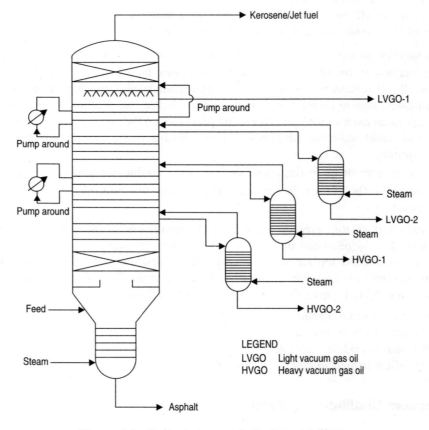

Figure 5.6 Typical vacuum distillation—lubes type

The operation of vacuum column is similar to that described for atmospheric column with refluxes, side strippers, pump arounds, steam injection, etc. except that the operation is under vacuum conditions.

Column simulation steps are also similar to those of atmospheric column.

The major operating parameters are in the ranges given below[1].

Top pressure	= 12 to 15 mmHg
Temperature at the top	= 225 to 250°C
Temperature of reduced crude (in the flash zone)	= 350 to 400°C
Pressure in the flash zone	= 30 to 40 mmHg
Steam rate	= maximum 5 kg per barrel
Overflash	= 2%

The residue obtained from vacuum column is suitable for delayed coking. Delayed coking is covered in detail in the next section.

5.2 COKING

Coking is a thermal cracking process. The coking operation may considered to be the last frontier in refining operations since the heaviest fraction (pitch or tar) is also converted into very useful products. Steel and aluminium industries are regular consumer of coke.

Coking is a process by which the vacuum residue obtained from vacuum column is processed to recover fractions of hydrocarbons and pure carbon in the form of coke. Coke is mainly used in making carbon electrodes. Different methods of coking are delayed coking, fluid coking, flexi coking and contact coking. The process of coking is complex because it produces solid, liquid and gaseous products and it combines batch and continuous operations.

5.2.1 Delayed Coking

Delayed coking which takes place in a large vessel at a slow pace (delayed) is the most important process among the various coking techniques (the precursor of delayed coking is the two-coil cracking process). It is capable of cracking all types of feed materials including solvent extracts. The process has a high decarbonizing efficiency (see Table 5.7).

The principle behind delayed coking operation is that heating is done in a furnace to initiate cracking and the chemical reactions are completed in huge and tall coke drums.

Several coke drums in series are operated in a cyclic manner. While one drum is getting filled, the other drums are in the process of coking and decoking. This way by orderly rotation of the drums the process may be contrived to work continuously. Minimum two drums are essential even for small capacity plants. However, a set of drums is desirable.

Heavy oils and light oils are recycled in different ratios to maximize the yield of either coke or distillates as per requirement. Table 5.1 shows the effect of parameters on yields[4].

Table 5.1 Effect of parameters on yields

Increased parameter	Gas yield	$C_5+(L)$ yield	Coke yield	HCGO quality
Pressure	Increases	Decreases	Increases	Improves
Recycle rate	Increases	Decreases	Increases	Improves
Coil temperature	Decreases	Increases	Decreases	Deteriorates
Velocity of medium	Decreases	Increases	Decreases	Deteriorates
Cycle time		Increases	Decreases	

The expected yield of coke may be about 30% for reduced crudes or 80% for tars and pitches. Coke from these units contains volatile matter up to 8–15% and the bulk density of the coke obtained may be around 9 kg per litre. CDE (Conradson Decarbonizing Efficiency) of the plant may be reaching up to 99.8%[1].

Usual conditions/parameters in delayed coking are given below[1]:

Heavy oil, discharge temperature	470–520°C
Coking temperature	450–470°C
Pressure in coke drums	5–6 atm
Drum diameter	4–5 m
Drum height	14–20 m
Thickness of drum walls	approx 40 mm

The most important technologies available are:

1. Foster Wheeler
2. ABB
3. CB&I Lummus
4. Conoco Philips
5. Exxon Mobile

Conventional delayed coking process is shown in the flow diagram below and the description aspects are generally based on the earlier ABB Lummus Delayed Coking process[5].

The Delayed Coker Unit is designed to produce the following products.

1. Coker gas
2. Coker LPG
3. Coker naphtha
4. Light coker gas oil (LCGO)
5. Heavy coker gas oil (HCGO)
6. Green coke

The Delayed Coker Unit consists of the following operations.

1. Feed Preheat
2. Coking
3. Primary Fractionation
4. Vapour Recovery
5. Coke Drum Steamout/Blowdown
6. Decoking
7. Green Coke Handling

Coke drums

See Figures 5.7 and 5.8. Fresh feed (vacuum residue from vacuum distillation unit of refinery) is preheated in preheat exchangers against HCGO product and HCGO pump around and passes through the fractionator boot (this is shown in Figure 5.8 only) where it gets heated by mixing with the heavy fractions, and the resulting heater charge enters the coker heater coil at a high velocity. Steam is introduced into the coil which prevents coke deposition in coil. The hot and partially vaporized mixture enters the coke drum. The coke drum is charged with hot mixture to half to two-third of the height of the drum.

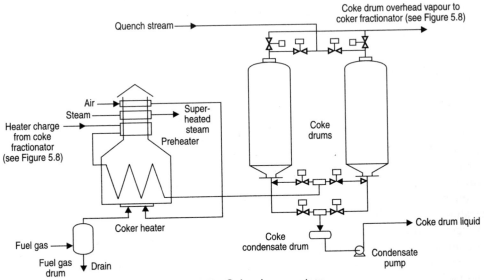

Figure 5.7 Coke drum system.

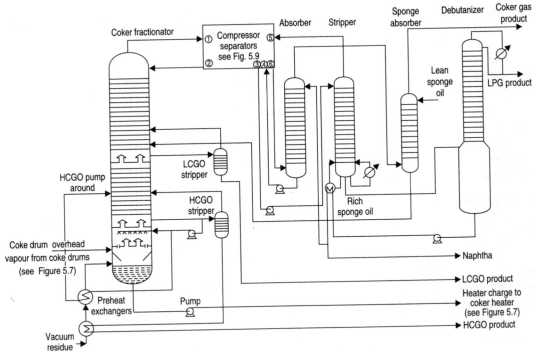

Figure 5.8 Typical coker fractionator system.

The level of hot-mix in drums is measured and controlled by cathode ray monitoring. Steam and volatiles escape from the coke drums and enter the coker fractionator. This fractionator was previously known as dephlegmator. Charging of a coke drum may require a time of 4 to 5 hours. Immediately after charging is over, the drum is isolated from the stream. Effluents of the coke heater are then switched to the second drum. Coking being slow, it usually takes a time of 10 to 16 hours. The time of charging coke drums must balance the time of coking and decoking operations.

The feed gets mixed with the stripped liquid (internal recycle) at the bottom of the coker fractionator. This recycled stream is heavier than HCGO and condenses in the wash section of the column. This mixture is called heater charge.

The introduction of relatively cool coker feed into the fractionator bottom reduces the tendency for coke formation in the tower bottoms. Fractionator bottoms liquid level is maintained by regulating the flow rate of coker feed from the coker feed drum. A side stream of fractionator bottom liquid is continuously circulated through the fractionator bottoms strainers by the fractionator bottoms recirculation pump, in order to remove the coke fines.

The coker heater charge is pumped by the coking heater charge pump, to the coker heater. The primary function of the coker heater is to quickly heat the feed to the required reaction temperature while avoiding premature coke formation in the heater tubes.

A steam coil above the process convection section in the coker heater superheats MP steam. This steam is utilized for stripping LCGO (Light Coker Gas Oil) and HCGO (Heavy Coker Gas Oil). The coker heater combustion air is also preheated against flue gas in the convection section in order to increase efficiency. The coker heater is fired with fuel oil or a fuel gas or combination of both fuels. A heater fuel gas drum separates any condensable liquids in the fuel gas before they reach the coker heater.

A two-heater configuration would allow offline decoking of one furnace without having to shutdown the entire unit.

The coke drum feed leaves the heater at approximately 506°C and 4.0 kg/cm². The coke drum inlet switch valve diverts the hot coker feed into the bottom of the filling or coking mode coke drum. In the coke drum, the hot feed cracks to form coke and cracked products.

Each coke drum is filled in a 24-hour period. An antifoam chemical is injected into the coke drum to prevent foam going into the coker fractionator.

The cracked product leaves the top of the coke drum at 450°C and 1.05 kg/cm², which is quenched to 426°C or less with HCGO to stop the cracking and polymerization reactions, and to thereby prevent coke formation in the vapour line to the coker fractionator.

Fractionation

See Figure 5.8. The fractionating column is provided with a distillations zone of typically 24 trays and a spray zone, divided into two major sections by the HCGO draw pan. The quenched coke drum effluent vapour flows upwards through the spray chamber, with some degree of cooling accomplished by contact with HCGO wash liquid. Heavy recycled liquid is condensed and collected on the wash section chimney and flows to the bottom sump to combine with fresh coker feed as already stated.

The net product vapour flows to the upper tower section through vapour risers in the HCGO draw-off pan. This vapour consists of the products and steam. Heat removal and fractionation are accomplished in the upper section of the coker fractionator.

HCGO production: HCGO draws are taken from the same draw tray and used for quenching, pump around and HCGO stripper feed. The HCGO quench, HCGO wash and HCGO pump around are pumped by common pumps. The heat in the pump around is also used to produce steam.

HCGO product draw from the coker fractionator flows under level control by gravity flow to the stripper. It is steam-stripped in the stripper and vapours are returned to the fractionator. The stripped HCGO product is pumped by the HCGO product pumps to heat the fresh feed as described earlier. Then, it is utilized in generating steam in the MP steam generator.

Finally, the HCGO product stream is cooled to 80°C in the HCGO product air cooler and is routed as (cold) HCGO product to the refinery fuels blending section.

LCGO production: Light Coker Gas Oil (LCGO) is typically withdrawn from the chimney tray below Tray 15 of the coker fractionator and it flows by gravity under level control to the light coker gas oil stripper. MP superheated steam is used for stripping LCGO product which is then successively cooled to 40°C and then goes to coalescers and salt driers, and then to storage/blending.

Fractionator overhead system

The fractionator overhead system separates the combined vapour and liquid overhead products from the coker fractionator into sweet coker gas, LPG and coker naphtha. It mainly consists of wet gas compression, absorber, stripper, sponge absorber and debutanizer.

Condensation: Overhead vapour from the coker fractionator is cooled by air cooler and condensed in the coker fractionator overhead condensers, and coker fractionator overhead trim condensers. It enters the coker fractionator overhead receiver where vapour, hydrocarbon liquid, and condensed water are separated. Vapour from the overhead receiver is routed to the vapour recovery section.

The fractionator reflux pumps are used to pump a portion of the hydrocarbon liquid back to the fractionator top tray as reflux under flow control reset by the tower overhead temperature. This indirectly controls the naphtha endpoint temperature. The fractionator overhead liquid pump is used to pump the net naphtha stream to the vapour recovery section. The sour water pumps are used to pump the condensed water on level control to the storage tanks. A portion of the sour water is recycled under flow control to the inlet of the compressor interstage cooler.

Wet gas compression: See Figure 5.9. The wet gas compressor is a two-stage high pressure steam turbine driven compressor (100% condensing type), that compresses wet gases from the coker fractionator receiver.

The first section discharges vapour at 3.6 kg/cm^2. The vapour is then cooled to 40°C in the compressor interstage cooler where vapour, hydrocarbon liquid, and water are separated. The condensed sour water recycles under its own pressure, back to the inlet of the coker fractionator condenser as wash water. The 'compressor interstage liquid pump' pumps the hydrocarbon liquid to the inlet of compressor discharge cooler, where it joins the second section discharge stream from the wet gas compressor.

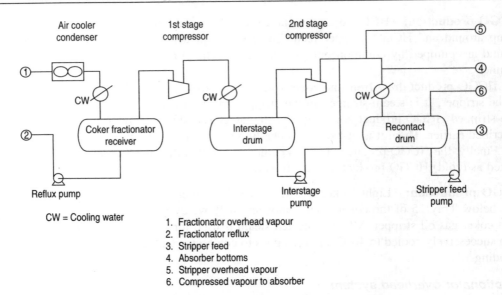

Figure 5.9 Wet gas compressor separators.

The second stage of the wet gas compressor compresses the vapour from the compressor interstage drum to 14.9 kg/cm² where the vapour combines with the interstage drum hydrocarbon liquid, the recycled sour water from the coker fractionator receiver, stripper overhead vapour from the stripper and absorber bottoms liquid from the absorber. The combined stream is cooled and condensed in the compressor discharge cooler and enters the recontact drum.

Absorber: Vapour from the recontact drum enters the bottom of the absorber. The absorber has approximately 30 trays. Cooled naphtha (lean oil) from the debutanizer bottoms is fed to top of the absorber, primarily to reduce the loss of LPG in the coker gas stream. Absorber overhead vapour is routed to the downstream sponge absorber. The absorber bottoms return the rich lean oil from the absorber bottom to the recontact drum via the compressor interstage cooler, under level control.

The liquid from tray 15 of the Absorber is pumped through the absorber intercooler into the absorber water separator. The hydrocarbon liquid flows from the top of the separator back to the column and enters on tray 17. Water from the bottom of the separator flows back to the coker fractionator overhead receiver.

Stripper: Liquid from the recontact drum goes to the stripper. The stripper column strips ethane and lighter components from the LPG and naphtha. The stripper has two reboilers. Liquid from the bottom sump of the column flows to the shell side of the stripper lower reboiler by thermosiphon action. The hot debutanizer bottoms stream is the heating medium. The MP steam flows to the upper reboiler.

Vapour from the top of the stripper column goes to the recontact drum via the compressor interstage cooler. Stripper bottoms liquid flows by its own pressure to the Debutanizer. A water trapout drum is provided towards the top portion of the stripper column to remove any free water.

Sponge absorber: The absorber overhead vapour flows to the bottom of the sponge absorber, which has about 21 trays. Cooled, lean sponge oil (Light Coker Gas Oil) from the light coker gas oil stripper is fed to the top of the sponge absorber to reduce the loss of naphtha in the absorber overhead vapour stream.

The sponge absorber overhead vapour is sour coker gas routed to refinery fuel system. The rich sponge oil from the sponge absorber bottoms is used to cool the lean sponge oil before it is returned to the coker fractionator.

Debutanizer: The debutanizer column is provided with typically about 40 trays. The unstabilized naphtha is separated into an overhead product of coker LPG and a bottom product of stabilized naphtha in this tower. The overhead vapour is totally condensed in the debutanizer overhead condensers and then routed to the debutanizer overhead reciever. From there, the debutanizer reflux pumps, pump the reflux back to the top tray.

A part of the reflux pump's discharge is drawn off as coker LPG product under receiver's level control, and is routed to LPG storage bullets or spheres. Pressure is controlled in the receiver by throttling the overhead condenser's hot vapour bypass control valve.

The debutanizer bottoms stream is pumped by the debutanizer bottoms pumps to supply stripper reboiler duty. Then the debutanizer bottoms stream is cooled to 40°C in the debutanizer bottoms air cooler followed by debutanizer bottoms trim cooler. It is finally routed as naphtha to the refinery tankage areas.

General coke drum operations

Coking is a semi-batch operation. Each drum will be on-line for 24 hours for coking including filling and off-line for 24 hours of decoking. The total cycle for each drum from oil-in to oil-in will thus be 48 hours.

The time taken for different operations is approximately given in Table 5.2.

Table 5.2 Time for coke drum operations

Operation	Hours
Coking	24.0
Switch Drums	0.5
Steamout to Coker Fractionator	0.5
Steamout to Blowdown System	1.0
Slow Water Cooling	1.0
Fast Water Cooling	4.0
Drain Coke Drum	3.0
Remove Top and Bottom Heads	1.0
Hydraulic Coke Boring/Cutting	4.0
Reheading/Pressure Testing	2.0
Drum Heat-up	5.0
Idle	2.0
Total	48.0

Switch-over of drums: Following the completion of a coke drum filling cycle, coker heater effluent is diverted to the other (empty) coke drum by means of the switch valve. Steam is then injected into the bottom of the coke-filled drum for about 30 minutes, with volatile light hydrocarbons purged to the coker fractionator. During the next 60 minutes, the steam rate is increased and the resultant vapours (mostly steam) are routed to the bottom of the blowdown tower.

Steamout and blowdown of drums: The blowdown tower cools and condenses the steam and hydrocarbon vapours evolved from the coke drum during the steaming, water cooling, and drum reheating operations. It has typically 12 special type trays to contact the incoming vapours from the coke drum with a stream of recirculating quench oil to condense the high-boiling-point hydrocarbons. The quench oil circulation rate is adjusted to maintain approximately 180°C as the blowdown tower overhead vapour temperature to prevent water condensation in the tower.

Effect of API gravity on delayed coker yields[1]

The effect of gravity, measured in degrees API, on delayed coker yields is enumerated in Table 5.3:

Table 5.3 Effect of API gravity on delayed coker yields

Charge API	Carbon residue (%)	Drum outlet temp in (°C)	Product yields in wt%				
			Gas (vol%)	Coke	Gasoline	Gas oil	Heavy gas oil
12.3	16	435	4.2	16.6	8.9	19.0	53.3
13.0	16	435	4.0	15.5	8.8	17.3	54.4
14.0	16	435	3.6	13.6	8.6	19.1	55.1

Effect of recycle ratio[1]

The effect of recycle ratio on product yields is given below in Table 5.4.

Table 5.4 Effect of recycle ratio on product yields

Recycle ratio in (vol. %)	Product yields				
	Coke	Gas	Gasoline	Gas oil	Heavy gas oil
0	15.5	4.0	13.0	21.5	46.0
20	18.0	5.5	15.0	26.0	36.0
40	20.0	6.0	15.5	28.5	24.5

Hydraulic jet decoking

Coke when set is very hard, and much difficulty is encountered in removing it. Decoking is a time consuming and arduous task. Earlier mechanical breaking with hammers was used, a method still prevailing in some countries that offer cheap labour. Drilling and mild dynamiting are also allowed.

Lot of developments have taken place in the direction of decoking operations. In one method, water jets at 150 to 200 atmosphere pressure are used to hit the coke formation. They break the coke into pieces.

Of late, an ingenious chain pulling technique consists of suspending strong chains in coke drums from the hooks fixed to the thick shell of the drum at the top. The other ends of the chains are free to fold and lie submerged under charge. After coking is over, steaming is usually done to drive off hydrocarbon vapours. The top flange is disconnected and the chains are pulled by cranes; this upward thrust shatters the coke to pieces, making it easy for removal.

Simulation/calculations

Simulation of coking fractionators is a little more complex than other simulations. While simulation is conducted, the number of theoretical trays may be fed (instead of the actual number of trays), a percentage of the actual number of trays for easy convergence which can be adjusted to real trays later on. The reason for this is simple. The number of real trays does not really affect the material and energy balance if you feed the pressure drop across the tower correctly. The number of theoretical trays affects the simulation. Pressure drop across trays should be accurately given. The simulation can be easily done using HYSYS or UNISIM.

A coker fractionator simulation is done using the following steps:

1. Create a unit set using FPS or MKS units along with common refining industry specific units.
2. Choose a property package, say PENG ROBINSON.
3. Select and add water, carbon dioxide, hydrogen, methane, ethane, propane, two butanes and two pentanes, hydrogen sulphide and any other components present which can be considered as free components.
4. Create hypothetical components (also called pseudo components).
5. Start with the hot overhead vapours from the coke drums as a feed to the coker fractionator. The wash section of the column should be treated as a separate scrubber through which the vapours pass and enter the fractionator. Model the fractionator as a refluxed absorber equipped with side stream operations. Fix the number of trays for the main column and the strippers. Configure each of the strippers as yielding a product. Converge it giving random results, add the side strippers and converge it again. With these results and assumed values, converge the overhead system and add it to the fractionator system. With assumed values, converge the stripper, two absorbers and debutanizer one by one. Finally, join the whole system together to give the required result.
6. Make a list of products and results you expect from your simulation. Run the simulation. See if you can match your expected figures after ensuring item by item that there are no bugs in your data fed.

Check quality and quantity of products obtained, and change the design parameters as required to obtain the required quality and quantity.

Case study example

A typical case of a plant for producing 500 tonnes per day of coke is simulated using the HYSYS or UNISIM software.

The simulation does not include the coke drums and the coking process. It starts from the overhead vapours from the coke drums going to the fractionation column. The reason is that it is not possible to get a complete analysis of CDU and VDU residues. A reasonable analysis starts from the coke drum overhead vapours only. However, user written programs can be made and used based on certain assumptions. Typical long chains of hydrocarbons can be assumed to be breaking up into smaller hydrocarbons with a certain portion ending up as coke. The propensity of cracking is an outright function of Conradson Decarbonizing Efficiency[1] (CDE).

220 Chemical Process Technology and Simulation

Apart from the above, Computational Fluid Dynamics (CFD) software may be used to analyse the flow of gases and liquids through the semicoke bed formed, as the coking proceeds.

A broad material balance is given below highlighting the production of various intermediate products obtained along with coke. While calculating production, a day of 22 hours is taken instead of 24 hours to accommodate fluctuations in flow rates during changeover of drums.

Daily production rate = 500 tonnes
Hourly production rate required = 55/22
= 22.73 tonnes
Design capacity (add 3.3% extra) = 23.5 tonnes per day

The simulation of the fractionation for the above plant gave the following results, shown in Tables 5.5 and 5.6.

Table 5.5 gives the inputs and output of coker heater.
Table 5.6 gives the outputs obtained from the complete system.

Table 5.5 Mass balance: inputs and heater streams

Stream	Coke drum overhead vapours to fractionator		Vacuum residue		Hot feed to heater	
Pressure	2.7		5		5	
Temperature	449		200		297	
Flow rate	68,768		85,000		89,250	
Component flow	Flow rate (kg/h)	%	Flow rate (kg/h)	%	Flow rate (kg/h)	%
H_2	38.5	0.056	0	0	0	0
H_2S	1180.15	1.716	0	0	0	0
CO	30.25	0.044	0	0	0	0
CO_2	30.3	0.044	0	0	0	0
N_2	0	0	0	0	0	0
Methane	1502.15	2.184	0	0	0.15	0
Ethane	1229.6	1.788	0	0	0.1	0
Ethylene	146	0.212	0	0	0.05	0
Propane	1021.65	1.484	0	0	0.15	0
Propylene	420.55	0.710	0	0	0.1	0
n butane	478.1	0.695	0	0	0.15	0
i butane	120.55	0.175	0	0	0.1	0
1 butene	514.6	0.748	0	0	0.05	0
n pentane	53.5	0.077	0	0	0.05	0
i pentane	430.15	0.625	0	0	0.15	0
1 pentene	74.55	0.108	0	0	0.1	0
Cyclopentanes	578.75	0.841	0	0	0.25	0
C_6+Residue	57865.95	84.146	0	0	4247.45	4.78
Vacuum residue	0	0	0	0	0	0
H_2O	3052.7	4.438	0	0	0	0
Residue	0	0	85000	100	85000	95.22
Coke	0	0			0	0
Total	**68,768**	**100**	**85000**	**100**	**89248.85**	**100**

Table 5.6 Mass balance: final product streams in the same order as in Table 5.5 (pressure, temperature, flow rate and components)

	HCGO		LCGO		Naphtha		LPG		Coker gas		Coke	
Pressure	6		5		6		13.7		14.5		1	
Temperature	80		40		40		39		43		30	
Flow rate	20956	0	26366	0	7433	0	2535.85	0	4206.65	0	23503	0
Component rate	kg/h	%	kg/h	%	kg/h	%	kg/h	%	kg/h	%	kg/h	%
H_2	0	0	0	0	0	0	0	0	38.5	0.941	0	
H_2S	0	0	0	0	0	0	224.95	8.870	955.05	22.704	0	
CO	0	0	0	0	0	0	0	0	30.25	0.719	0	
CO_2	0	0	0	0	0	0	0	0	30.25	0.719	0	
N_2	0	0	0	0	0	0	0	0	0	0	0	
Methane	0	0	0	0	0	0	0	0	1502	35.704	0	
Ethane	0	0	0	0	0	0	1.8	0.071	1227.7	29.184	0	
Ethylene	0	0	0	0	0	0	0	0	146	3.471	0	
Propane	0	0	0	0	0.05	0.0006	948.55	37.406	72.9	10733	0	
Propylene	0	0	0	0	0	0	359.2	14.165	61.3	1.457	0	
n butane	0	0	0	0	47.85	0.6438	397.2	15.663	32.7	0.777	0	
i butane	0	0	0	0	2.35	0.0316	115.75	4.565	2.4	0.057	0	
1 butane	0	0	0	0	16.65	0.224	484.25	19.096	13.55	0.322	0	
n pentane	0	0	0	0	50.5	0.679	0.1	0.004	2.65	0.063	0	
i pentane	0	0	0	0	377.2	5.095	3.5	0.138	47.5	1.129	0	
1 pentane	0	0	0	0	67.25	0.904	0.4	0.0158	6.55	0.156	0	
Cyclopentanes	0	0	0.15	0.0006	571.7	7.690	0.15	0.006	3.75	0.0892	0	
C_6+Residue	20945	99.948	26334	99.88	6299.5	84.750	0	0	12.15	0.288	0	
Vacuum residue	0	0	0	0	0	0	0	0	0	0	0	
H_2O	11	0.0524	31.5	0.1194	0	0	0	0	21.5	0.511	0	
Residue	0	0	0	0	0	0	0	0	0	0	0	
Coke		0		0		0		0		0	23503	0
Total	20956	100	26366	100	7433.05	100	2535.85	100	4206.7	100	23503	100

Overall check by manual calculation

Manual calculation can be done for coking fractionators too, as it can be done for crude and vacuum distillations. The method will depend on graphs for separations based on OVERLAP and GAP. These charts are given in R.N. Watkin's book titled *Petroleum Refinery Distillation, Second Edition*[3], Figures 4.2 and 4.3. These charts are common and can be used for FCCU fractionator and coker fractionator.

5.3 OTHER TYPES OF COKING

5.3.1 Fluid Coking

W.L. Nelson predicted that continuous coking may revolutionize the coking methods[6]. However, it has not yet replaced batch operating coke drums very significantly.

Search for continuous coking has inducted the concepts of fluidization with necessary modifications. Fluid catalytic crackers use catalysts to aid cracking and here cracking and coking are catalysed by coke particles. Explicitly, it is nothing but the formation of excessive coke on the surface of particles; as a special case, these happen to be on the coke particles themselves.

Coke particles produced in the same unit assume more or less spherical shape and act as heat carriers while travelling from burner (regenerator) to reactor, and coke carriers in reverse travel. Some portion of steam stripped coke is burnt and the remaining coke is taken out. The hot coke particles are in a state of fluidization caused by incoming vapours. Thus, the effective continuous circulation of coke seeds, namely coke particles, is unavoidable. A close look at Figure 5.10 reveals the process sequence. About 20% of coke produced is consumed to maintain coking reaction[1].

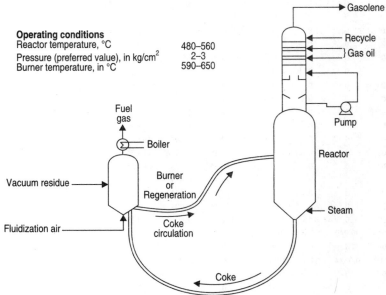

Figure 5.10 Fluid coking.

5.3.2 Flexi Coking

The feed consisting of heavy residues is fed into a scrubber fractionator reactor. Here thermal cracking takes place. Steam is admitted from the bottom. The coke fines circulate through the heater where further coke formation takes place and then passes on to gasifier where it encounters a stream of air and steam. Coke can be withdrawn between the heater and the gasifier.

5.3.3 Contact Coking

In contact coking, coke circulates always between the reactor and the heater. A part of coke is always essential for supplying thermal energy and the remaining portion is separated in the disengager. This method gives great flexibility in operation and control.

5.3.4 Comparison of Methods of Coking[1]

A comparison of the methods of coking discussed is shown in Table 5.7.

Table 5.7 Comparison of methods of coking

	Delayed coking	Fluid coking
Gravity API	15	15
Conradson Carbon Residue	9	9
Sulphur	1.2	1.2
Products		
C_3 and lighter fractions %	6.0	5.5
Coke %	22.0	11.0
CDE%	99.8	91.2
	Flexi coking	*Contact coking*
Gravity API	18.9	18.9
Conradson Carbon Residue	11.7	11.7
Sulphur	0.6	0.6
Products		
Lighter fractions %	7.5	14.9
Coke %	13.0	20.0
CDE%	99.3	99.3

Note: CDE indicates Conradson Decarbonizing Efficiency

5.4 CATALYTIC CRACKING

5.4.1 Fluid Catalytic Cracking Unit

The most important catalytic process unit in refineries is the fluid catalytic cracker unit commonly known as FCCU. High molecular weight portions of crude oil are converted to gasoline and to high demand olefinic gases which can be used as raw material for petrochemical production. High boiling point molecules are thereby converted to high demand, low molecular weight substances.

The feedstock for FCCU consists of heavy gas oil with initial boiling point around 340°C and average molecular weight ranging between 200 and 650.

Basically, there are two different configurations for FCC units. One is the "stacked" type where the reactor and the catalyst regenerator are contained in a single vessel with the reactor above the catalyst regenerator, and the "side-by-side" type where the reactor and catalyst regenerator are in two separate vessels. The side-by-side design is used by several companies whereas the stacked configuration is mainly used by Kellogg Brown & Root only[7].

Typical process description

See Figure 5.11. The high-boiling petroleum feedstock, consisting of long-chain hydrocarbon molecules, is preheated to anywhere between 315 and 430°C. Then it is combined with recycled slurry oil from the bottom of the distillation column and injected into the catalyst riser of the reactor. It is vaporized and cracked into smaller molecules by contact with hot powdered catalyst recycled from the regenerator. This is where all the cracking takes place. The upcoming vapours fluidize the powdered catalyst and the mixture of vapours and the catalyst flows upward to enter the reactor at a temperature of about 535°C and a pressure of about 1.7 kg/cm^2.

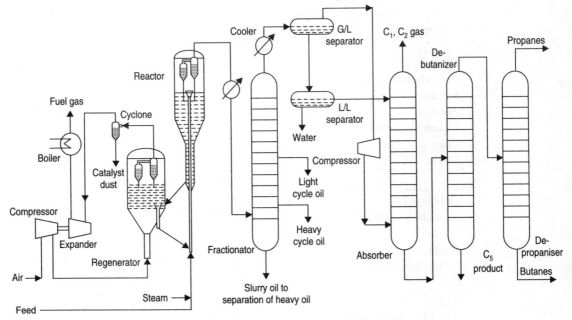

Figure 5.11 Fluid catalytic cracking unit.

The reactor is in fact merely a vessel in which the cracked product vapours are:

(i) separated from the spent catalyst by flowing through a set of two-stage cyclones within the reactor and

(ii) the spent catalyst flows downward through a steam stripping section to remove any hydrocarbon vapours before the spent catalyst returns to the catalyst regenerator. The flow of spent catalyst to the regenerator is regulated by a slide valve in the spent catalyst line.

In the early stages of development the catalyst used was an aluminium trichloride solution. Later, the catalyst used was amorphous silica alumina. It had superior properties like high thermal and attrition stability, high activity and optimal pore structure. Now the catalysts used are zeolite-based which are even better since they are more active, more stable, form less coke, and have higher acidity.

Since zeolite crystals are too active for practical use in the reactors, the zeolite is diluted with porous silica-alumina material. The dilution also has the benefit of increasing the pore diameter which allows bigger-sized molecules to enter the pores and get cracked[8].

The cracking reactions produce some carbonaceous material or coke. This material deposits on the catalyst and very quickly reduces the catalyst reactivity. The catalyst is regenerated in the regenerator by burning coke with air blown into the regenerator. The regenerator operates at a temperature of about 715°C and a pressure of about 2.4 kg/cm^2. The exothermic chemical reaction produces a large amount of heat that is partially absorbed by the regenerated catalyst, which in turn provides the heat required for the vaporization of the feedstock as well as the heat required for catalytic cracking. This makes the FCCU system a largely heat-balanced system.

The hot catalyst at 715°C leaves the regenerator and flows into the catalyst well where any entrained combustion gases are allowed to escape to the upper part of the regenerator.

Then the regenerated catalyst flows to the feedstock injection point below the catalyst riser. This flow is regulated by a slide valve in the regenerated catalyst line. The hot flue gas exits the regenerator after passing through a multiple sets of two-stage cyclones which remove the entrained catalyst from the gas.

The amount of catalyst circulating between the regenerator and the reactor is about 5 kg per kg of feedstock[7].

Distillation section: The distillation section consists of the main fractionating column, oil water separators, absorber, debutanizer, depropanizer and other columns similar to other refinery fractionation sections. Cracked naphtha, sulphur compounds, butane and butylenes, propane and propylene, ethane and ethylene and methane are recovered and used in the refinery itself or sold to petrochemical manufacturers.

The bottom product oil from the main fractionator contains the residual catalyst particles which were not completely removed by the cyclones in the top of the reactor and hence the bottom product is referred to as slurry oil. Part of that slurry oil is recycled back into the main fractionator above the entry point of the hot reaction product vapours so as to cool and partially condense the reaction product vapours as they enter the main fractionator. The remainder of the slurry oil is pumped through a slurry settler. The bottom oil from the slurry settler contains most of the slurry oil catalyst particles and is recycled back into the catalyst riser by combining it with the FCC feedstock oil. The decant oil is withdrawn from the top of slurry settler for use as a heavy fuel oil blending component, or as a carbon black feedstock.

Pollution control: The flue gases from the FCCU are highly polluting since they contain sulphur oxides and nitrogen oxides. Hence, suitable sulphur recovery and effluent treatment measures are required as per regulations.

Effect of oil contaminents on catalyst

Modern FCC catalysts are fine powders with a bulk density of 0.80 to 0.96 g/cc. They have a particle size distribution ranging from 10 to 150 micrometre and an average particle size of 60 to 100 micrometre. The design and operation of an FCC unit is dependent upon the chemical and physical properties of the catalyst.

The matrix component of an FCC catalyst contains amorphous alumina which provides catalytic activity sites resulting in larger pores that allow entry for larger molecules than does the zeolite. The binder and filler components provide the physical strength and integrity of the catalyst. Nickel, vanadium, iron, copper and other metal contaminants to the catalyst are present in FCC feedstocks in minute quantities. They all have detrimental effects on the catalyst activity and performance. Nickel and vanadium are specifically very detrimental towards catalyst activity.

Some of the methods for mitigating the effects of contaminants are:

1. Avoiding feedstocks with high metals content.
2. Allowing feedstock feed pretreatment for removing some of the metals and sulphur.
3. Increasing the addition of the fresh catalyst.
4. Demetallization: There are commercial proprietary processes which remove nickel and vanadium from the withdrawn spent catalyst.

5. Metals passivation: Certain materials can be used as additives which can be impregnated into the catalyst or added to the FCC feedstock in the form of metal-organic compounds. These passivate the contaminants by forming less harmful compounds. Antimony is effective in passivating nickel. Tin can passivate vanadium. A number of proprietary passivation processes are available which can be used.

Design and simulation

The simulation of an FCCU unit is somewhat complicated due to the presence of many catalytic reactions. The hints given in respect of the simulation of coker fractionators in the previous section are all applicable to the simulation of FCCU fractionation section as well.

5.4.2 Hydrocracking

FCCU and hydrocracking are not competing technologies but complementary technologies.

Hydrocracking involves the cracking of heavy oils in the presence of hydrogen using fixed-bed catalytic reactors since immediate regeneration of the catalyst is not required as in FCCU. Regeneration is required only once in two years. Hydrocracking uses more aromatic feeds such as FCC cycle oils along with atmospheric and vacuum gas oils. It will be gaining more acceptance in the future due to the decreasing consumption of heavier oils in the markets around the world.

A typical hydrocracking process flowsheet is shown in Figure 5.12. Here the feed is mixed with recycled and fresh hydrogen and first stage hydrocracking is done in the first reactor and the products obtained are sent to separators to separate the unreacted hydrogen. The separated liquid is distilled in a fractionator which produces C_1, C_2, C_3, C_4, gases at the top, naphtha and middle distillates as side draws. The bottoms are heated and sent to the second hydrocracker whose cracked products are also mixed with those of the first cracker.

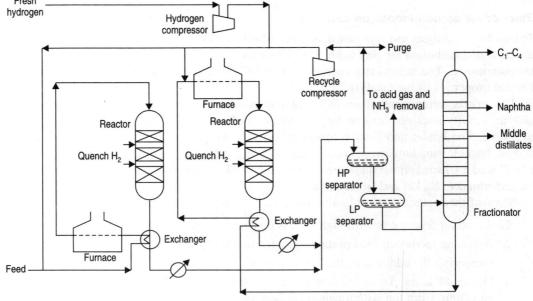

Figure 5.12 Two-stage hydrocracking process.

5.4.3 Catalytic Reforming

The catalytic reforming process is a method of producing high octane gasoline from modern refineries. It is done by reforming naphtha. The reforming process mainly consists of rearranging the molecules (with or without the addition or substraction of hydrogen atoms).

Some of the processes that take place during reforming are as follows:

(i) Naphtha consists of C_6 to C_{11} straight chain hydrocarbons. One of the chemical reactions is dehydrogenation of some of the molecules in naphtha, for example, methyl cyclohexane being converted to toluene.
(ii) Cyclization of molecules such as hexane to form methyl cyclohexane.
(iii) Isomerization of molecules such as heptanes converted into isoheptanes.

The above changes have the advantage of increasing the octane number of gasolene made from naphtha. They also have an additional advantage of forming large quantities of aromatics, which have a high demand from petrochemical companies.

Raw materials and products

The preferred raw materials for catalytic reforming are the following:

1. Straight-run naphtha
2. Coking naphtha (naphtha from coking plant)
3. Catalytic naphtha obtained from catalytic cracking.

Typically, the product will contain 45 to 60% aromatics suitable for high octane number gasoline or for extraction and sale of individual aromatics such as benzene, toluene or xylene.

Catalysts and reaction kinetics

The most desirable catalyst in terms of selectivity and activity is platinum. However, platinum is also very costly. Since loss of platinum occurs during the process, other catalysts such as chromium or molybdenum oxides or cobalt molybdate on silica-alumina base catalysts have also become competitive.

The chemical reactions taking place are endothermic and hence the heat of reaction is provided by heating the gas to a high temperature before it enters the catalyst bed. Similarly, low pressure is favoured by the reaction. Previously, high pressures were used but new plants operate at low pressures, such as 4 kg/cm^2. Since the catalyst gets poisoned, regeneration is essential.

Process description

There are three main processes of production based on the catalyst regeneration method employed. These are:

1. The fixed-bed process (not preferable)
2. Semi-continuous catalytst regeneration process
3. Moving bed continuous catalyst regeneration process.

The arrangement shown in Figure 5.13 is common to semi-continuous and continuous catalyst regeneration processes.

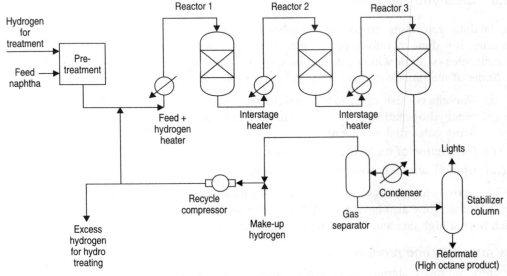

Figure 5.13 Catalytic reforming process.

The first step is the pretreatment of the feed, the requirement of which depends on the process adopted, catalyst used, catalyst poisons and undesirables present in the feed. The catalyst can be adversely affected by lead or arsenic, hydrogen sulphide and sulphur compounds, ammonia and organic nitrogenous compounds. These are largely removed during pretreatment.

Then the treated feed consisting of naphtha mixed with hydrogen enters three sets of heaters and reactors. In the heaters, the feed is preheated to approximately 500°C. The effluent from the reactor is again heated to 510°C and 520°C respectively for the second and third reactors.

In the earlier semi-regenerative fixed-bed technologies, high pressures of the order of 35 to 45 kg/cm^2 were being used. In the new continuous regeneration moving bed catalytic processes, low pressures, about 4 to 5 kg/cm^2, are being used.

The effluent obtained from the third reactor is sent to a separator where the hydrogen gas is separated and recycled and also used for other refinery operations such as hydrotreating.

The liquid product from the third reactor is fed to a stabilizer column, where lights are separated at the top and the product, called reformate, is taken out at the bottom.

Catalyst regeneration

The arrangement of the moving bed continuous catalyst regeneration process is shown in Figure 5.14.

Catalyst regeneration is a continuous operation in the case of moving catalyst beds. The catalyst moves from the top of the reactor to the bottom at a slow rate, and enters the outlet hoppers. It also moves from the first reactor to the second, and so on. On reaching the bottom of the last reactor, it is regenerated in the catalyst regeneration section and fed back to the top of the first reactor. Spherical catalysts are being used for easy movement through the equipment.

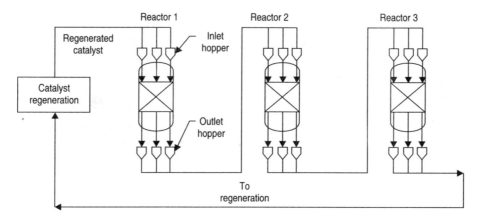

Figure 5.14 Continuous catalyst regeneration.

REVIEW QUESTIONS

1. What are the nine salable products from refineries and how are they used?
2. Why do we use distillation curves for petroleum products?
3. What are the refluxes used in the atmospheric distillation column?
4. What is the disadvantage in giving more than two pump arounds in a column?
5. What are the steps for simulating a refinery distillation column?
6. What is flash zone in an atmospheric distillation column? What is its significance?
7. What are the major operating parameters of a vacuum distillation column?
8. How are the hot vapours from coke drum converted into HCGO, LCGO, naphtha, LPG and coker gas?
9. How is the coke removed from the drums?
10. Describe the FCCU method for obtaining light end hydrocarbon products. What is the effect of metal contaminants in FCCU feed?
11. Briefly describe the hydrocracking process.
12. What is the benefit of catalytic reforming? How is it done?
13. In which countries is the use of high octane gasoline prevalent? What are its grades and compositions? (for further reading)

Chapter 6

OIL AND GAS (UPSTREAM)

6.1 PHASE SEPARATION

Oil and Gas industry (also called 'Upstream' to differentiate it from refinery operations called 'Downstream' as per industrial practice) includes the extraction, preprocessing and distribution of petroleum and natural gas. Due to the relative unpredictability of petroleum reservoirs, a facility handling oil and gas has to be built such that it is flexible for accepting different cases with variations in flow rates and properties.

Separators for oil and gas production facilities are different from those of ordinary separators used in other chemical industries. This is due to the fact that these separators have to deal with oil, water (brine), gas, emulsions and sand. Moreover, the gas and oil from wells emanate at somewhat unpredictable rates, pressures and temperatures. Here we discuss the various configurations of separators in oil and gas production, also the different types of phase separators, and how one can decide as to which type and design of separator should be used.

6.1.1 Four Types of Reservoirs

The reservoir can be seen as a storehouse of hydrocarbon fluid. Production operation is essentially the withdrawal of this fluid in a usable form. There are four kinds of reservoirs. One is the Black Oil Reservoir, so called since the oil is usually darker in colour as well as thicker. The second type is the Volatile Oil Reservoir, where the gas to oil ratio (GOR) is higher than that of the black oil reservoir. In the third type called Gas Condensate Reservoir, condensation occurs as governed by pressure–temperature conditions, and gas composition in the well bore changes due to falling back of the formed liquid. The flow rate may also be sometimes erratic due to flow getting blocked by formed liquid. The fourth type of reservoir is known as the True Gas Reservoir. In this type, during operation, the temperature may drop to a point where some liquid is formed.

A good estimate of the quantity of water which is likely to accompany oil is necessary for separator design as well as an indication of sand or paraffin wax contamination. Two-phase separators are used when no water is expected and three-phase separators have to be used in case water is to be separated.

The complete nature of production operations and the changes occurring in reservoir pressure in course of time dictate the various considerations in separator design.

Water (brine) is usually present in reservoirs. It is almost well known that petroleum was formed by the burial of marine life-forms, whereas coal was formed by the burial of ancient land-based flora.

Sand removal pans are required to be incorporated at the bottom of the separators in case sand is expected.

Horizontal versus vertical separators

Separators can be either horizontal or vertical. In the gravity settling section, droplets fall down against the direction of gas flow in vertical separators whereas the droplets fall perpendicular to the direction of flow in horizontal separators. Wherever very high or very low GOR (gas to oil ratio) is specified, including for scrubbers, vertical separators are preferred. For all other operations where separation is more difficult, horizontal separators are preferred.

The horizontal separators have the following advantages over the vertical separators:

1. Liquid droplets are more efficiently separated from gas since they fall perpendicular to the flow of gas.
2. Gas bubbles can break out of the liquid more easily because of the large gas–liquid interfacial area.
3. Top platform of the separator is more easily approachable for operation/maintenance.

The horizontal separators nonetheless have certain disadvantages over the vertical separators such as:

1. The horizontal separators require more area than the vertical ones. Therefore, where area is critical, such as offshore locations, the vertical separators may be preferred.
2. Sand separates easily in vertical separators since it falls to the bottom of the dish from where it can be removed. In horizontal separators, sand pans have to be installed with water flushing facility to remove the accumulated sand.
3. Level control in horizontal vessels is more complicated than in vertical vessels.

6.1.2 Two-Phase Separators

A two-phase separator for separating gas and liquid must consist of the following elements:

1. A separation section to remove bulk liquid from gas.
2. Sufficient liquid capacity to handle surges from the headers.
3. Mist eliminator.
4. Level control.

The separator can be further subdivided into scrubbers and separators.

A scrubber is a simple separator. It is not an elaborate design and does not have enough liquid capacity to handle surges. It can be said that a scrubber handles only small quantities of liquid compared to a separator. A scrubber is not installed as the first separation equipment out of any well. This is due to the fact that even dry wells produce liquid at times.

Separators usually have facilities to remove liquid particles up to 10 microns. Particles of size less than 10 microns are referred to as particulates and are not handled in separators.

A typical horizontal separator is shown in Figure 6.1. The inlet stream first hits an inlet distributor, or inlet momentum device, where particles of liquid more than 500 microns are removed. The larger particles get thrown into the target surface by centrifugal action caused by a change in direction of the fluid. The particles coalesce on this surface and fall down to the bottom of the vessel or fall to the liquid layer. The gas then passes through the gravity separation section. In this section, particles of diameter more than a definite size (100 to 150 microns) are removed by the action of gravity. Finally, the gas passes through a demister pad where particles more than a definite size (10 microns) are removed by coalescing into larger particles and falling to the liquid layes.

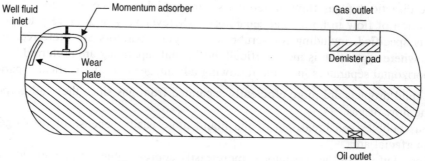

Figure 6.1 Two-phase separator.

The retention time of oil in the separator should be more than the minimum values recommended in API 12J standard[1].

Oil relative density	Residence time
Below 0.85	1 min
0.85 to 0.93	1–2 min
0.93 to 1.0	2–4 min

The given residence time, however, should be even more depending upon the presence of emulsions, foaming tendency, etc. Since level control is provided for liquid, a certain amount of liquid volume should be provided for effective level control.

6.1.3 Three-Phase Separators

Three-phase separators handle two immiscible liquids (oil and water in our case) and a gas.

The design of the liquid–gas separation portion is same as that for two-phase separators except that in this case the densities of both liquids (oil and water) have to be considered. Usually, a certain separation of oil from the water layer and water from the oil layer is to be supplied, to act as a guide to the design of the liquid–liquid separation portion. A certain size of droplets of the distributed phase (500 microns) in the continuous phase is assumed. Calculations are done to see whether the droplet will reach the interface layer in the time that will be taken for the travel using settling/buoyancy phenomena. Calculations are also cross checked using certain minimum residence times required for oil and water to achieve the required separation. API 12J standard recommends the following residence times based on the relative density of oil:

Oil relative density	Residence time
Below 0.85	3–5 min
Above 0.85—100°F and above	5–10 min
80—100°F	10–20 min
60—80°F	20–30 min

However, residence times are usually fixed using more rigorous calculations taking settling/buoyancy effects into consideration as explained above and also using better configuration of separators and use of internals. A typical three-phase horizontal separator is shown in Figure 6.2.

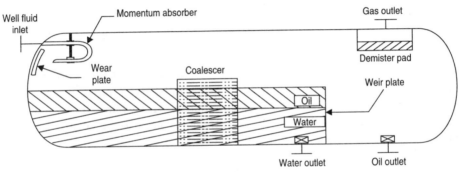

Figure 6.2 Three-phase separator.

Demister pads

Demister pads or mist eliminators may be either of mesh type or of vane type. The most popular demisters are wire meshes knitted in the form of a pad. The knitted pad has about 97 to 98% void space and it collects particles primarily by impingement. A liquid particle striking the wire surface, which it does not wet, flows downwards where the adjacent wires provide some capillary space. At these points, liquid collects and continues to flow downward. Surface tension tends to hold these drops on the lower face of the pad until they are large enough for the downward force of gravity to exceed that of the upward gas velocity and surface tension.

The efficiency of the pad is a function of pad thickness, wire diameter, closeness of weave, etc. The thickness of the pad is not very critical since droplet coalescence is completed in the first one or two inches and the extra thickness is mainly for catching the re-entrained material. Thickness usually provided is between 100 mm and 300 mm.

A vane type demister consists of a number of parallel corrugated plates. When gas passes between the plates it has to change direction a number of times. The heavier particles are thrown out and get collected in the pockets. Coalescence of small particles into bigger particles is provided by agitation and surface contact. The surface of the element is usually wet and particles striking it are absorbed into the liquid film. The efficiency of the vane type demister is a function of pressure drop since closer plates introduce more pressure drop.

Coalescers

Coalescers are used in three-phase separators to make the separation of water and oil more efficient. They essentially provide surface contact area by means of surface elements arranged inside the liquid medium.

The particles of the dispersed phase contact the surfaces by the action of gravity/buoyancy. The surface provides a place where small particles join together and become big and when these come out of the coalescer they reach the liquid–liquid interface level easily because of their bigger size.

Other internals

In long horizontal vessels, it is desirable to install wave breakers, which are simply vertical baffles with large holes punched into them, spanning the gas–liquid interface perpendicular to the flow.

Foam occurs at the vapour–liquid interface as gas is dispersed in the liquid. Foam can be broken by the addition of anti-foaming chemicals. It is also possible to break the foam by passing the foam through a series of inverted angle sections or tubes that coalesce the tiny bubbles into larger bubbles or by providing a slotted horizontal plate which will restrict the movement of foam and break it.

High, very high and low, very low levels for instrumentation and control

Figure 6.3 shows the positions of high and low levels to be fixed during the design stage. The tripping or emergency shutdown of the separators takes place where HLSD and LLSD values occur. Alarms occur at HLA and LLA values, sufficiently prior to the occurrence of tripping so that operator can take the necessary corrective action.

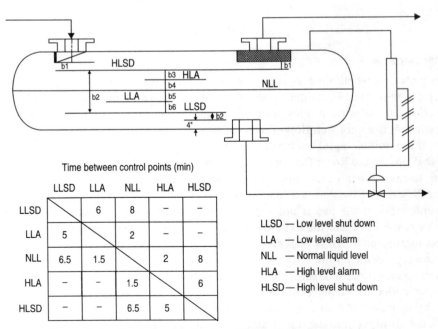

Figure 6.3 Guide for low level and high level alarms and trips.

The table given along with Figure 6.3 provides a rule of thumb guideline for the residence time in minutes between levels. The values to the right of diagonal are for surging flow or feed to a fired heater. The values to the left are for normal operation[2].

Separator selection

API 12J recommends the following procedures for selecting a separator for a particular application:

1. Determine which shape fits the particular installation best, considering space, mounting and ease of access for maintenance. Both present and future operating conditions should be considered.
2. Determine whether any unusual well stream conditions (foam, sand, etc.) would make the vessel selected difficult to operate or maintain.
3. Check whether the cost of the vessel is reasonable.
4. Check whether any heating coil is required against contamination by solid paraffin wax/ check hydrate formation.
5. Check three-phase requirement for water removal.
6. Check any possible liquid slugging of separator. Slugs are large lumps of liquid flouring along with gas.

6.1.4 Development of Separator Sizing

Separator sizing is always done after determining the residence time required for separation. The residence times in turn depend upon the properties of fluid and the conditions of flow.

Development of a program for separator sizing

Separator sizing program[3,4,5] can be developed in spreadsheet or in any programming language. C++ is preferred, because its object-oriented features are helpful to make programs with clarity, extensibility and ease of maintenance. For the initial development of the program, spreadsheet is more useful, as it is easy to make. The logic and various formulae used in the calculation can be easily checked using spreadsheet. Once it is found acceptable, a more dependable, faster and user-friendly program can be written in C++.

Two-phase oil–gas separator design

The diameter and length of the separator are calculated based on gas capacity and liquid capacity. The gas capacity is calculated based on the following assumptions:

1. Separator is one-half filled with liquid.
2. The time required for the droplet to fall to the gas–liquid interface and the residence time of the gas are equal.

The gas capacity provided shall be capable of removing 100 to 150 micron liquid drops from the gas. It must also provide sufficient retention time to allow the liquid to reach equilibrium.

Liquid capacity constraint can be used to calculate the various combinations of vessel internal diameter and effective length of vessel, using the liquid retention time desired. A reasonable combination of vessel diameter and length that satisfies both gas capacity constraint and liquid capacity constraint can be selected. Usually a slenderness ratio of 3–4 is used to select the diameter and length of horizontal two-phase separators.

Three-phase oil–gas separator design

Various combinations of vessel diameter and effective length are calculated based on gas capacity constraint and retention time constraints of oil and water. In the case of gas separation the concepts pertaining to two-phase separation described above are valid. The maximum diameter of the separator is calculated based on the need to settle typically 500 micron water droplets from the oil.

The retention time of oil and water provided determines the amount of water content in oil and oil content in produced water. The oil capacity provided should ensure that the oil reaches equilibrium with flashed gas being liberated, and free water has time to coalesce into droplet sizes sufficient to settle. The water capacity provided should ensure that most of the large droplets of oil entrained in water have sufficient time to coalesce and rise to the oil–water interface.

An acceptable combination of diameter and length is arrived at on the basis of the above constraints. A slenderness ratio between 3 and 5 is used for horizontal three-phase separators.

Sizing parameters and guidelines

In today's market, even if an application only requires a two-phase separator (gas/oil), the majority of customers see the need to look ahead to when water production usually becomes a reality. Hence, provision nowadays is usually made for the facilities to be added for three phases (oil/water/gas).

In three-phase separators there are three main functions to be considered in sizing:

1. Provision of enough residence time for the liquid so that degassing can occur.
2. Provision of enough free cross-sectional area above the liquid so that the gas velocity is low enough to allow liquid particles to be separated.
3. Provision of enough residence time for liquid–liquid separation to occur.

A separator sizing will be either liquid or gas controlling, which is determined by the gas/oil ratio.

Water in oil can be quantified in terms of percentage and particle size.

Obviously, very tight emulsions will cause problems. In general terms, a three-phase separator with a residence time of 3 minutes will produce an outlet oil quality of between 3 and 8% water in oil.

If the residence time is extended, resulting in an increase in price due to a larger vessel size, the oil quality can in most cases be improved. Other factors such as heating prior to separation, addition of demulsifying agents, etc. can also greatly assist with outlet oil quality improvement (in most cases).

Gas in outlet oil does not usually raise a problem, unless the gas is controlling, due to a very high gas/oil ratio. Normally, a three-phase separator is liquid controlling in terms of sizing. Hence, on letdown in pressure to the next stage of separation (be it another separator, free water knockout, or flow tank), liquids should not liberate any more gas than would theoretically exist under equilibrium conditions.

The general requirement for sales oil is 15 psia Reid vapour pressure, easily measured as per ASTM standard D323. Generally, the same parameters which dictate water in oil quality are relevant to oil in water quality. In addition, if only a very small quantity of water is present in the formation, much tighter emulsions would tend to form. With larger quantities, an interface can be established much more quickly.

As a guideline for settling time requirement the following chart can be followed.

Crude	Outlet water in oil
Heavy crude < 30° API	5000 ppm — for 3 min, 1000 ppm for 30 min
Medium crude 30°–45° API	3000 ppm — for 3 min, 1000 ppm for 10 min
Light crude > 45° API	1500 ppm — for 3 min

An interpolated chart for oil in water for various residence times may look as in Table 6.1.

Table 6.1 Oil in water for various residence times

Residence time (min)	Density < 30°API	Density 37.5°API	Density 45°API
3	5000 ppm	3000	1500
6	3000	1800	1000
10	1800	1080	500–1000
20	1080	500–1000	
30	500–1000		

Special considerations should be given to the sizing of separators under the following conditions.

- Degassing of very light >40° API dry oil
- Degassing and dehydration of waxy crude
- Handling of foaming crude

Sometimes there is a problem of identifying crude that is likely to foam prior to specifying surface facilities. In general, a crude oil is likely to foam if the crude density is less than 40° API, the temperature is less than 170°F, and the crude has very low viscosity.

Other special separators (i) double barrel and (ii) spherical

A double barrel separator is a special type of horizontal separator (see Figure 6.4) consisting of two horizontal cylinders connected by downcomers. The liquids are drained to the lower cylinder, thereby preventing gas re-entrainment due to high gas flow velocities. They are used for high GOR applications.

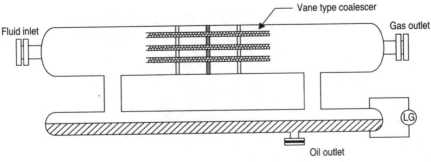

Figure 6.4 Double barrel separator.

Spherical separators are spherical in shape and take advantage of reduced wall thickness required, as a consequence. A spherical separator can be used instead of a double barrel separator if the cost and application justify the same.

6.1.5 Free Water Knockouts

Free water knockouts are usually specified downstream of the degassing or first stage separator. They are normally employed to remove large amounts of water from oil streams, i.e. 70% to 90% of water.

In North America, especially Canada, free water knockouts are usually specified upstream of the crude oil dehydration (or heater treater), the purpose being to remove bulk amounts of water prior to electrostatic treating.

The free water knockout is essentially a separator with sizing completely controlled on liquid–liquid separation rather than gas.

The inlet fluid enters and impinges on an inlet diverter as with a standard separator. After the velocity has been slowed in this way, the liquid enters the vessel. This vessel is sized to give a large interface area for gravity settling of water/oil.

Usually, to aid separation, coalescing plates are installed to enhance separation time. Depending on the gas volume expected, the vessel will be either liquid packed or a small gas volume will be allowed at the top of the vessel, or if larger gas volumes are present, a gas dome can be provided in the top of the shell for degassing.

Free Water Knockout (FWKO) special considerations

The free water knockout, as its name implies, is designed to remove free water, and not emulsions from oil (see Figure 6.5). The use of demulsifying chemicals is usually a must to help in separation, and even then the outlet quality depends on the extent and tightness of the emulsion. As a guide, the following typical outlet ppm qualities of oil in water and water in oil emulsions will give an indication of achievable separation.

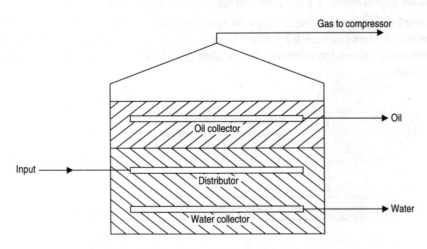

Figure 6.5 Free water knockout tank.

Oil in water emulsion

Heavy crude < 30 API	– 150 ppm outlet
Medium crude, 30 API to 45 API	– 100 ppm outlet
Light crude, > 45 API	– 50 ppm outlet

Water in oil emulsion

Heavy crude, < 30 API	– 5 to 6%
Medium crude, 30 API to 45 API	– 3 to 5%
Light crude, > 45 API	– 2 to 3%

The FWKO is usually sized on a residence time between 15 and 30 minutes.

PPM water in oil chart

Crude	Outlet water
Heavy crude < 30 API	5000 ppm — 3 min
	500 ppm — 30 min
Medium crude 30–45API	3000 ppm — 3 min
Lighter crude > 45API	1500 ppm — 3 min

Generalized ppm oil in water chart for different crudes

Minutes	< 30° API	< 37.5° API	< 45° API
3	5000	3000	1500
6	3000	1800	1000
10	1800	1080	500–1000
20	1080	500–1000	
30	500–1000		

As an example from the above, by regression analysis, to achieve less than 250 ppm oil in water, the residence time will be about 125 minutes which can be considered unviable. For very tight specifications, other methods such as centrifugation or use of hydrocyclones will be more appropriate.

6.1.6 Three-Phase Separator—A Case Study

A three-phase separation is to be carried out for the well fluid which consists of 110 MMSCFD of gas with 20,000 barrels per day (BPD) of oil and 5000 BPD of water at given specified properties and conditions.

The calculation principle is that a farthest particle of oil should join the oil phase within a specified time due to its vertical upward velocity. Similarly, for the water too, the farthest water particle should join the water layer[6, 7].

Notations used

- d : vessel inside diameter, inch
- GLEFF : vessel effective length, feet for gas flow
- GLSS : vessel seam-to-seam length, feet for gas flow
- LLEFF : vessel effective length, feet for liquid flow
- LLSS : vessel seam-to-seam length, feet for liquid flow
- T : operating temperature, degree R
- Z : gas compressibility
- Q_g : gas flow rate, MMSCFD (million standard cubic foot per day)
- Q_w : water flow rate, BPD
- Q_o : oil flow rate, BPD
- P : operating pressure, psia
- ρ_g : density of gas, lb/ft^3
- ρ_l : density of liquid, lb/ft^3
- μ : viscosity, cP
- C_d : drag coefficient
- d_m : liquid drop diameter, microns
- $(t_r)_w$: water retention time, minutes
- $(t_r)_o$: oil retention time, minutes
- ΔSG : difference in specific gravity
- A_w : area of cross section of water flow, ft^2
- h_o : oil pad thickness
- h_w : water pad thickness
- A : area of cross section of total flow, ft^2

Procedure for vertical separator

The required steps are given below:

1. Calculate the minimum diameter from the requirement for water droplets to fall through the oil layer. Use 500-micron droplets if no other information is available.

$$d^2 = \frac{6690\, Q_o \mu}{(\Delta SG) d_m^2}$$

For 500-micron, $$d^2 = \frac{Q_o \mu}{\Delta SG}$$

2. Calculate the minimum diameter from the requirement for oil droplets to fall through gas. Use 100-micron oil droplets if no other information is available.

$$d^2 = \frac{5040\, TZQ_g}{P}\left[\frac{\rho_g}{\rho_l - \rho_g} \times \frac{C_d}{d_m}\right]^{0.5}$$

3. Choose the larger of the two as $d_{\min}$.
4. Select $(t_r)_o$ and $(t_r)_w$, and solve for $h_o + h_w$ for various values of d.

$$h_o + h_w = \frac{(t_r)_o Q_o + (t_r)_w Q_w}{0.12 d^2}$$

5. Estimate the seam-to-seam length using the larger value.

$$\text{LSS} = h_o + h_w + \frac{76}{12} \quad \text{or} \quad h_o + h_w + d + \frac{40}{12}$$

6. Select a size of reasonable diameter and length. Slenderness ratios (12LSS/d) of the order of 1.5 to 3 are common.

Procedure for horizontal separator

The required steps are given below:

1. Select $(t_r)_o$ and $(t_r)_w$.
2. Calculate $(h_o)_{max}$ (maximum oil pad thickness). Use 500-micron droplets if no other information is available.

$$(h_o)_{max} = \frac{1.28 \times 10^{-3} (t_r)_o (\Delta SG) d_m^2}{\mu}$$

For 500-micron droplets, $(h_o)_{max} = \dfrac{320 (t_r)_o (\Delta SG)}{\mu}$

3. Calculate A_w/A:

$$A_w/A = \frac{0.5 (Q_w (t_r)_w)}{(t_r)_o Q_o + (t_r)_w Q_w}$$

4. Determine h_o/d using the relation between areas and sector lengths given in mathematics section of Perry[8].
5. Calculate d_{max}.

$$d_{max} = \frac{(h_o)_{max}}{h_o/d}$$

Note: d_{max} depends on Q_o, Q_w, $(t_r)_o$ and $(t_r)_w$.

6. Calculate combinations of d, LEFF for d less than d_{max} that satisfy the gas capacity constraint. Use 100-micron droplets if no other information is available.

$$d, \text{LEFF} = \frac{420 (TZQ_g)}{P} \times \left(\frac{\rho_g}{\rho_l - \rho_g} \times \frac{C_d}{d_m} \right)^{1/2}$$

7. Calculate the combinations of d, LEFF for d less than d_{max} that satisfy the oil and water retention time constraints.

$$d^2, \text{LEFF} = 1.42 [(t_r)_o Q_o + (t_r)_w Q_w]$$

8. Estimate the seam-to-seam length.

$$\text{LSS} = \text{LEFF} + \frac{d}{12} \quad \text{for gas capacity}$$

$$\text{LSS} = \left(\frac{4}{3} \right) \text{LEFF for liquid capacity}$$

9. Select a reasonable value of diameter and that of length. Slenderness ratios (12LSS/d) of the order of 3 to 5 are common.

242 Chemical Process Technology and Simulation

For separators other than 50% full of liquid, equations can be derived using the same principles.

The above calculations are shown with an example in a worksheet format in the following pages.

Calculation using spreadsheets—a case study for designing a horizontal separator

The advantage of using spreadsheets is that various options can be tried out very quickly.

Three-Phase Horizontal Separator

	Operating conditions		
1.	**Inputs**		
QG	Gas flow rate	MMSCFD	110
QW	Water flow rate	BPD	5000
QO	Oil flow rate	BPD	20000
G	Specific gravity of gas	From HYSYS	0.799
°API	Oil specific gravity	°API (calculated from the density of oil)	42.2
SGW	Water specific gravity	–	1.03
P	Operating pressure	PSIA	1074.57
T	Operating temperature	R	570
DM	Diameter of water droplet to be separated from oil	microns (as per specification)	500
DML	Diameter of liquid (water and oil) droplet to be separated from gas	microns (as per specification)	150
TRW	Water retention time	min (as per selection)	5
TRO	Oil retention time	min (as per selection)	5
LVIS	Viscosity of oil liquid	CP (centipoise)	1.433
2.	**Calculation of Separator Sizing**		
Note 1: For separator sizing the gas and liquid capacity constraints are to be calculated to find what governs the separator sizing. These constraints are calculated at various diameters as tabulated in Table 1 and Table 2 below. It is evident from the tables that the liquid capacity constraints govern the separator sizing in this case. A slenderness ratio between 3 and 5 is ideal. By trial and error, a standard diameter of 102 inches and seam-to-seam length of 30 ft are selected and their suitability checked.			

Oil and Gas (Upstream) 243

The equations given below are based on *Surface Production Operations*, Volume 1 By Ken Arnold & Maurice Stewart Jr.[7] which may be referred as required.

M	Molecular weight	(28.97*G)	23.147
SGO	Specific gravity of oil	(141.5/(API+131.5))	0.815
LGO	Average density of oil and water	(SGO*QO+SGW*QW)/(QO+QW)	0.858
LDEN	Liquid density	(62.4 * LGO)	53.520
DSGWO		(SGW-SGO)	0.215
GDEN	Gas density	(2.7*P*G)/(Z*T)	4.62155076

2.1 Gas capacity constraints

DHG	Vessel I.D.	
CKF	K factor (see 2.5 below)	
Z	Compressibility factor (see 2.3 below)	
GLEFF	Vessel effective length	420*CKF*Z*T*QG/(P*DHG)
GLSS	Vessel seam-to-seam length	4/3 *GLEFF
HSPAT	Slenderness ratio	12*GLSS/DHG

Table 1

HSPAT	DHG(IN)	GLEFF(FT)*	GLSS(FT)**	*effective length. **seam-to-seam length.
3.26	50	10.1973	13.60	
2.12	62	8.22363	10.96	
1.49	74	6.89007	9.19	
1.10	86	5.92867	7.90	
0.85	98	5.20271	6.94	
0.78	102	4.99868	6.66	
0.53	124	4.11182	5.48	

2.2 Liquid capacity constraints

AY	(for iteration)	(0.5 *QW*TRW)/(QO*TRO+QW*TRW)	0.1
YTWO	(for iteration)	((ACOS(2*AZ))-4*AZ*((0.25-AZ^2)^0.5))/3.14	0.099776282
DY	(for iteration)	(ABS(AY−YTWO))	0.00
AZ	Z coefficient	(by iteration)	0.4306
HOMAX	Max. allowable oil pad thickness	0.00128*TRO*DSGWO*DM^2)/LVIS	240.4768591

DMAX	Max. vessel I.D.w.r.t HOMAX	(HOMAX/AZ) inches		558.4692501
RATIO		LLEFF/LLSS ratio		0.75
DHL	Vessel I.D			
LLEFF	Vessel effective length	1.42*(QW*TRW+QO*TRO)/(DHL^2)		
LLSS	Vessel seam-to-seam length	LLEFF*RATIO		
HSPAT	Slenderness ratio	12*LLSS/DHL		

Table 2

HSPAT	DHL(IN)	LLEFF(FT)	LLSS(FT)	
22.72	50	71	94.66666667	
11.91635058	62	46.1759	61.56781131	
7.00846939	74	32.4142	43.21889457	
4.465015659	86	23.9995	31.99927889	
3.017450212	98	18.4819	24.64251007	
2.67619543	102	17.0607	22.74766116	(L/D = 2.7) Selected
1.489543822	124	11.544	15.39195283	

2.3 Calculation of Z (compressibility factor)

PPC	Critical presssure	(709.604−58.718*G)	662.688318
TPC	Critical temperature	(170.491+307.344*G)	416.058856
PR	Reduced pressure	(P/PPC)	1.621531527
TR	Reduced temperature	(T/TPC)	1.369998479
QI		(0.27*PR/(ZII*TR))	0.363150297
Z	Compressibility factor		0.88

2.4 Calculation of gas viscosity (with GDEN,M,T and based on regression analysis constants)

KONE		(9.4+0.02*M)*(T^1.5)	134220.4533
KTWO		(209+19*M+T)	1218.79357
X		(3.5+986/T+0.01*M)	5.461294861
Y		(2.4−0.2*X)	1.307741028
GDENI		(GDEN/62.4)	0.074063313
K		(KONE/KTWO)	110.1256657
VISICI		(K*EXP(X*GDENI^Y))	132.050699
VISC		(VISCI/10000)	0.01320507

Oil and Gas (Upstream)

2.5 Calculation of K Factor — Initial VT 1.7

DEN	Density difference	(LDEN−GDEN)	48.899
REN	Reynolds number	(0.0049*GDEN*DML*VT/VISC)	175.1015992
CD	Drag coefficient	(24/REN+3/REN^0.5+0.34)	0.703776165
VT	Terminal settling velocity	(0.0119*(DEN*DML/(GDEN*CD))^0.5	0.565106968
CKF	K factor	(GDEN*CD/(DML*DEN))^0.5	0.02105796

2.6 Calculation of length required for settling

DS	Diameter of separator	ft	8.5
ROG	Density of gas	lb/ft^3	0.227
VFG	Volume flow of gas	ft^3/s	70.092
VT	Terminal settling velocity	ft/s (iterated to REN)	0.565106968

(a) Liquid/Gas interface %	(b) Height above the L/G interface (ft) (DS*(1−(a))	(c) Area of sector (ft^2)	(d) Gas velocity (ft/s) (VFG/(c))	(e) Time taken for settling ((b)/VT)	(f) Length reqd for settling (ft) ((e)*(d))
45.00%	4.68	31.96409631	2.19	8.27	18.14
50.00%	4.25	28.358	2.47	7.52	18.59
55.00%	3.83	24.75545999	2.83	6.77	19.16
60.00%	3.40	21.18492676	3.31	6.02	19.91

Note 2:

(a) The gas velocity is less than the liquid droplet settling based on droplet size of 150 microns

(b) The calculated length required for settling of entrained oil from gas is less than the length of the separator.

2.7 Calculation of maximum operating level

When sudden surging occurs, levels can go high. Its effect on gas velocity is given in the table below.

(x) Liquid/Gas interface %	Height above the L/G interface (ft) (DS*(1-(x))	Area of sector (ft^2)	Gas velocity (ft/s) (VFG/(c))
45.00%	4.68	31.96409631	2.19
50.00%	4.25	28.358	2.47
55.00%	3.83	24.75545999	2.83
60.00%	3.40	21.18492676	3.31
65.00%	2.98	17.6926323	3.96
70.00%	2.55	14.3105834	4.90
75.00%	2.13	11.08856835	6.32

Note 3:				
The maximum operating level (HHLL) should not be greater than 65% because in the low pressure mode the gas velocity above this level is greater than the terminal velocity.				
Note 4:				
Operation under upset condition				
The level may rise up to 75% during upset condition. But considering the length of the vessel there will be no appreciable change in liquid carryover by gas.				

2.8 Oil and water pad thickness and rettention time calculation				
VFO	Volume flow of oil	ft^3/m	132.60	
VFW	Volume flow of water	ft^3/m	89.70	
DS	Shell diameter	ft	14.25	
LEFF	Effective length	ft	43.50	
LSS	Seam-to-seam length	ft	60.00	
LLL	Low liquid level	ft	1.00	
HLL	High liquid level	ft	7.46	

Partial volumes in ellipsoidal heads (2 :1 horizontal)				
DS			ft	14.250
RATIOLL	(LLL/DS)			0.070
RATIO-HL	(HLL/DS)			0.523
KLL	Coeffnt from Table for RATIOLL=0.070175439			0.014
KHL	Coeffnt from Table for RATIOHL=0.523391813			0.535
VLL	Volume wrt LLL in one head (0.2618*DS^3/2*KLL)		ft^3	5.302885809
VHL	Volume wrt HLL in one head (0.2618*DS^3/2*KHL)		ft^3	202.6459934
VLLHL	Volume between LLL and HLL (VHL–VLL)		ft^3	197.3431076

Liquid/Gas interface %	Wier height (ft)	Oil pad thickness (ft)	Water pad thickness (ft)	Cross sectional area of total pad (ft^2)	Cross sectional area of oil pad (ft^2)	Cross sectional area of water pad (ft^2)
45.00%	6.41	3.48	2.93	69.61	45.98	23.63
50.00%	7.13	3.48	3.64	79.74	47.57	32.18
55.00%	7.84	3.48	4.36	89.88	48.58	41.29
60.00%	8.55	3.48	5.07	99.91	49.07	50.84

Volume of water pad (ft³)	Volume of oil pad (ft³)	Retention time for water (min)	Retention time for oil (min)	Volume of oil pool for LLL (ft³)	Volume of oil pool for HLL (ft³)	Retention time for oil pool between LLL and HLL (min)
1027.90	1999.97	11.46	15.08	86.59	1596.55	11.39
1399.69	2069.1	15.6	15.6	86.59	1596.55	11.39
1796.33	2113.39	15.94	15.94	86.59	1596.55	11.39
2211.67	2134.53	16.1	16.1	86.59	1596.55	11.39
Note 5:						
The holdup time between LLL and HLL is well above the minimum required time of 3 minutes						

Results and conclusion

From the above discussion and calculations it is found that a diameter of 102 inches and a length of 30 ft are suitable for the separator. A weir of variable height with range between 45% and 60% of the diameter, will give extra flexibility in operations. If conditions require, this can be incorporated which will take care of changing gas–liquid flow scenarios.

With reference to oil water separation, if emulsion is a possibility and consequently five minutes residence time is insufficient, then either the residence time can be increased or a coalescer pad consisting of parallel plates can be designed and fitted into the separator so that the same effectiveness is obtained. This may work for particles up to about 50% size of what we have assumed for separation. For smaller particles, a reliable method will be to fit proprietary coalescers as recommended by vendors based on particle size distribution.

The advantage of spreadsheet-based calculation is that we get many options at one go with trade-offs between length and diameter from which a desired one can be selected.

Increase in diameter can sometimes result in a step change of the 'shell thickness of vessel', which increases the cost of the vessel. In the case of offshore platforms, it leads to increase in the cost in the platform structure itself. The designer in this case should optimize the weight at the same time and it should result in utilizing the space, or result in a utilizable length on the platform. This means that in smaller platforms it would be advisable to use either the entire length or the entire width of the platform, leaving only enough space for operation and maintenance.[9]

While locating a separator on an offshore platform, the general relationship between hazard and layout should be strictly followed as shown below[10].

    Make the initial layout.
    Visualize hazard events
    Study the release and dispersion of gases.
    Consider chances of fire/explosion.
    Consider chances of effect on facility.
    Modify the layout to minimize damage.

If the separator is to be located on floating platforms such as an FPSO (Floating Production Storage and Offloading) unit, the sizes have to be considerably larger to take care of the natural rolling and pitching movements of the FPSO unit. It may be noted that FPSO units are production platforms located at sea but not firmly supported from the sea bed.

6.2 NATURAL GAS TRANSPORTATION NETWORK

Natural gas is obtained from gas reservoirs. First, it is processed in gas processing units to remove natural gasoline and heavier fuels and then supplied to consumers through pipelines.

It consists of some or all of the following chemicals.

1. Nitrogen
2. Helium (traces)
3. Carbon dioxide
4. Hydrogen sulphide (traces)
5. Methane
6. Ethane
7. Propane
8. Butanes
9. Pentanes (minor quantities)
10. Heavies (traces)

(Trace quantities of carbonyl sulphide, carbon disulphide and mercaptans have also been reported in rare cases.)

The operation of a modern natural gas network (gas grid) using a centralized control room is definitely a challenge akin to a 'process plant' operation and the same is dealt with here.

The pressure which a pipeline can withstand depends upon the yield strength of the material used in the pipeline and the wall thickness. In the early days, as only low strength materials of construction were available, the pipes could not withstand high pressures and, therefore, gas was transported at low pressures through such pipelines. As better materials with higher yield strength came into the field, the gas pressure used has consequently increased and thus costs have come down. From the year 2000 onwards, gas is being transported at 200 barg pressure, thereby decreasing pipe diameters. Materials used have been API SL X65 (where 65 stands for minimum yield strength of 65,000 psi), X70, X80, and in the future X100, and further X120 are also likely to be used. Also it is likely that steel FRP composite materials will be used for gas pipelines in the future[11].

Centrifugal compressors driven by gas turbines for gas pipelines and centrifugal pumps driven by electric motors for oil supply provide the motive energy for gas/oil transportation. Diesel engine drives are also common for pumps. Modern pipelines are almost always controlled by a Supervisory Control and Data Acquisition (SCADA) system located in a centralized control centre. All supervisory and operational as well as safety and shutdown operations are controlled from this centre. Measurement of gas and liquid flows, and totalization for custody transfer, has come a long way from the orifice plates used in the initial stage. P.D. meters, turbine meters, ultrasonic flow meters and coriolis meters have been developed to various degrees of accuracy for accurate measurement and totalization. Microprocessor-based flow measurement has also been developed with many automatic features built into it.

Sophisticated metering facilities with filtering equipment to filter unwanted constituents have been developed, too.

Advanced softwares like the 'Shell Systems' which capture all aspects of pipeline management and monitoring are also aiding pipeline transportation.

6.2.1 Operation of Gas Pipelines

Operating a gas pipeline involves balancing the gas supply from suppliers with gas delivered to consumers. Gas from different suppliers generally gets mixed in the gas pipeline, however, this is not a problem as long as customers receive gas of quality as specified by the contract.

Compressors are used for pressurizing the gas to a pressure that is less than the design pressure of the pipe at all operating conditions, except during surging. Limits are fixed for various parameters, primarily pressure, flow and power after which point the line shuts down as a safety measure. The pressure is limited by design pressure; flow would be limited by erosional velocity and power by the design power consumption of the compressor.

A fire or gas leak or other safety issues also cause the unit to shut down. A central control room usually controls all the operations of the pipeline including shutdown.

The gas line also serves as a storage of gas. Line pack in a gas pipeline is the amount of gas the line is holding in Nm^3 (normal cubic metres) at any point of time. It depends upon variables such as pressure, temperature and composition of the gas and pipeline geometry.

If the total supply is more than the total consumption, then the line pack tends to increase consequent to pressure increase and vice versa. However, the operators in the central control room would always act to keep the supplies and receipts equal as far as possible. Central control keeps a materials balance check and sends daily reports to both suppliers and consumers regarding the gas quantities and equivalent energies supplied and received by each party. In addition, more detailed weekly reports are also given.

In all gas lines, sectionalizing valves, also called block valves, are provided along the main line at intervals up to a maximum of 32 km depending upon the location and the area class through which the gas line passes. These sectionalizing valves close automatically when they sense either low pressure or abnormal pressure loss between the different sections of pipe, indicating that a leak has occurred. Thereby, the gas in other sections of pipeline is safeguarded by positive isolation of the affected section. At the same time the central control room checks on material balance in the pipeline. It is to be noted that walking the pipeline is an important field activity, which can detect even small leaks by means of sophisticated instruments. Early detection and correction of leaks proves very useful in providing an efficient supply of gas to consumers. The local people are also encouraged to report any suspected leaks using the phone numbers located at different points on the pipeline route.

Pigging, cleaning and corrosion monitoring are some of the other important activities planned and carried out by the central control room. Typical oil/gas pipeline details including launchers and receivers are shown in Figure 6.6.

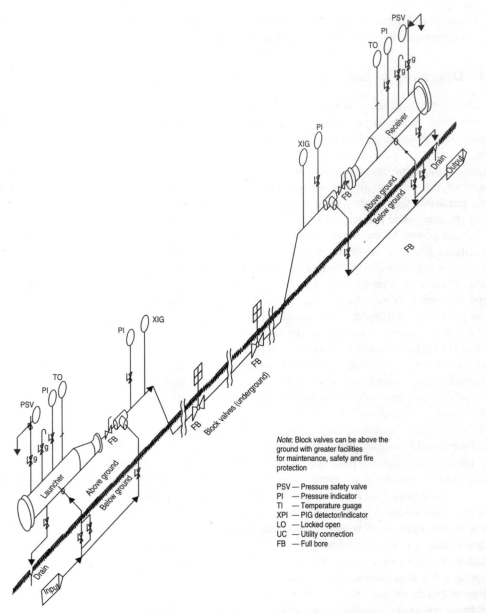

Figure 6.6 Typical oil/gas pipeline showing input, output, launcher, receiver and block valves.

Optimization of pipelines

Optimization of pipelines[9] is conducted by minimizing CAPEX (capital expenditure) and OPEX (operating expenditure) for the design life of the pipeline. Design is done in a manner so as not to overshoot the maximum erosional and noise velocity considerations[11]. One such erosional consideration is given in API RP 14E, which suggests that velocities should not exceed the relation given by

$$\text{Velocity (ft/s)} = E / \sqrt{\text{(Density)}}$$

where E, an empirical constant in FPS units, can vary from 100 to 200, based on continuous, non-continuous or corrosive service and density in lb/ft^3. For a two-phase flow, a minimum velocity consideration also exists which is often quoted as 3 m/s, below which slugs will create flow instability. There are no minimum velocities for single-phase flow.

Transient analysis

A transient analysis of pipelines is important to fix the opening and closing times of valves so that high pressure in excess of maximum allowable operating pressure is not experienced in pipelines. In this case, Maximum Allowable Operating Pressure (MAOP) is defined as the pressure which the weakest link in the system will be able to withstand. The US Department of Transportation (DOT) standards can be used to calculate the MAOP of a system.

Before making a transient analysis, a steady state analysis is to be made. Pressure drops for varying flow rates are calculated for steady state conditions and tabulated. Further transient scenarios are built up, such as pump, compressor failure, closure of emergency shutdown valves, opening of valves from pressurized header, etc. The results are found out using manual calculation and more often using softwares such as Pipesim/Pipeline Studio Tlnet & Tgnet/Pipenet/OLGA and others.

Importance of SCADA

Supervisory Control And Data Acquisition (SCADA) systems are extensively used in the operation of pipelines. SCADA consists of the following sections:

- RTU to transfer data
- Telecommunication network to aid communication
- Centrally located SCADA Host Computer

SCADA is similar to a Distributed Control System (DCS) but differs to the extent that it is designed to gather data from geographically remote field locations. All data is gathered to a centralized location for display and overall control. The main difference between SCADA and DCS is that SCADA is slower because data has to traverse large distances. Telecommunication systems are much in use in SCADA. Telecommunication is usually done using fibre optic cables, radio networks, satellite networks, IP (Internet Protocol) networks and others. Displays, alarms and their annunciation, supervisory controls, report generation, historical data views, data archiving are the activities performed by SCADA.

HIPPS

High Integrity Pressure Protection System (HIPPS) is a method of protecting any system from high pressure by isolating the source of high pressure. Normally, systems are protected from high pressures by safety valves. However, safety valves discharge gases to atmosphere, or to flares which in turn discharge gases to atmosphere. Hence, HIPPS can be used on condition that it is made as safe as or safer than safety valves.

In pipeline systems, it is to be ensured that when a pipeline is connected from a higher pressure system to a lower pressure system, under no circumstances should the lower pressure system be overpressurized. HIPPS should be used in such cases. HIPPS must be operated, tested and maintained throughout the life of the pipeline. Once a testing frequency is determined, it must be followed for the full life of the pipeline at specified intervals.

HIPPS basically consists of a number of pressure transmitters which transmit pressure readings to a control station from where the source of pressure is isolated in case of any increase beyond the limit set for the downstream pipeline.

Compressor systems and costs[12,13]

Compressor systems provide the motive force for the flow of gases through pipelines.

Centrifugal compressors use a series of rotating impellers to impart velocity head to the gas. This is then converted into pressure head as the gas is slowed in the compressor case. Sizes of compressors range from 1000 hp to 20,000 hp. Being small in size they are popular in Offshore Process Platforms too.

The drive is mostly either an electric motor or a gas turbine. Lower energy costs justify the higher investment required for gas turbines, apart from easy availability of gas. Turbines also have the advantage of being coupled with other drives such as electric motors and expanders.

Centrifugal compressors and their drive turbines rotate at very high speeds of more than 20,000 rpm which makes the design of its parts very critical. The centrifugal compressor occupies less space and produces more work per unit weight. However, it is less efficient than the reciprocating compressor. Further, the centrifugal compressors are less flexible and need to be used at or near design conditions.

In reciprocating compressors, pressure increase is achieved by decreasing the volume inside a cylinder by means of a piston. The ratio of outlet pressure to inlet pressure rarely exceeds 6 for a single stage. The volumetric efficiency decreases and mechanical stress limitations increase, as this ratio increases. Usually, a stage is added for every compression ratio of 4. The total power requirement is minimum when the compression ratio in each stage is the same. Inter-stage coolers and separators are provided to achieve near isothermal operation for better efficiency.

The cost of compressors plays an important role in pipeline operation. The fixed cost goes into the CAPEX and the operating cost goes into the OPEX. The fixed cost can be reasonably computed from market surveys and operating cost from the cost of gas. The most realistic value needs to be assigned to gas for optimizing a gas pipeline.

Corrosion mitigation and related design optimization [14]

There are different ways to mitigate corrosion, such as providing proper coatings and providing current cathodic protection. Both methods are used in combination.

Polyethylene coating is applied to the surface of a pipeline, which is very effective except that it cannot withstand high temperatures. Fusion bonded epoxy (FBE) coating is excellent but impact resistance is low and moisture absorption high. Polyolefin coating, too, has excellent properties. For large diameter pipes, composite coatings are also used.

Cathodic protection (CP) is usually applied to coated pipes to ensure that those minute areas which get exposed are also protected. Insulation provides the basic protection against corrosion. When the installation is new, the insulation is good and less of pipe is exposed to soil. Hence, there is not much of CP current flow. Later, as the area exposed increases, this current value becomes high. However, as the voltage for the protection is very less, the energy consumption is well within limits.

The better the coating, the lesser the power consumed by cathodic protection. There are essentially two methods for cathodic protection. Both function by causing current to flow from anode to pipeline. One is an impressed current method and the other is called sacrificial anode method. In the first method, an anode bed is created near the pipeline using coke breeze and zinc (or some other type of) electrodes. The anode is made slightly positive and the pipe is made slightly negative. The potential difference is only between 0.95 and 1.5 volts. The anode bed gets corroded based on the current flow and prevents the cathode from getting corroded. This is called conventionally "cathodic protection". The second method employs a more electro-positive element affixed in contact with the soil (or sometimes inside for exposure to water flowing in the line along with the product). The more electro-positive element now acts as an anode. The anode first has to fully get corroded before start of corrosion to the main pipeline. The placement of this sacrificial anode is very strategic and is based on soil resistivity, pipe thickness and material of pipeline.

Hydrate formation check

Hydrates are multiple cage-like clathrate formation of water molecules in which hydrocarbon molecules reside and result in a solid mass after due growth. See Figure 6.7.

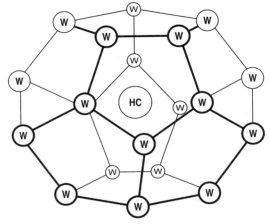

W = Water molecule
HC = Hydrocarbon molecule

Note: A pentagonal face can be common to other hydrocarbon molecules.

Figure 6.7 A single cage of hydrate formation.

Hydrates are formed in natural gas lines under certain conditions of temperature and pressure based on the composition of gas. No hydrate will form if there is no water at all. However, small quantities of water are enough to form hydrates and plug the lines at critical points. There are different ways of calculating whether hydrates will form. Parrish and Prausnitz[15] developed an algorithm based on a statistic model of van der Waals and Platteeuw [16] that can predict the incipient of hydrate formation for pure water. Later, this model was improved by Ng and Robinson [17] et al. The gas gravity method is very simple for predicting the gas hydrate conditions. The gas gravity method was conceived by Katz [18] of the GPSA Data Book. Also, the gas gravity method has served the gas processing industry well, as an initial estimate for a long period of time. Based on the GPSA data book, hydrate equations were developed for gases where specific gravity was known. The available correlations for a specific gravity method to calculate the hydrate formation conditions are those of Sloan, Berge, Motie and Hammerschmidt [19].

A designer can easily check hydrate formation using simulation software. Most simulation software, such as HYSYS or UNISIM, have a hydrate check utility which can be used. We can either change the operating conditions or add inhibitors to mitigate hydrate formation.

6.2.2 Methodology for Hydraulic Simulation

Suppose, a company is expanding its gas grid to meet the increasing demand for gas from new as well as existing customers. This will put pressure on the existing pipeline network. The capacity of the present pipeline network is not equipped to handle the flow rates that are being expected in the years to come. In this context, the study is aimed at finding the optimum solution to the above problem by conducting a hydraulic simulation study of the existing grid and to suggest the optimum solution by which the capacity of the existing grid can be increased to meet the growing demand.

The one-dimensional time dependent continuity, momentum equation, equation of state of gas, gas viscosity models with appropriate boundary conditions need to be solved to capture the flow conditions in the pipeline. Sophisticated and recognized fluid dynamic simulation codes have to be used to determine the same.

Any of the suitable softwares can be used for this study. The software, Pipeline Studio (TGNET), Energy Solutions International, United Kingdom is one of them.

The correlations used in the thermodynamic module to estimate fluid properties and friction factor are those of Bennedict–Webb-Rubbins, with Starling modification (BWRS) equation of state and Colebrook–White, respectively. The viscocity is calculated using the empirical correlation of Lee, Gonzalez and Eakin which also takes into account the temperature variation of fluid along the pipeline.

Solution technique

A very stable numerical solution technique with fast error-free convergence to the solution is used. The solution technique also captures the second order effects in the flow.

The pipeline network model is made for the entire grid comprising all pipelines in operation, and pipelines planned. In actual practice, parts of pipeline will be commissioned while other parts are being built. This also leads to throughput increase requirement. The pipeline throughput capacity can be increased by two methods.

1. Looping the pipeline.
2. Adding compressor stations.

Looping the pipeline

Looping a gas pipeline is equivalent to increasing the pipe diameter, and hence results in an increased throughput capacity. This method of increasing the pipeline capacity by looping involves initial capital investment. However, the operating cost is negligible compared to that of compressor stations.

When looping is provided to the existing pipeline appropriately, the supply pressure from various sources to the network is itself adequate to transport the required quantity of gas to all delivery points at the required delivery pressure.

The advantages and disadvantages of looping a gas pipeline are given below:

Advantages

Low capital investment compared to overall cost of compressor stations. The operating cost is negligible compared with the compressor operating cost.

Disadvantage

The time period for the project completion is comparatively higher. Looping is generally used for 'capacity increase' purpose.

Compressor stations

Compressors compress the natural gas and raise its pressure (and temperature) to the level that is required to ensure that the gas is transported to delivery points at the required flow and pressure. By compressing gas, high volume of gas can be transported through a small diameter pipeline. Though the initial capital cost of the compressor is low, the operating cost is higher and can be a major disadvantage.

With supply pressure (which is fixed) available at supply points, gas can be transported for a short distance, after which due to pressure drop in the pipeline the volume of gas increases considerably, then at this stage further transportation of gas at reduced pressure will not be possible due to the constraint imposed by the pipeline capacity (i.e. cross-sectional area of pipeline). To overcome this problem, intermediate compressor stations in the pipeline network at strategic locations are recommended.

Advantages

Comparatively, the project can be implemented faster with minimum problems to the existing system. The capital investment is also low.

Disadvantage

Gas cannot be compressed beyond the pressure for which the weakest part of pipeline is designed for. The annual operating cost of compressors is low; however, for the entire design life the operating cost of compressors is comparatively higher compared with the cost of looping option in some cases. To arrive at the optimum network configuration, both looping compressor stations need to be considered. This approach has the following advantages.

1. The supply pressure is utilized to maximum extent to transport gas.
2. By looping the compressor upstream section, the pressure drop in the upstream section of the compressor is reduced. Gas will be available at a higher pressure resulting in lowered compression ratio and as a result reduced fuel consumption at compressor driver.

Pigging

Pigging can be defined as the process of pushing a solid object through a pipeline. As already shown in Figure 6.6, one end of any pipeline should have a launcher and the other end should have a receiver.

The necessity for pigging can arise due to many reasons such as:

1. Removing debris, wax and dirt from the pipelines. Removing condensed liquids from gas pipelines.
2. Checking the thickness of lines for measuring corrosion by sending instrument pigs through the pipeline.
3. Special pigs are used for providing multipurpose coatings to the inner wall of pipelines.
4. If two or more products (mainly liquids such as petrol, diesel, kerosene) are transported batch-wise, then the pig acts as a separating agent.

6.2.3 Pipeline Network—A Case Study

An example of a pipeline network configuration is given below.

Natural gas networks are structured on geographical and demand considerations. In the present example, a network to be positioned in South India is considered. See Figure 6.8.

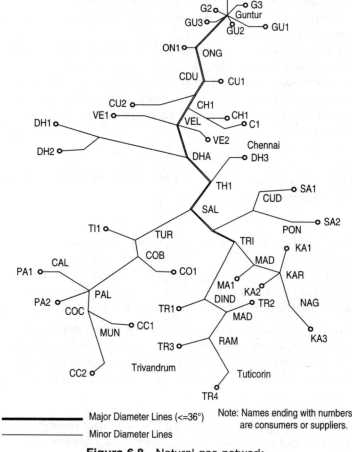

Figure 6.8 Natural gas network.

It is assumed that gas is supplied from a place called Guntur (located at the turning point of the shoreline of Bay of Bengal which is rich in gas deposits) and proceeds southwards supplying to the cities of Bangalore and Chennai and further proceeds to reach Trivandrum and Tuticorin. In between one more gas supply point is considered in Chennai, it being a port city. Only arbitrary values of distances between places (not actual values) are taken for the simulation.

Supply details considered

(Note: Flow units are in MMSCMD — Million Standard Cubic Metres per Day). Three different suppliers supply gas at Guntur and one at Chennai. See Table 6.2.

Table 6.2 Gas suppliers at Guntur and Chennai

Description	*Supply pressure*, Barg	*Flow*, MMSCMD
Guntur 1 Supply G1	91	9.0
Guntur 2 Supply G2	91	5
Guntur 3 Supply G3	53	9.0
Chennai Supply C1	53	12.0

Delivery details considered

Twenty-eight consumers are considered as per Table 6.3.

Table 6.3 Details of consumers

Consumer description	*Minimum delivery pressure*, Barg	*Flow*, MMSCMD
GU1	30	2
GU2	25	2
GU3	30	0.3
ON1	30	5 or 2
CU1	25	3
CU2	25	0.5
CH1	40	1.0
VE1	25	0.7
VE2	30	1.2
DH1	30	0.6
DH2	25	0.1
DH3	25	0.05
SA1	20	0.5
SA2	20	0.5
TR1	30	1.2
TR2	25	0.5
TR3	25	0.6
TR4	25	0.55
MA1	20	0.2
KA1	35	9–13
KA2	35	1.5
KA3	20	1.0–1.5
TI1	20	1.0
CO1	30	4.0
PA1	25	0.5
PA2	25	1.0
CC1	30	1.5
CC2	25	1.0

Simulation cases

This simulation consists of two studies—one about the selection of pipeline configuration and another about the flow rate changes based on a particular configuration.

For illustration purposes, two alternative cases are examined for each study to find out which one is most profitable.

Configuration selection

The costs indicated are only by way of illustration and not the actual costs.

There is an extra demand by the addition of a client at point TRI. To meet this demand various options are considered and configurations in terms of pipe diameters and use of compressors were techno-economically analysed in terms of capital and operating expenses, also called Capex and Opex[12]. The configurations are given below.

The existing major diameter line from CHI to TRI is 36 inch. As per simulations done if we increase the size to 42 inch in a selected stretch between the points CHI to TRI, the extra capacity required at TRI is then automatically taken care off.

As an alternative we can also decide not to increase the diameter of the existing major diameter line. As per simulations done in this case, a centrifugal compressor will have to be installed at a place called SAL and to be run continuously with a power requirement of 10.3 MW. In this case, there is both a capital cost as well as a fixed cost. Since the total cost is the sum of capital and operating costs, an analysis is done for obtaining results as given below.

Configuration 1:
Simulation case 1:

Case: 42 inch line in a selected stretch between CHI and TRI.

Case name diameter increase	Capital cost, in crores	Extra operating cost for 10 yrs, in crores	Total cost, in crores
Config 1	2310	Negligible	2310

Total cost for the portion between CH1 and TR1 = ₹ 2310 crores

Configuration 2:
Simulation case 2:

Case: 36 inch line from CHI to TRI + One Compressor at SAL of 10.3 MW.

Compressor at SAL	Capital cost, in crores	Extra operating cost for 10 yrs, in crores	Total cost, in crores
Config 2	1910	746	2656

Total cost for configuration 2 = ₹ 2656 crores

From the above analysis we find that the cost of pipeline laying is much less in the case of configuration 2 and the extra cost for operating the compressor is ₹746 crores. This, in actual case will be even higher, taking inflation into account for oprating cost estimation.

Since the total cost is low for Configuration 1, this configuration can be selected based on Capex and Opex.

Flow-rate changes and effect on delivery pressures

In this flow rate analysis, two simulation cases are considered based on the two alternative requirements of one of the clients. The gas pipeline network configuration in the present case is mainly decided by the gas requirement of the client. Simulation cases are developed based on flow to client ON1 (5 MMSCMD or 2 MMSCMD). The rest of the customers' requirement is constant.

In actual cases, several variations in pressure and flow are possible for any of the customers. For each simulation case, options are developed with different line sizes, out of which one optimum case is selected which is feasible in terms of economics as well as the one which offers better flexibility for future expansion.

Details of sample calculations are given below for illustration. Two options of differing consumption for a particular consumer are dealt with.

Now let us consider by way of illustration, the case of decreasing the flow rate to customer ON1 from 5 MMSCMD to 2 MMSCMD. See Tables 6.4 and 6.5. The pressures and other conditions at various points will change only to a small extent and hence are within acceptable limits.

Table 6.4 Delivery output data up to TR1—option 5: MMSCMD

Delivery point	Flow, in MMSCMD	Pressure, in Barg	Temperature, in °C
GU1	2	85.15	27.99
GU2	2	85.36	28.00
GU3	0.3	85.36	33.96
ON1	5	80.64	31.07
CU1	3	78.77	28.00
CU2	0.5	75.88	28.01
CH1	1.0	75.89	28.00
VE1	0.7	75.89	30.39
VE2	1.2	81.40	33.99
DH1	0.6	68.85	27.46
DH2	0.1	68.80	27.60
DH3	0.05	67.70	27.98
SA1	0.5	66.52	26.82
SA2	0.5	64.85	28.00
TR1	1.2	64.65	27.87

Table 6.5 Delivery output data up to TR1—option 2: MMSCMD

Delivery point	Flow, in MMSCMD	Pressure, in Barg	Temperature, in °C
GU1	2	85.06	27.99
GU2	2	85.27	28.00
GU3	0.3	85.27	33.96
ON1	2	80.37	31.07
CU1	3	80.24	28.05
CU2	0.5	75.88	28.06
CH1	1.0	77.13	28.10
VE1	0.7	77.15	30.50
VE2	1.2	82.63	34.01
DH1	0.6	69.87	27.46
DH2	0.1	69.83	27.70
DH3	0.05	68.68	27.98
SA1	0.5	67.45	26.92
SA2	0.5	65.8	28.10
TR1	1.2	65.6	27.98

Analysis

There will be a minimum inlet pressure requirement for each consumer. These figures have to be obtained from the consumer.

As per the inlet pressure figures given in the tables above for each consumer for both cases, option should be selected considering the minimum supply pressure requirement of any of the consumers downstream of ON1, so that the consumers supply pressure requirement is not violated.

Dynamic analysis

The hydraulic analysis should preferably include dynamic analysis. In dynamic analysis, two aspects are very important. The customers will not appreciate sudden changes in flow or pressure when

1. the compressors trips
2. the valves are opened and closed at either end.

We will have to define the events which are likely to occur, such as startups, shutdowns, tripping of one or many compressors, restarting of compressors, sudden stoppage of consumption, pipe rupture, etc.

A typical effect of two successive compressor trips at a client's receiving point will be as shown in Figure 6.9.

A forty-eight hour cycle in which important valves are closed and opened is considered. The first trip occurs after about 10 hours but is restarted before expiry of the 24 hours period. But again in the 24–48 hours period of time, it trips as in the previous case. The effect on flow rate

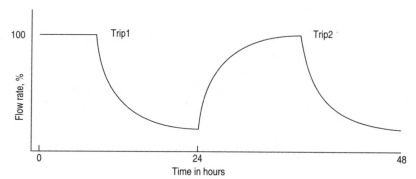

Figure 6.9 Flow rate change for compressor tripping.

is shown in Figure 6.9. The graphs of the following parameters can also be obtained for various conditions, and risks can be studied and eliminated.

1. Pressure
2. Temperature
3. Velocity
4. Flow fluctuations

The above example illustrates that complex natural gas network units can easily be solved using modern simulation software with all necessary flow, pressure and temperature results for taking quick decisions. The other types of dynamic analysis are covered in Chapter 8.

6.2.4 Shipping of Natural Gas

Natural gas is shipped in the form of Liquified Natural Gas (LNG) for which LNG carrier ships with attached refrigerations facility are used. On reaching the destination port the LNG which is at −160°C is stored in tanks provided with refrigeration.

REVIEW QUESTIONS

1. Describe the design features of two-phase and three-phase separators.
2. Describe the procedure for sizing Free Water Knockout Tank.
3. What is pigging in pipelines?
4. What is hydrate formation? How is it avoided?
5. What retention time will you use for light, medium and heavy crudes in the design of a three-phase separator?
6. What is dynamic analysis of a gas pipeline and how is it applied to gas pipelines?
7. What features are important in the design of gas transportation networks?
8. How are pipelines laid under the sea? (for further reading)
9. How is liquid natural gas stored? How is it regasified? (for further reading)

Chapter 7: NANOTECHNOLOGIES

7.1 POLYSILICON

Ultra-pure silicon in the form of polysilicon is the major building block of the second and third generation information technologies. Some of the properties of silicon and its forms are given below:

Atomic number	14
Molecular weight	28.09
Protons, electrons, neutrons	14 each
Density	2330 kg/m^3
Hardness	7 Moh
Melting point	1413°C
Boiling point	2878°C
Structure (crystalline)	diamond cubic structure
Structure (amorphous)	as powder

Intel Corporation was founded in California in 1968 by Gordon E. Moore (of Moore's law fame, a chemist and physicist) and Robert Noyce (a physicist and co-inventor of the integrated circuit). Robert Noyce left Intel to start Fairchild Semiconductor. A number of other Fairchild employees went on to participate in other Silicon Valley companies.

Intel's third employee was Andy Grove, a chemical engineer, who ran the company through much of the 1980s and the high-growth 1990s. Grove is now remembered as the company's key business leader and strategist of that period. By the end of the millennium, Intel was one of the largest and most successful businesses in the world.

Polysilicon with the crystal structure is extensively used in electronics, telecommunication and computer industries. The requirement of polysilicon for semiconductors has seen tremendous growth. Similarly, since it is also used for making solar panels, this becomes another huge area of growth for polysilicon.

7.2 POLYSILICON BY TRICHLOROSILANE ROUTE

The main raw material for the production of polysilicon is the metallurgical grade silicon (MG-Si).

7.2.1 Production of MG-Si

Silica is the main ingredient of all rocks and sands. Quartz and quartzite are pure forms of silica. While processing silica, the main product obtained is the metallurgical grade silicon. The by-products of these operations comprise various grades of silicon carbide and other intermediate materials.

The MG-Si is produced in large quantities since it is also used in the aluminium industry. Silica is converted to MG-Si, which takes place in submerged electrode arc furnaces. The molten silicon flows out of the arc funace, which is solidified and converted to powder. The greatest demand (60%) for MG-Si is in the aluminium industry where it is added in small amounts to improve machinability, castability and erosion resistance. The polysilicon industry has the next highest demand (25%) for MG-Si. The combined demand is several lakh tonnes per year, which includes polysilicon for solar panels.

MG-Si is produced by the carbothermic reduction of quartzite in submerged electrode arc furnaces. A typical capacity of a plant is about 10,000 tonnes per year. The following overall chemical reaction takes place during the reduction process.

$$SiO_2 + 2C \rightarrow Si + 2CO$$

The carbon is supplied as coke. It is also mixed with some wood chips to give porosity for easy escape of gases. The MG-Si obtained is refined using chlorine or oxygen to obtain a product with about 99% silicon. The carbon monoxide produced is used as a fuel.

7.2.2 Production of Polysilicon from MG-Si

Fluidized bed reactor

See Figure 7.1. The raw material, MG-Si, is ground in ball mills such that 75% of the particles are less than 40 microns in size. Then it is fed to a fluidized bed reactor (FBR). The bed is fluidized by HCl gas fed at the bottom of the reactor. The silicon in the MG-Si is converted to trichlorosilane (TCS) by the following reaction.

$$Si + 3HCl \rightarrow SiHCl_3 + H_2$$

Filtering and purification by distillation

Distillation is an all important step since ultra-pure TCS quality is achieved using distillation columns with more than hundred trays in most columns. All purification steps aim to avoid any type of contamination since the product to be achieved is to be very pure. In all distillation columns, the reboilers and condenser tubes are made of double concentric tubes with a buffer liquid between them to minimize leakage and contamination of process streams with water or other utilities.

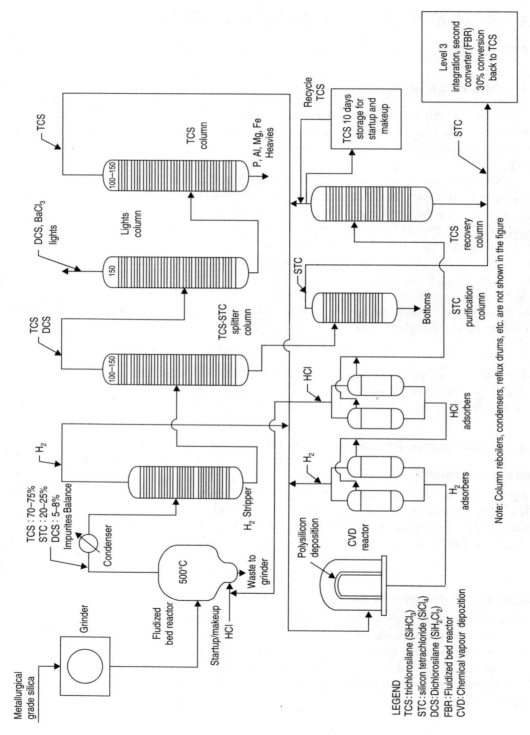

Figure 7.1 Flowsheet for the production of polysilicon.

The gases are passed through a particle filter to remove solid impurities such as $AlCl_3$ particles. Then the gases are sent to a hydrogen stripper where hydrogen is produced at the top.

The bottom product is sent to a TCS-STC splitter column where the silicon tetrachloride (STC) is removed at the bottom, and the TCS and some dichlorosilane (DCS) are removed at the top. The STC obtained at the bottom is further sent to an STC purification column where impurities in STC are removed before being sent to reconversion to TCS.

The TCS obtained at the top of the TCS–STC splitter column is sent to a lights column where the DCS and the light ends such as $BaCl_3$, etc. are removed. The bottom product of lights column is sent to the TCS column where the heavies such as compounds of phosphorus, aluminium, magnesium, iron, etc. are removed at the bottom and pure TCS is sent to the CVD reactor.

CVD reactor

The trichlorosilane entering the chemical vapour deposition (CVD) reactor dissociates directly on the resistance heated filaments inside the reactor chamber at a temperature of around 1100°C. The gas residence time in the reactor is between 5 and 20 seconds. The reactor consists of a bell jar sealed at the bottom to a base plate with inlet and outlet nozzles. The bell jar itself is made of quartz and has another housing of metal with startup heating rods. The rods either can be the polysilicon itself or can be made of metal conductors. New research suggests that specially shaped silicon conductors can provide higher areas for deposition[1]. As the deposition proceeds, the rod grows and hence the power supply to rods is increased to maintain the temperature by automatic temperature control.

The bell jar is made of quartz and is surrounded by a heater shroud. Between the bell jar and the heater shroud are the seed-rod preheaters which surround the bell jar. If the bell jar is made of metal, it is usually a double-walled bell jar using a water jacket for cooling. The auxiliary systems for the reactor are a power source, a system to cool the bell jar and a system to remove the grown silicon rods.

A typical Siemens type quartz bell jar reactor is shown in Figure 7.2. Rogers–Heitz polysilicon reactors are also similar wires in lieu of the seed rods and the seed rod heaters. The latter reactor can be either a metal or a quartz bell jar system.

Reactor bell jars vary in size from 18 inches to 10 feet in diameter and from 48 inches to 10 feet in height.

The capacity of a typical unit will be 40 tonnes per year. A number of CVD reactors are kept in parallel. At the end of the run of one set of reactors, the reactors are turned off, vessels purged with nitrogen, rods are removed, and the reactors are made ready for the next run.

Silicon is produced by the following chemical reaction:

$$2SiHCl_3 \rightarrow SiCl_4 + Si + 2HCl$$

This reaction gives the desired silicon but also poses a challenge, i.e. how to utilize the STC produced? The simplest option is to sell it. Selling of intermediate products is not good economics. Recycling of intermediates is the answer. All chemical industries try for maximum conversion of all intermediates to the desired product. For this the level 2 and level 3 integration systems are made, where progressively higher levels of system efficiency are achieved.

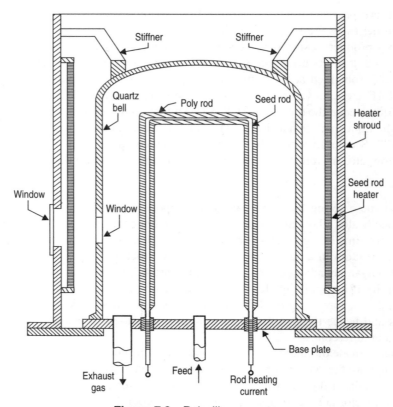

Figure 7.2 Polysilicon reactor.

Physical properties and critical constants

Most of the data for SiH_4, SiH_2Cl_2, $SiHCl_3$, $SiCl_4$ and Si is available, and so also the data for HCl and H_2 from literature.

Vapour pressure is given by the relation given below.

$$\log p = a + b/t + c \log T + dT + eT^2$$

Free energy is taken from TANAE tables.

Materials of construction

Union Carbide tested various materials of construction under chlorosilane conditions. Samples of carbon steel and SS316 were exposed to liquid chlorosilane for 600 hours. The carbon steel showed a uniform erosion rate of 2 μm per year. The SS 316 showed no detectable loss in weight but did exhibit pitting. Carbon steel was thus considered suitable.

Two studies have examined corrosion rates of materials of construction placed in a fluidized bed of silicon used to hydrogenate $SiCl_4$ to $SiHCl_3$. In each case, a reactor was operated at a temperature of 773 K and with the $H_2/SiCl_4$ molar ratio of 2.

Level 2 integration

In this integration step, the STC (SiCl$_4$) is converted back to TCS (SiHCl$_3$). The on-line practical single-pass conversion efficiency from STC to TCS is stated to be 18% and 32% at near 500°C, and at 73 psig and 300 psig, respectively[2]. In both conversions the residence time required is nearly sixty seconds at a mole ratio of 2.8 for H$_2$/SiCl$_4$[3].

In Level 2 integration, higher amount of recycling and more number of recovery distillation columns provide maximum production of polysilicon.

Level 3 integration

Here a second fluidized bed reactor is introduced and a better hydrogen and chlorine by-product recovery is made. This design reuses STC as in Level 2 integration to produce additional TCS and, in addition, uses the by-product HCl to react with MG-Si in a second FBR to produce both TCS and STC. There are several design variations of the level-three vertical by-product integration.

7.3 ALTERNATIVE ROUTES

7.3.1 Polysilicon from Magnesium Silicide

Magnesium silicide, produced by combining magnesium and silicon, is used to produce larger quantities of silane (SiH$_4$) via the chemical reaction given below.

$$Mg_2Si + 4NH_4Cl \rightarrow SiH_4 + 2MgCl_2 + 4NH_3$$

Since silane can be more easily purified to a higher degree than chlorosilanes, it can be decomposed in a Siemens-type reactor to produce very high purity polysilicon. Compared to cholorosilane, lower silicon deposits occur due to the lower temperature and pressure that must be used to present the gas-phase nucleation of silane in the reactor.

A low-cost method has been developed to produce large quantities of silane (SiH$_4$), for decomposition either in a Siemens-type reactor or in a fluidized bed reactor[3]. The Union Carbide process uses several of the chemical reactions developed (shown below) and involves a high recycling of chlorosilanes. The process is represented by the following sequence of chemical reactions by which SiCl$_4$ is converted to SiH$_4$ in four steps and finally to silicon. The first reaction uses copper as catalyst. The next three chemical reactions are also catalysed reactions.

$$3SiCl_4 (g) + 2H_2 (g) + Si (s) \rightarrow 4SiHCl_3 (g)$$
$$2SiHCl_3 (g) \rightarrow SiH_2Cl_2 (g) + SiCl_4 (g)$$
$$2SiH_2Cl_2 (g) \rightarrow SiH_3Cl (g) + SiHCl_3 (g)$$
$$2SiH_3Cl (g) \rightarrow SiH_4 (g) + SiH_2Cl_2 (g)$$
$$SiH_4 (g) \rightarrow Si(s) + 2H_2 (g)$$

The last reaction is for the decomposition of silane and deposition of silicon.

7.3.2 Silicon Production from Hydrosilisic Acid Produced from Superphosphate Plant

Ethyl Corporation has developed a large-scale process to produce SiH_4 for a polysilicon plant based on fluidized bed technology[4]. Their starting material is H_2SiF_6, a by-product of the superphosphate fertilizer industry. The flurosilisic acid is reacted with concentrated H_2SO_4 to liberate SiF_4 by the following chemical reaction.

$$H_2SiF_6 + H_2SO_4 \rightarrow SiF_4 + 2HF$$

The SiF_4 can be easily converted to silicon. This process plant can be placed adjacent to fertilizer plants.

Feedstock selection

Feedstock selection is as important as the process route selection. As already seen, the four most-used polysilicon feedstocks are:

1. Silane (SiH_4)
2. Dichlorosilane (DCS)
3. Trichlorosilane (TCS)
4. Silicon tetrachloride (STC)

The selection is difficult but has to be brought into practice. Feedstock selection is also the business of safety, quality, price and dependability. At any period in history, the importance of these factors will also depend on social and economic conditions as well as availability of technology.

Several criteria need consideration: They are: (i) Product purity, (ii) Manufacturing Cost, (iii) Safety, (iv) Alternative sources, (v) Transportation, (vi) Storage, (vii) By-product recovery, (viii) By-product use, (ix) Reactor deposition rate, (x) Construction methods, (xi) Reactor choices.

As of now, trichlorosilane continues to be the best feedstock for polysilicon. It is also the most practised technology.

Capital and operating costs

(Illustrative only, costs based on the year 1990)

With Level 3 chemical by-product recovery and reuse, the initial capital investment to build a polysilicon plant increases. The cost of production of polysilicon decreases because of higher efficiencies.

Table 7.1 indicates the approximate capital cost for a 2000 tonnes/year polysilicon plant. A plant using a Level 3 closed loop hydrogen and chlorine recovery system is considered.

Table 7.2 indicates the expected operating costs for this plant. A plant using a Level 3 closed loop hydrogen and chlorine recovery system is considered. These estimates are based on the use of bell jar reactor technology and the recovery and on-site reuse of nearly all reactor by-products.

Table 7.1 Capital cost of a TCS polysilicon plant[1]

Item	Cost (₹ crores)
Plant and equipment (including recovery systems)	4500
Safety systems Enviornmental systems	600
Installation and construction	525
Engineering, design, start-up	75
Total	**5700**

Table 7.2 Operating costs and cost of production of a TCS polysilicon[1]

Item	Cost (₹/kg silicon)
Raw materials	1095
Utilities	337.5
Labour	360
Operating supplies	172.5
Waste treatment	30
Depreciation (10 year st. line)	975
Total	**2970** (say ₹ 3000 per kg)

7.4 PILOT PLANT AND MINI PLANTS

7.4.1 Pilot Plant

Experimental study of TCS production[5]

Experimental studies are an important method of conceptual design validation and further development of plant designs. An abstract of such an experimental study is as follows.

The aim of the experiment was to study the chemical reaction between silicon and hydrogen chloride gas to produce, primarily, trichlorosilane in a fluidized bed condition. The experimental data was obtained first in the fixed bed condition and then in the fluidized bed condition. It was also required to determine the reaction rate constants for the chemical reaction of HCl gas on metallurgical grade silicon. The reactor used for the purpose was a laboratory reactor of inside diameter 26.6 mm and height 470 mm, made of Stainless Steel 316L. The reaction products were condensed at −78°C after coming out from the fluidized bed reactor. Uncondensed trichlorosilane vapours, if any, from the condenser were trapped in caustic soda solution, forming silica that was washed and weighed to calculate the content of uncondensed trichlorosilane. The experiments were carried out in the temperature range 316–326°C, using the average silicon particle size of 88–208 microns and flow rates of HCl gas of 1.6–3.2 litres per minute at 1 atmospheric pressure. Experimental conversions of HCl were evaluated. Fluidized bed models of Kunii and Levenspiel, and Kato and Wen were analysed using the experimental data to determine the overall rate constant and conversions. The experimental conversions of HCl were compared with the estimated values of Kuni and Levenspiel model as well as from those from the Kato and Wen model.

The experimental setup made up of Stainless Steel SS316L consisted of a reactor having 26.6 mm ID and 470 mm height.

Refer to Figure 7.3. The reactor had perforated plate distributor over the conical bottom through which HCl was supplied. It also had a flanged joint at the top and a pressure gauge. Approximately 0.056 kg oven dried silicon powder of size 208 microns or as per requirement was introduced from the top to obtain an initial height of bed at about 10 cm. Product outlet line was connected to a condenser and it had a sample valve. The reactor was heated by an electric resistance coil with an on-off controller. The reactor temperature was measured by a thermocouple located in the lower middle portion. The reactor was well insulated with glasswool. Heat generated due to the chemical reaction was removed by air flowing through a copper coil brazed around the reactor. A silicon bin was used for adding silicon powder to the bed in the reactor, and a condenser was used for condensing trichlorosilane. Trichlorosilane, a chemical reaction product, was condensed and collected in a small flask and weighed. The vapour of uncondensed trichlorosilane was reacted with NaOH and analysed.

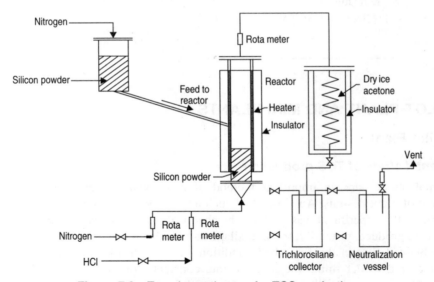

Figure 7.3 Experimental setup for TCS production.

Under the fixed bed conditions when the superficial gas velocity was lower than the minimum fluidization velocity for a particular size (438 microns) of silicon particle, trichlorosilane generated was not condensed but reacted directly with caustic soda solution. Silica formed by the chemical reaction of trichlorosilane with NaOH was precipitated, washed, dried and weighed to estimate the conversion of HCl to trichlorosilane. The reaction temperature was 321°C at atmospheric pressure. The flow rate of HCl was maintained at 1.75 litres per minute. The HCl conversion obtained by reacting unconverted HCl with NaOH was used to calculate the rate constant.

Unconverted HCl gas from the reactor and uncondensed trichlorosilane from the condenser were neutralized in caustic soda solution. The following chemical reactions occurred:

$$HCl + NaOH \rightarrow NaCl + H_2O$$
$$SiHCl_3 + 3NaOH \rightarrow SiO_2 + 3NaCl + H_2 + H_2O$$

The silica generated was dissolved in warm (due to chemical reaction of escaping TCS vapour with NaOH) caustic soda solution and then precipitated by 35.4% hydrochloric acid as per the following chemical reactions.

$$SiO_2 + 2NaOH \rightarrow Na_2SiO_3 + H_2O$$
$$Na_2SiO_3 + 2HCl \rightarrow 2NaCl + SiO_2 + H_2O$$

The boiling point of the produced trichlorosilane was 32°C, which matched well with the literature value. The specific gravity was found to be 1.37, which was slightly higher than the value given in the literature.

The chemical reactions were done for various silica particle sizes too, and results obtained. Conversion of silica increased with reduction in particle size, hence more conversion of HCl was obtained. A model was also built for the prediction of conversion based on the Kunii and Levenspiel model, and it was also compared with the other models.

7.4.2 Mini Plants

A mini plant is a new concept. It can be less costly and more useful than a pilot plant but manpower requirements are higher due to more qualified staff required. The objective of a mini plant is similar to that of a pilot plant, but before building a mini plant, computer simulations and mathematical modelling of the plant are done. These are then validated by a mini plant. It has the dual advantages of observing the theoretical as well as the practical aspects. It uses only minimum quantities of raw materials when compared to a pilot plant and consequently the wastes are also less.

Before making a mini plant, the kinetic data, equilibrium data, mass and heat transfer coefficients, etc. are determined in the bench-scale itself. Then the mini plant is built which reproduces the configuration as in flow sheet. This is done at the smallest scale possible but with maximum instrumentation so as to reproduce the operating conditions.

The mini plant is operated continuously for continuously operating plants and batchwise for batchwise operating plants.

The mini plants can be run for thousands of hours and kept monitored. This can result in direct construction of plants from the mini plant stage.

Mini plants can be designed and built based on new ideas and can be built for any continuous/batch process plants including trichlorosilane and polysilicon plants.

7.5 SEMICONDUCTORS AND SOLAR PANELS

7.5.1 Semiconductors

Semiconductors are used in electronics and computer industry for making chips and in the energy industry for making solar panels. The semiconductors for both these uses are the same but in computer industry the purity of the semiconductor material should be 99.9999999% whereas in solar panels it needs to be much less, say 99.9999%. Hence, the basic processing is the same but the quality requirement is different. If we build a big unit to produce solar energy grade silicon semiconductor along with a portion of it to produce semiconductors for computer industry, the economy of scales will benefit both the industries.

Several levels of intermediate industry exist for the manufacture of raw materials (precursors), polysilicon, single crystal ingots and wafers for use as silicon substrates in device manufacturing. Companies have different levels of integration across these industries.

Crystal manufacture

The manufacture of ploysilicon ingots is discussed in this chapter. Silicon crystals are made from these ingots. Crystals have been made by man from ancient times. But never has man made crystals of such perfection as silicon crystals. There are different methods for the same such as the Czochralski method, the Float Zone method, etc. The Czochralski method is described here since it is the most popular method.

Once the polysilicon of desired purity is obtained it is charged into the crucible of the Crystal Pullers, so called because the crystal is pulled out of the silicon melt as described below. See Figure 7.4.

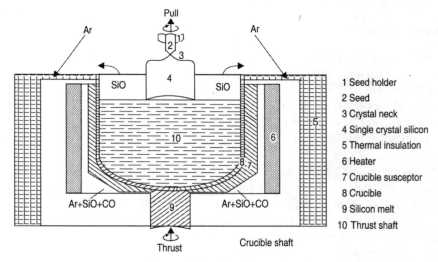

Figure 7.4 Crystal growth.

The silicon charge is melted to a liquid. A crystal seed is dipped into the melt. The freezing of a layer of melt occurs at the bottom of the seed. Further layers will grow on this layer as the crystal is slowly pulled out from the melt at a particular speed which depends on the heat transfer rates.

The latent heat of solidification is released by the melt during crystal formation. Under a steady-state condition the heat input to a system should be equal to heat output. The overall heat balance is given by the following equation.

$$Q_h + Q_l = Q_m + Q_c$$

where

Q_h = heat supplied by heaters
Q_l = heat of solidification
Q_m = heat loss from melt
Q_c = heat loss from crystal.

When the solid crystal meets the liquid melt, there will be temperature gradients in the solid as well as the liquid in the direction of heat flow which can be denoted by dt/dx. A theoretically evaluated equation for this process is the following differential equation.

$$d^2t/dx^2 = B * T^4$$

where $B = 2k_r e/k_s R$,
with

e = emissivity,
k_s = thermal conductivity,
k_r = radiation constant,
R = radius of crystal,
x = distance along the ingot from the cold end.

The above is only a simple example. There are many other advanced equations with different approaches, for which literature may be referred[1].

Doping theory

A silicon atom has fourteen protons in its nucleus with an equal number of electrons surrounding it, which makes the atom electrically neutral. Two electrons are in the first orbit, eight in the second orbit, and four more in the outermost orbit. This last orbit is not saturated since eight electrons are required for saturation. This is the case of a single atom. When there are many atoms as in pure silicon material with crystal arrangement, the material becomes stable by sharing four electrons of an atom with the neighbouring four atoms.

When an atom gets heated, an electron may move out of the atom and what is called a hole is created. The holes can move in a semiconductor producing a current, in addition to the movement of electrons, but in the opposite direction. The holes are equivalent to positive charges. We can increase or decrease electron availability by doping semiconductors with materials which have either five electrons in their outermost orbit or three electrons in their outermost orbit.

The *n*-type and *p*-type semiconductors are formed by doping. Pentavalent atoms such as arsenic, antimony and phosphorus are added to get extra electrons. Silicon doped in this manner is called the *n*-type semiconductor. To get extra holes instead of electrons, trivalent atoms such as boron or aluminium are added into the silicon. The number of holes will be equal to the number of atoms added. This type is called the *p*-type semiconductor.

A crystal becomes a semiconductor only because of the doped substances. The dopant concentrations ranging from 10^{14} to 10^{19} atoms/cm are used to meet the needs of various device applications. As already mentioned, there are two types of dopants—group 5 elements for *n*-type conduction and group 3 elements for *p*-type conduction. It is necessary to stir the contents for complete dissolution. The deposition of dopants can be done by many methods, of which the chemical vapour deposition is one of the popular methods.

The segregation coefficient in thermodynamic equilibrium gives the concentration of dopants in the growing crystals and that in the melt. It is usually less than one because the impurity atoms in excess stay in the melt. The crystal is always cleaner than the liquid melt. Hence, the impurities in the melt increase and so the impurities in the crystal. This increases as the crystal grows because of the impurities in the melt. Hence, the last portion of melt is to be discarded.

Silicon crystals should not contain any dislocations. It should not also contain lattice defects.

Wafer production

Wafer production is not done by a simple cutting process since with such a process the stringent requirements of the silicon industry cannot be met.

The first major operation is crystal-shaping consisting of cropping, grinding, fat grinding and etching. The second operation consists of wafering using a machine with annular diamond blade, heat treatment, edge rounding, lapping and wafer etching. The third operation is mainly polishing, cleaning and marking.

We will briefly describe the wafer cutting process. See Figure 7.5. The cutting is done by an inside diameter (ID) nickel-coated saw blade with diamond particles deposited on the inside edge. (Diamond is one of the hardest materials, about 4 times harder than corundum in the absolute scale of hardness.)

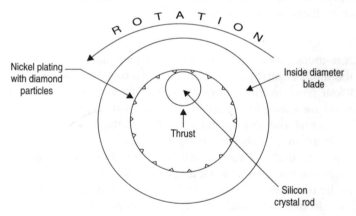

Figure 7.5 Wafer cutting.

The heart of the ID slicing machine is the blade made of high strength stainless steel with nickel plating and deposition as mentioned. The silicon crystal is sliced into thin wafers when the silicon crystal rod is pressed against the inside edge of the rotating blade.

Edge contouring and lapping are other important operations. Lapping is done to achieve very precise uniformity of thickness, flatness and parallelism. The polishing is mostly done by using a colloidal dispersion of silica in water. Finally, mechanical and chemical cleaning and heat treatment are required. The wafers produced should be traceable. Hence, they are marked with their history. Laser marking with chemical etching is also done.

Unlike other polished metals, polished silicon wafers have a perfect surface. The crystal just ends followed by less than 2 nanometers of silicon oxide which forms quickly in air and protects the wafer from chemical attacks. The defects in the product should generally be less than the detection limit of the best analytical tools. Also, no atomic or molecular impurities should be present on the surface.

Epitaxy

In Greek, *epitaxy* means something like "above (*epi*) in an ordered form (*taxi*)".

In the process of epitaxy a monocrystalline film (with a required dopant) is deposited on a monocrystalline substrate. The deposited film is known as epitaxial layer. Integrated circuits are made by repeating the operation several times. A simple integrated circuit made from crystal wafer chips by etching and vapour deposition is shown below. (A single VLSI (Very Large Scale Integration) chip may consist of millions of such transistors, which are fabricated using computer programs and sold at a reasonable price.) See Figure 7.6.

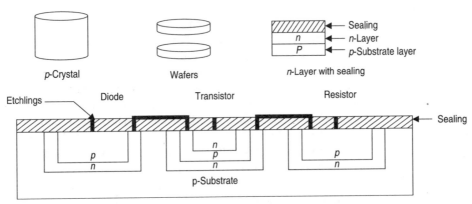

Figure 7.6 Steps in the manufacture of IC chips.

Because the substrate acts as a seed crystal, the deposited film takes on a lattice structure and an orientation, identical to those of the substrate. If the film is deposited on a substarte of the same composition, the process is called homoepitaxy, otherwise heteroepitaxy.

With homoepitaxy you can grow a layer which is more pure than the substrate and also layers having different doping levels. Heteroepitaxy is used for films of materials for which single crystals cannot be grown and also to produce compound material layers. Epitaxy is particularly important for making compound semiconductors.

Photolithography

The wafers are coated with a light sensitive liquid and cured by heating and then soft baked. Using ultra violet rays, masks are made between layers, and light is passed through selected areas so that a required pattern is transferred to the wafer surface.

New developments

In a rapidly changing field such as semiconductors it would be unreasonable not to mention something on new developments. A small list is given below.

1. Increase in crystal diameter will be a clear trend.
2. A decrease in average oxygen content may be achieved using magnetic fields.

3. Bulk defect reduction using hot zone engineering.
4. New crystals with high density of interstitials in the outer area and lower density of vacancies in the centre.
5. Crystals grown under near equilibrium conditions will be required for clustering of point defects to larger agglomerates.
6. Advanced single side and double side polishing of wafers.
7. Annealing at high temperatures and rapid thermal processing.
8. New epitaxial wafer types.
9. New generation photolithographic techniques.

7.5.2 Solar Panels

The big driver of solar panels is crystalline silicon which is getting more and more efficient, with thinner wafers and cheaper silicon materials. It is a kind of Moore's law for solar photovoltaics (PV). One example is using lasers in solar PV production. This will revolutionize the way solar cells are made, since the efficiency can be improved and we can go to thinner wafers to save on materials. Weber has indicated that due to the potential of 'concentrated photovoltaics' (which the Fraunhöfer Institute used in 2009 to scoop the world record for conversion of sunlight into electricity with 41.1% efficiency) new developments will occur[3]. The current commercial solar PV cells range between 10 and 20% efficiency.

Efficiency is lost because conventional PV cells can only absorb a small bandwidth of light. Concentrated PV broadens this spectrum using the 'triple junction' solar cells made of three different materials—gallium indium phosphide, gallium indium arsenide, and germanium. Each material absorbs different colours in the sunlight's spectrum. Special lenses amplify the energy intensity further by concentrating the beams up to 500 times.

Markets, however, are reacting differently to new technologies. This may result in a geographical split in solar PV research and development. Europeans are generally focusing on low-cost, high-volume production of silicon cells. Americans are trying to get ahead by developing the next generation photovoltaics.

The aim of increasing efficiency is producing results. A team at the California Institute of Technology lab tested a silicon wire design which they said had an efficiency of 85% in plain sunlight. At certain wavelengths, it had 95% efficiency. The researchers used tiny silicon wires one-millionth of a metre long, instead of conventional wafers. They are encased in a flexible polymer that can be rolled or bent.

More developments are also taking place. The University of Texas scientist, Xiaoyang Zhu, reported in June 2010 that semiconductor nanocrystals, or 'quantum dots', could capture the sun's energy currently lost as heat in conventional solar cells. Theoretically, this will boost performance to more than 60% efficiency.

Solar energy storage

The necessity of concentrated solar power (CSP) with energy storage facility arises since solar panels produce power only when the sun is shining but not when it is dark or when it is cloudy.

The biggest advantage of CSP over Solar PV is the ability to store solar energy at night or on sunless days. To quote an example, the Andasol solar power plant in Spain heats molten salt

In the evening, as the salt cools, it emits heat to make steam, giving the plant 7.5 hours of extra generating time even without sunlight.

Storage would open up new markets very quickly. If we have cost effective storage for solar PV, as we have for solar thermal, it would be a real breakthrough.

The other storage methods under investigation include using solar power generated during the day to drive pumps to compress air underground so that at night, the air could be released for generating power.

Despite promising research, CSP still challenges photovoltaics when it comes to efficiency. Most CSP systems have a back-up power installed, usually a natural gas turbine, which helps guarantee electricity production during cloudy weather. But concentrating solar power with mirrors demands a large land area and a water footprint. CSP plants also have to be built in areas with very good solar resources that often tend to be far away from large population centres. But as with photovoltaics, the CSP industry is also locked in an exciting technology race. The leading design is still the parabolic trough that uses mirrors to concentrate sunlight and heats synthetic fluids that then heat water to run steam turbines. But these 'transfer fluids' are costly and cannot be heated beyond 380°C, which limits efficiency and energy output.

The International Energy Agency's Concentrated Solar Power Roadmap reports, "The challenge is to enable the next generation of trough plants to produce steam at temperatures close to 500°C, thereby feeding the state-of-the-art turbines without continuous backup from fuel". The Agency recommends replacing the transfer fluids with water and thereby using 'direct steam generation' in trough plants.

Another way of increasing the heat is to use solar power towers. In these systems, hundreds or even thousands of mirrors focus sunlight on the tip of a tower where it heats a liquid in a central tank to create superheated steam.

Researchers are currently trying to increase temperatures (and thereby efficiency) to over 530°C in the newly planned CSP plants in California.

REVIEW QUESTIONS

1. Describe the process of manufacture of trichlorosilane and polysilicon with MG-Si as raw material.
2. What is a CVD reactor? How does it work?
3. Write a short note on doping.
4. What is crystal growth? Explain with a sketch.
5. How are silicon wafers produced?
6. Explain the term epitaxy and describe the process.
7. Compare the different methods of utilizing solar power.
8. How can we increase the efficiency of solar panels?
9. What geographical conditions/locations help in converting light into electrical energy? (for further study)

Chapter 8: MODELLING AND SIMULATION

8.1 REFORMER MODELLING

Mathematical modelling along with simulation is a useful and important tool for configuring process plant systems and subsystems employing computers. The results and observations obtained by modelling and simulation are widely accepted in process plant design as well as for optimizing plant operations.

A typical example of modelling a hydrogen plant reformer subsystem has been worked out as an illustration in this chapter to highlight the procedures required for the same.

There are three types of steam hydrocarbon reformers in vogue, these being top fired (typically Kellogg design), side fired (typically Haldor Topsoe design) and bottom fired also called terrace walled (typically Foster Wheeler design). The profiles of flame or flue gas temperatures are to be established as required for the particular case. This model can be adapted to all three types provided that the flame and flue gas temperatures are correctly modelled to suit the relevant case and provided that all characteristics of the catalysts are taken into consideration.

8.1.1 Equations for Model[1,2,3]

Let us consider a single tube of the reformer packed with catalyst. The feed gas, which is chemically reacting, is flowing through this tube from top to bottom at mass velocities sufficiently high that one can neglect the radial diffusion both of heat and mass in comparison with the main transport in the direction of flow. External heat is being supplied to this reacting system proportional to the temperature difference between the inside tube wall and the mean mixed bulk of the gas at any cross-section. In developing the model, the important reactions of the steam–methane reforming are considered.

The three independent reactions are:

$$CH_4 + H_2O \rightarrow CO + 3H_2 \tag{1}$$

$$CO + H_2O \leftrightharpoons CO_2 + H_2 \tag{2}$$

$$2CO \rightarrow CO_2 + C \tag{3}$$

The last reaction (3) is very very small and hence can be neglected. This will not affect the heat balance or the mole balance considerably. The model assumes equal distribution of feed in all tubes as well as equal exposure of radiation on all tubes.

The main reaction of reforming, reaction (1), is highly endothermic and consequently the appropriate rate of this reaction has to be considered for correct heat balance at any point inside the reactor. It is assumed that the water gas reaction (2) is at equilibrium under the normal operating temperatures. This fact is obvious from the available water–gas shift reaction data.

Heat balance

If one considers the differential element of the reactor of length dz as shown in Figure 8.1, then the amount of heat being brought into this element will be as given by the following expression (4),

$$\pi(R_i)^2 \cdot (G\, c_p\, t_B) + 2\pi R_i\, dz\, q_i'' \qquad (4)$$

where
  R_i = inside tube radius, ft
  G = mass velocity, lb/h-ft^2
  c_p = mixture, specific heat capacity, Btu/lb-°F
  t_B = gas bulk temperature, °F
  q_i'' = heat flux on the inner tube surface, Btu/h-ft^2

The first term is the sensible heat of the gas, and the second term is the external heat. Here the diffusion of heat, due to molecular conduction and eddy conduction, is considered negligible both radially and axially in comparison with the main transport of heat in the direction of gaseous flow.

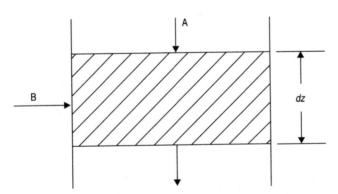

Figure 8.1 Catalyst in tube.

The heat leaving the differential element is given by the following expression (5),

$$\pi(R_i)^2(G\, c_p\, t_B) + \pi(R_i)^2\, G \frac{\delta}{\delta z} z(c_p\, t_B)\, dz - F \frac{\delta}{\delta z}(x \cdot H_{R_1})dz - F \frac{\delta}{\delta z}(y \cdot H_{R_2})dz \qquad (5)$$

where x is the moles of methane reacted per mol of feed by reaction (1), y is the moles of CO reacted per mol of feed by reaction (2), H_{R1} is the heat of reaction (1) in Btu/mol of CH$_4$, H_{R2} is the heat of reaction (2) in Btu/mol of CO. H_{R_1} and H_{R_2} are expressed as positive for exothermic reactions. They are dependent on the bulk temperature t_B at any location. F is the total feed flow rate in moles/h.

For the steady state case, one obtains from heat balance, Eq. (6) as follows:

$$\pi R_i^2 \cdot G \cdot \frac{d}{dz}(c_p \, t_B) = F \frac{d}{dz}(x \cdot H_{R_1}) + F \frac{d}{dz}(y \cdot H_{R_2}) + 2\pi R_i q_i \tag{6}$$

The heat flux, q_i, on the internal tube surface is related to the flux on the external surface q_o by Eq. (7) given below.

$$q_i \, R_i = q_o \, R_o \tag{7}$$

The heat flux on the outer tube wall surface is expressed by Eq. (8) in terms of radiation and convection from the furnace flue gas to the outer tube wall as

$$q_o = \sigma \, \mathcal{E}^{(z)}_m \cdot \{T_F^4 - T_{ho}^4\} + h_c \cdot (T_F - T_{wo}) \tag{8}$$

where

σ = Stephen-Boltzmann constant
$\mathcal{E}^{(z)}_m$ = effective emissivity of flue gas at distance z
T_F = flue gas temperature, °R (Absolute)
h_c = free convective heat transfer coefficient of flue gas, Btu/h-ft^2-°F
T_{wo} = outside tube wall temparature, °R (Absolute)

T_F is established based on the type of firing of the reformer as well as the type of flame. For side-fired reformers, T_F will be uniform. $\mathcal{E}^{(z)}_m$ is the product of the radiation view factor and the net emissivity between the radiant gas and the tube wall.

Tube wall temperature, T_{wo}, is related to the inside tube wall temperature, T_{wi}, and the gas bulk temperature, T_B, by Eqs. (9) and (10) for the cylindrical geometry of the tube wall.

$$T_{wi} = T_{wo} - q_o(R_o/k) \ln(R_o/R_i) \tag{9}$$

$$T_B = T_{wo} - q_o \, R_o * (1/(h_i \, R_i) + \ln(R_o/R_i)/k) \tag{10}$$

where

R_o = outside tube radius, ft
R_i = inside tube radius, ft
k = thermal conductivity of tube metal, Btu/h-ft^2-°F
h_i = internal heat transfer coefficient, Btu/h-ft^2-°F

The internal heat transfer coefficient from the tube wall to the bulk gas is simulated from the generalized Leva correlation as follows:

$$2h_i R_i/k_g = 0.813 \, e^{(-3*dp/Ri)} * (d_p G/\mu)^{0.9}$$

which reduces to

$$h_i = 0.813(k_g/2R_i) \, e^{(-3*dp/Ri)} * (d_p G/\mu)^{0.9} \tag{11}$$

where

k_g = conductivity of the gas, Btu/h-ft-°F
μ = viscosity of gas, lb/ft-h

The transport properties such as c_p, μ, k_g for use in Eqs. (6) and (11) are expressed as functions of temperature and pressure for each component i present and weighted for composition at the location in the reaction

$$c_p = (1/W)\Sigma_{(i=1 \text{ to } 6)} M_i(x, y) c_{pi} t_B \tag{12}$$

where $M_i(x, y)$ are moles of component i flowing per hour, c_{pi} the molar heat capacity, Btu/mol-°F and W total mass flow rate. c_{pi} is further expressed as

$$c_{pi}(t_B) = a_i + b_i t_B + c_i t_B^2 \tag{13}$$

where a_i, b_i, and c_i are coefficients dependent on pressure. Similarly, the other properties are expressed as follows.

$$\mu = \frac{1}{F(1+2x)} \sum_{i=1 \text{ to } 6} M_i(x,y) \mu_i t_B \tag{14}$$

$$k_g = \frac{1}{F(1+2x)} \sum_{i=1 \text{ to } 6} M_i(x,y) k_{gi} t_B \tag{15}$$

The term $(1 + 2x)$ accounts for the molecular expansion.
The properties μ_i and k_{gi} are expressed as

$$\mu_i(t_B) = a'_i + b'_i t_B + c'_i t_B^2 \tag{16}$$

$$kgi(t_B) = a''_i + b''_i t_B + c''_i t_B^2 \tag{17}$$

Mole balance and reaction kinetics

The proper stoichiometric relations among the reacting components and the accounting of the moles consumed and produced by the chemical reactions are considered as shown in the Table 8.1.

Table 8.1 Mole balance

Component	Moles in feed	Moles reacted by reaction (1)	Moles reacted by reaction (2)	Moles present
CH_4	f_1	xF	0	$f_1 - xF$
CO	f_2	$-xF$	yF	$f_2 - (y-x)F$
CO_2	f_3	0	$-yF$	$f_3 + yF$
H_2	f_4	$-3xF$	$-yF$	$f_4 + (y+3x)F$
H_2O	f_5	xF	yF	$f_5 - (y+x)F$
N_2	f_6	0	0	f_6
Total	F	$-2xF$	0	$(1+2x)F$

where f_1–f_6 is the feed flow rate of component i (i from 1 to 6).
The generation of new species within the differential element dz can be determined from the methane mole balance:

$$F\left(\frac{dx}{dz}\right) = \rho_c \pi R_i^2 (1 - \epsilon) r \tag{18}$$

where
ρ_c = catalyst density, lb/ft^3
r = rate of reaction (1), moles CH_4/h-lb of catalyst
z = tube length, ft

The net reaction rate in the forward direction, r, for the rate controlling reaction (1) is given in terms of partial pressures p_i as

$$r = k_r \left[p_{CH_4} - \left(\frac{p_{CO} p_{H_2}^3}{p_{H_2O} KE_1} \right) \right] \tag{19}$$

where
k_r = reaction rate constant, mol CH_4/h-lb of catalyst at atmospheric pressure
KE_1 = equilibrium constant of reaction (1)

The specific reaction forward rate constant, k_r, is expressed in terms of temperature, pressure, specific geometric surface per unit volume in inch^{-1}, a_s, and the activity of the catalyst, a_c, as follows:

$$k_r = \left(\frac{3.02}{P} \right) a_s a_c e^{(-E/(R_g T_B))} \tag{20}$$

where, E is the activation energy and R_g is the gas law constant.

The effect of diffusion on-reaction rate constant is not discussed. The relationship between x and y is obtained from the equilibrium statement of reaction (2) as follows:

$$KE_2 = ([f_3 + yF] [f_4 + (y + 3x)F])/([f_2 - (y - x) F] * [f_5 - (y + x)F]) \tag{21}$$

(representing, CO_2, H_2, CO, H_2O).

The dependence upon temperature of equilibrium constants KE_1 and KE_2 for reactions (1) and (2) is given by the relations

$$KE_1 = g_1 e^{\beta_1/T_B} \tag{22}$$
$$KE_2 = g_2 e^{\beta_2/T_B} \tag{23}$$

where $g_1, g_2, \beta_1, \beta_2$ are constants.

The expression for dy/dz required in Eq. (6) is obtained in terms of dx/dz and dt/dz from Eqs. (21) and (23) respectively.

Reaction heats HR_i are also expressed as functions of temperature.

$$HR_1 = g_3 + g_4 t_B + g_5 t_B^2 \tag{24}$$
$$HR_2 = g_6 + g_7 t_B + g_8 t_B^2 \tag{25}$$

This completes the basic equations of this model which is to be developed into a complete model except the pressure drop through the bed.

Pressure drop through bed

The pressure drop through the reactor can be determined by the following Eq. (26), developed by Ergun, which takes into account both the kinetic energy losses as well as the frictional losses.

$$-\frac{dp}{dz} = \frac{-\gamma G}{\rho_g d_p g_c} \frac{1-\epsilon}{\epsilon^3} \left(\frac{150(1-\epsilon)\mu}{d_p} + 1.75 G \right) \tag{26}$$

where
γ = a constant.
ρ_g = gaseous density given by

$$\rho_g = \frac{P}{K_g T_B} \cdot \frac{W}{F(1+2x)} \tag{27}$$

G = mass velocity, lb/h-ft^2
γ = Ergun equation constant
d_p = catalyst particle diameter, ft
ϵ = void fraction of tube catalyst bed
μ = velocity of gas, lb/ft-h

However, it is better to use more suitable equations for pressure drop, based on a particular catalyst manufacturer's recommendation.

A flowchart of the reformer simulation program is given, See Figure 8.2. This is only an illustrative version and the pressure drop calculation for each centimetre is not included.

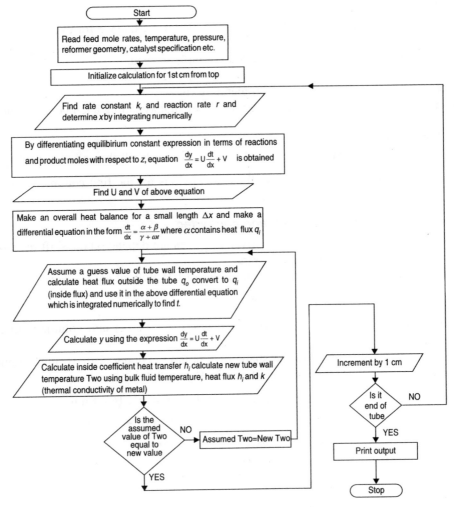

Figure 8.2 Flowchart for reformer simulation program.

8.1.2 Development of Model

The development of a model is illustrated along with an example to give a better perspective of the calculations. Only the first centimetre calculations are shown.

Manual calculation is shown for the beginning portion.

The following hydrocarbon plus steam mixture at temperature and pressure conditions given below is considered at the inlet to the reformer tube:

CH_4 = 398.2 kgmol/h
CO = 1 kgmol/h
H_2 = 8.1 kgmol/h
H_2O = 1388.9 kgmol/h
Temperature = 615°C
Pressure = 23 atm

The program will be required to work in the following sequence for every step, 1 cm being a unit of calculation for a step.

1. Find the moles of methane converted per mole of feed.
2. Find the moles of CO converted per mole of feed.
3. Find the heat transfer rate and temperature rise.
4. Steps 2, 3 will be done together iteratively by first assuming a temperature and changing it until convergence is obtained.
5. Find the pressure drop.
6. Proceed to the next centimetre and so on downwards till the end of the tube is reached.

The logic of the development as well as calculations for the first centimetre are discussed below.

1. To calculate the moles of methane converted

The methane conversion reaction is rate driven and does not depend on the shift reaction. Hence, this is calculated first for the first centimetre.

To find the reaction rate constant (k_r) in moles of CH_4/h/lb of catalyst/atm

Use Eq. (20), $k_r = (3.02/P) \, a_s \, a_c \, e^{(-E/R_g T_B)}$

where P is in atmospheres (take 23 atm), a_s is catalyst surface per unit volume in inch^{-1} and a_c is catalyst activity.

Consider a catalyst particle in the shape of a hollow cylinder with 16 mm length, 16 mm outside diameter and 2 mm thickness as shown in Figure 8.3.

$$\begin{aligned}
\text{Total area} \quad &= \pi/4(16^2 - 4^2)*2 + \pi*16*16 + \pi*16*4 \text{ mm}^2 \\
&= 0.785*(16^2 - 4^2)*2 + \pi*16*(16 + 4) \text{ mm}^2 \\
&\quad (\pi/4 \text{ taken as } 0.785, \text{ in all cases}) \\
&= 1381 \text{ mm}^2/(25.4)^2 \text{ inch}^2 \\
&= 2.14 \text{ inch}^2
\end{aligned}$$

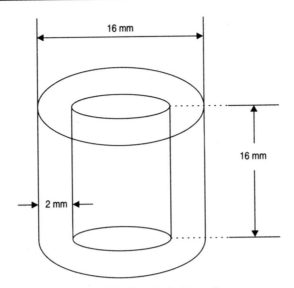

Figure 8.3 Catalyst dimensions.

$$\text{Total volume} = \pi/4 * 16^2 * 16 - \pi/4 * 4^2 * 16 \text{ mm}^3$$
$$= 0.785 * 16 * (16^2 - 4^2) \text{ mm}^3$$
$$= 0.1839 \text{ inch}^3$$

$a_s = 2.14 \text{ inch}^2/0.1839 \text{ inch}^3$
$\quad = 11.64 \text{ inch}^{-1}$

$a_c = 5$ (catalyst activity)

E = activation energy (Btu/mol)
$\quad = 44{,}000 \text{ kJ/kgmol}$

R_g = Universal gas constant = 8.314 kJ/kgmol-K or 35 Btu/lbmol-°R

T_B = bulk temperature = 615°C = 888 K or 1598.4°R

$E/R_g T_B = 44{,}000 \text{ kJ/kgmol}/(8.314*888) = 5.96$

∴ $\quad k_r = 3.02/23 * 11.64 * 5 * e^{-(5.96)}$
$\quad\quad = 0.0197 \text{ atm}^{-1} \text{ inch}^{-1}$ or (lbmols CH_4)/(h-lb catalyst-atm)

To find r, the rate of chemical reaction in lbmol CH_4 reacted /(h-lb of catalyst)

$$r = k_r \, (p_{CH_4} - (p_{CO} * p_{H_2O}^3)/(p_{H_2O} * KE_1))$$

where p is the partial pressure in atm.

Taking $KE_1 = 1.139 \text{ atm}^{-2} = (1/(1.139 * 1.139)) = 0.7708 \text{ atm}^2$

$r = 0.0197 * ((398.2/1815.2) * 23 - ((1/1815.2) * 23 * (8.1 * 23/1815.2)^3))/$
$\quad ((1388.9/1815.2) * 23 * 1/(1.139 * 1.139))$

$$= 0.0197 * (5.045 - (0.1267 * .001081^3)/(17.598 * 0.7708))$$
$$= 0.0197 * (5.045 - 0.000000886)$$
$$= 0.09938 \text{ atm}^{-1} \text{ inch}^{-1}$$

To find x, we use
$$\rho_c = \int dx = \int [(\rho_c * \pi R_i^2 * (1 - \epsilon) * r)/F] * dz \tag{18}$$

First value $= (\rho_c \, \pi \, R_i^2 * (1 - \epsilon) * r/F) * (0.03278 \text{ ft})$ [dimensionless] ($\because dz = 1$ cm $= 0.03278$ ft)

where $\rho_c = 68.685$ lb/ft^3, density of catalyst

$R_i = 0.164$ ft, inner radius of tube

$\epsilon = 0.5$

$F = 1815.2$ kgmol/h $= 3993.44/96$ lbmol/h per tube (number of tubes $= 96$)

$$x = (68.685 * \pi * (0.164)^2 * 0.5 * 0.09938 * 0.03278)/(3993.44/96)$$
$$= 0.000227 \text{ lbmol CH}_4 \text{ converted/lbmol of feed.}$$

$dx/dz = 0.000227/0.01$

$= 0.0227$ lbmol CH$_4$ converted/lbmol feed/metre length of tube.

$= 0.0227$ m^{-1} (also called 'w' in the next section)

(If you can get F as a function of x, integrate for 1 cm using the Runge-Kutta method.)

2. Relation between dy/dz, dt/dz and dx/dz

The conversion of CO by water gas reaction is an equilibrium reaction and depends on the temperature as well as the composition of reactants and products. Here dy/dz is the kgmoles of CO converted per kgmole feed, per metre of tube length. In the conversion of CO to CO$_2$, KE_2 is the dimensionless equilibrium constant.

$$KE_2 = ((f_3 + yF) * (f_4 + (y + 3x) * F))/((f_2 - (y - x) * F) * (f_5 - (y + x) * F))$$

LHS $= g_2 * e^{\beta_2/t}$

($g_2 =$ constant, $\beta_2 =$ association constant.) $\tag{23}$

Differentiating both sides with respect to z,

$d/dz(\text{LHS}) = g_2 e^{\beta_2/t} * (- \beta_2/t^2) * (dt/dz)$

$= ((-g_2 * \beta_2)/t^2) * e^{\beta_2/t} * dt/dz$

RHS $= d/dz \, (u/v) = (1/v^2) \, [v * (du/dz) - u * (dv/dz)]$ (formula for differentiation by parts)

where,

$u = (f_3 + yF) * (f_4 + (y + 3x) * F)$

$v = (f_2 - (y - x) * F) * (f_5 - (y + x) * F)$

$du/dz = d/dz * [(f_3 + yF) * (f_4 + (y + 3x) * F)]$

$= (f_3 + yF) * d/dz[f_4 + (y + 3x) * F] + (f_4 + (y + 3x) * F) * dy/dz(f_3 + yF)$

$= (f_3 + yF) * (dy/dz + 3dx/dz) * F + (f_4 + (y + 3x) * F) * (dy/dz)F$

$= 3(f_3 + yF) * F * (dx/dz) + (f_3 + yF) * F * dy/dz + (f_4 + (y + 3x) * F) * F * (dy/dz)$

$$dv/dz = d/dz * [(f_2 - (y - x) * F) * (f_5 - (y + x) * F)]$$
$$= (f_2 - (y - x) * F) * d/dz \, (f_5 - (y + x) * F) + (f_5 - (y + x) * F) * d/dz (f_2 - (y - x) * F)$$
$$= (f_2 - (y - x) * F) * (-F(dy/dz) - F(dx/dz)) + (f_5 - (y + x) * F) * (-F(dy/dz) + F(dx/dz))$$
$$= (f_2 - (y - x) * F) * (-F) * (dy/dz) + (f_2 - (y - x) * F) * (-F) * (dx/dz) + (f_5 - (y + x) * F) * (-F)$$
$$\quad * (dy/dz) + (f_5 - (y + x) * F) * F * (dx/dz)$$
$$= dx/dz[(f_2 - (y - x) * F) * (-F) + (f_5 - (y + x) * F) * F] + dy/dz[(f_2 - (y - x) * F) * (-F) +$$
$$\quad (f_5 - (y + x) * F) * (-F)$$

RHS $= 1/v^2 \, [v * (du/dz) - u * (dv/dz)]$

Take $H = (f_2 - (y - x) * F)^2 * (f_5 - (y + x)F)^2$

$G = ((-g_2 * \beta_2)/t^2) * e^{(\beta 2/t)}$

$$G * H * (dt/dz) = v * (du/dz) - u * (dv/dz)$$
$$= (f_2 - (y - x) * F) * (f_5 - (y + x) * F) * \{3 * (f_3 + yF) * F * (dx/dz)$$
$$\quad + [(f_3 + yF) * F + (f_4 + (y + 3x) * F) * F] * dy/dz\}$$
$$\quad - (f_3 + yF) * (f_4 + (y + 3x) * F) * \{dx/dz[(f_2 - (y - x) * F) * (-F)$$
$$\quad + (f_5 - (y + x) * F) * F] + (dy/dz)[((f_2 - (y - x) * F) * (-F) + (f_5 - (y + x) * F) * (-F)]\}$$
$$= (f_2 - (y - x) * F) * (f_5 - (y + x) * F) * 3 * (f_3 + yF) * F * (dx/dz)$$
$$\quad + (f_2 - (y - x) * F) * (f_5 - (y + x) * F[(f_3 + yF) * F + (f_4$$
$$\quad + (y + 3x) * F) * F] \, (dy/dz) - (f_3 + yF) * (f_4 + (y + 3x) * F) \, [(f_2$$
$$\quad - (y - x) * F) * (-F) +$$
$$\quad (f_5 - (y + x) * F) * F] \, (dx/dz) - (f_3 + yF) * (f_4 + (y + 3x) * F)[(f_2$$
$$\quad - (y - x) * F) * (-F) + (f_5 - (y + x) * F) * (-F)] \, (dy/dz)$$
$$= (P - Q) * (dy/dz) + (R - S) * (dx/dz)$$

where
$$P = (f_2 - (y - x) * F) * (f_5 - (y + x) * F)\{3 * (f_3 + yF) * F + (f_4 + (y + 3x * F) * F\}$$
$$= (f_2 - (y - x) * F) * (f_5 - (y + x) * F) * (f_3 + yF + f_4 + (y + 3x) * F) * F$$
$$= (f_2 - (y - x) * F) * (f_5 - (y + x) * F) * (f_3 + f_4 + 2yF + 3xF) * F$$

$$Q = (f_3 + yF) * (f_4 + (y + 3x) * F) \, [f_2 - (y - x) * F + f_5 - (y + x) * F] * (-F)$$
$$= (f_3 + yF) * (f_4 + (y + 3x) * F) * (-f_2 + (y - x) * F - f_5 + (y + x) * F) * F$$
$$= (f_3 + yf) * (f_4 + (y + 3x) * F) * (-f_2 - f_5 + 2yF) * F$$

$$R = (f_2 - (y - x) * F) * (f_5 - (y + x) * F) * 3 * (f_3 + yF) * F$$

$$S = (f_3 + yF) * (f_4 + (y + 3x) * F) * (f_2 - (y - x) * F - f_5 + (y + x) * F) * (-F)$$
$$= (f_3 + yF) * (f_4 + (y + 3x) * F)(-f_2 + (y - x) * F + f_5 - (y + x) * F) * F$$
$$= (f_3 + yF) * (f_4 + (y + 3x) * F) * (f_5 - f_2 - 2xF) * F$$

Now,

$$GH\,(dt/dz) = (P - Q)*(dy/dz) + (R - S)*(dx/dz)$$

or $GH/(P - Q)\,(dt/dz) = dy/dz + ((R - S)/(P - Q))*(dx/dz)$

Call, $W = dx/dz$

$$(GH/(P - Q))\,(dt/dz) = dy/dz + (R - S)*W/(P - Q)$$

Relation obtained:

$$dy/dz = [GH/(P - Q)]\,(dt/dz) + [-(R - S)*W/(P - Q)]$$

Putting $U = GH/(P - Q)$ and

putting $V = -(R - S)W/(P - Q)$

$$dy/dz = U*(dt/dz) + V$$

Dimensional check: $\dfrac{1}{M} = \dfrac{1}{T}*\dfrac{T}{M} + \dfrac{1}{M}$ (all terms are in M^{-1}). The preceding relation is therefore dimensionally consistent.

Values of U and V for the first cm

$$G = (-g_2*\beta_2)/t^2 * e^{\beta_2/t}$$

where g_2 and β_2 are found by regression of data

$g_2 = e^{-4.14484} = -0.01585$

$\beta_2 = 4466.272$

$G = (-0.01585 * 4466.272)/(888)^2\ e**(\ 4466.272/888)$

$\quad = -0.01372\ (K)^{-1}$

$H = (f_2 - (y - x)*F)^2 * (f_5 - (y + x)*F)^2$

$\quad = f_2^2 * f_5^2$

$\quad = 1^2 * (1388.9)^2/96^4$

$\quad = 0.02268$

$P = (f_2 - (y - x)*F)*(f_5 - (y + x)*F)*(f_3 + f_4 + 2yF + 3xF)*F$

$\quad = (1*1388.9*(11.3 + 8.1)*1815.2)/96^4$

$\quad = 0.57585$

$Q = (f_3 + yF)*(f_4 + (y + 3x)*F)*(-f_2 - f_5 + 2yF)*F$

$\quad = (11.3*8.1*(-1 - 1388.9)1815.2)/96^4$

$\quad = -2.7188$

$R = (f_2 - (y - x)*F)*(f_5 - (y + x)*F)*3*(f_3 + yF)*F$

$\quad = (1*1388.9*3*11.3*1815.2)/96^4$

$\quad = 1.00626$

$S = (f_3 + yF)*(f_4 + (y + 3x)*F)*(f_5 - f_2 - 2xF)*F$

$\quad = (11.3*8.1*1388.9*1815.2)/96^4$

$\quad = 2.7169$

$$U = (-0.01372 * 0.02268/(0.57885 - (-2.7188))$$
$$= -0.00009436 \text{ K}^{-1}$$
$$V = -(1.00626 - 2.7169) * W/(0.57885 - (-2.7188))$$
$$= -(1.00626 - 2.7169) * 0.0227/(0.57885 - (-2.7188))$$
$$= 0.0117 \text{ m}^{-1}$$

3. To calculate heat balance, CO converted and temperature rise

(a) *Overall heat balance over infinitesimal length of tube.*
It can be done in (Btu/h per ft of tube) or (J/h per m of tube)
$$\pi R_i^2 * G * d/dz \, (c_p t_B) = F * (d/dz) * (x \, HR_1) + F * (d/dz) * (y \, HR_2)) + 2\pi R_i \, q_i$$

Calculation in SI system:
c_p is available in J/kgmol K
$$\text{LHS} = (\pi R_i^2 / \pi R_o^2) * (F/96) * d/dz(c_p t_B) \quad ((F/96) \text{ is flow through 1 tube})$$
$$= (\pi R_i^2 / \pi R_o^2) * (F/96) * (dc_p/dz * t + dt/dz * c_p)$$
$$= (\pi R_i^2 / \pi R_o^2) * (F/96) * (dc_p/dt * dt/dz * t + dt/dz * c_p)$$

By regression analysis, $c_p = (a_i t + b_i)$, hence substituting for c_p,
$$\text{LHS} = (\pi R_i^2 / \pi R_o^2) * (F/96) * t * dt/dz * a_i + (\pi R_i^2 / \pi R_o^2) * (F/96) \, dt/dz * (a_i t + b_i)$$

[Units are kgmol/h * K * K/m * J/kgmol-K-K = J/m-h]
$$= (F/96) \, t \, dt/dz \, a_i + (F/96) \, t \, dt/dz \, a_i + (F/96) \, dt/dz \, b_i \quad [\text{kgmol/h °K/m J/kgmol °K}]$$
$$= 2 (F/96) \, t \, dt/dz \, a_i + (F/96) \, dt/dz \, b_i$$
$$\text{RHS} = (F/96) \, HR_1 \, dx/dz + (F/96) \, HR_2 \, dy/dz + 2\pi R_i \, q_i$$

[Units kgmol/h * J/kgmol * 1/m = J/h-m] [m * J/h-m^2 = J/h-m]

Equating LHS = RHS
$$2(F/96) \, t \, dt/dz \, a_i + (F/96) dt/dz \, b_i = (F/96) \, HR_1 \, dx/dz + (F/96) \, HR_2 \, dy/dz + 2\pi R_i q_i$$

Substituting dy/dz by $U \, dt/dz + V$ and dx/dz by W,
$$dt/dz(2(F/96) \, t \, a_i + (F/96) b_i) = (F/96) \, HR_1 \, W + 2\pi R_i q_i + (F/96) \, HR_2(U \, dt/dz + V)$$
$$= (F/96) \, HR_1 \, W + 2\pi R_i q_i + (F/96) \, HR_2 \, u \, dt/dz + (F/96) \, HR_2 \, V$$

i.e. $dt/dz(2(F/96) \, t \, a_i - (F/96) \, HR_2 \, U + (F/96)b_i) = (F/96) \, HR_1 \, W + 2\pi R_i q_i + (F/96) \, HR_2 V$

Substituting
$$\text{Alpha} = (F/96) \, HR_1 \, W + 2\pi R_i q_i$$
$$\text{Beta} = (F/96) \, HR_2 \, V$$
$$\text{GAM} = (F/96) \, b_i - (F/96) \, HR_2 \, U$$
$$\text{OMEG} = 2(F/96) \, a_i$$
$$dt/dz \, (\text{GAM} + \text{OMEG} * t) = \text{ALPHA} + \text{BETA}$$

we have
$$dt/dz = (\text{ALPHA} + \text{BETA})/(\text{GAM} + \text{OMEG} * t)$$

Units:
ALPHA = kgmol/h * J/kgmol * 1/m = J/m-h
BETA = kgmol/h * J/kgmol * 1/m = J/m-h
GAM   = kgmol/h * J/kgmol K = J/h-K
OMEG * t = kgmol/h * J/kgmol K * K/K = J/h-K
dt/dz = (J/m-h)/(J/h-K) = K/m

(b) *Overall heat flux equation*
$$q_o = \sigma \epsilon_m \{T_F^4 - T_{wo}^4\} + h_c(T_F - T_{wo})$$
Units are J/h-m^2 = (σ * 3600) J/(h-m^2 (K)^4) ϵ_M (K)^4 + (h_c * 4196)J/(h-m^2-K)K
where $\sigma = 5.68 * 10^{-8}$ J/(s-m^2-K^4)

(c) *Convection coefficient by Nusselt equation*
For flow outside tubes—Properties of flue gas have been assumed.
[Units dimensionless = (kcal/h-m^2-K) * (m/kcal) * (h-m^2-K/m)]
$h_c D_e/k = 0.023 * (D_e G/\mu)^{0.8} * (c_p \mu/K)^{1/3}$

D_e = equivalent diameter
    = (4 * Free Area)/Perimeter = (4 * 9 * 9.4)/2 * (9 + 9.4) + (π * 121 * 96)
    = 4.6 m

$D_e G/\mu$ = 4.6 * (61292/0.785 * (4.6)^2) * 1/(0.02 * 3.6)
         = (4.6 * 61292)/(0.785 * (4.6)^2 * 0.02 * 3.6)
         = 235,745

$c_p \mu/k$ = ((0.02 * 3.6) * 135)/0.12 (kg/h-m) * (kcal/kg-°C)
         = 81 kcal/(h-m-°C)

$h_e = (k/D_e * 0.023) * (NRe)^{0.8}(NPr)^{0.33}$ = (kg/h-m) * (kcal/kg-°C) * (h-m-°C/kcal)
    = 0.023 * (0.12/4.6) * (235745) ** 0.8 * (81) ** 0.33

(Units of k/D_e = kcal/(h-m-°C) * m)
          = 50.8 kcal/(h-m^2-K)

(d) *Heat flux and temperature calculation*
Assume T_{wo} (guess value) = 800°C = 1073 K
q_o = 5.68 * 10^{-8} * 3600 * $\epsilon_m\{(1623)^4 - (1073)^4\}$ + 50.8 * 4196(1623 − 1073)
    = (5.68 * 10^{-8}) * (3600 * 0.375) * (5.613 * 10^{12}) + (50.8 * 4196) * 550
    = (0.43 * 10^9) + (0.117 * 10^9)
    = (0.547 * 10^9) J/(h-m^2)

q_i = q_o * (121/100) = (0.6623 * 10^9) * J/(h-m^2)
ALPHA = (F/96) * HR_1 W + 2$\pi R_i q_i$
Now
  (HR_1 = −206,000 J/kgmol, HR_2 = 41,000 J/kgmol)
∴ ALPHA = (1815.2/96) * ((−2,06,000) * 0.0227) + (2π * 0.05) * (0.6623 * 10^9)

$$= -88419 + 0.20806 * 10^9$$
$$= 0.2079 * 10^9 \text{ J/h-m}$$

$$\text{BETA} = (F/96) * HR_2 * V$$
$$= (1815.2/96) * 41,000 * 0.51875$$
$$= 402,156.6 \text{ J/h-m}$$

$$\text{GAM} = (F/96) \, b_i - (F/96) \, HR_2 \, U$$

Units = (kg mol/h) * (J/kgmol-K) − (kgmol/h) * (J/kgmol) * 1/K i.e. J/h-K

$$\therefore \text{GAM} = (1815.2/96) * 33899.5 - (1815/96) * 41000 * (-0.00009436)$$
$$= 640983.05 + 73.1518$$
$$= 641,056.2 \text{ J/h-K}$$

$$\text{OMEG} = 2 * (F_i/96) * a_i \quad \text{Units: (kgmol/h)} * \text{(J/kgmol-K-K)}$$
$$= 2 * (1815.2/96) * 7.497$$
$$= 300.529 \text{ J/h-K-K}$$

$$\text{OMEG} * t = 300.529 * 888$$
$$= 266,869.79 \text{ J/h-K}$$

$$\Delta T = ((\text{ALPHA} + \text{BETA})/(\text{GAM} + \text{OMEG} * t)) * \text{K/m} * 0.01 \text{m}$$
$$= ((0.2079 * 10^9) + 402,156.6)/(641,056.2 + 266,869.79)) * 0.01 \text{ K}$$
$$= 2.29426 \text{ K}$$

$$dt/dz = 2.29426/0.01 = 229.426 \text{ K/m}$$

(e) *CO converted*

$$dy/dz = U * (dt/dz) + V$$
$$= -0.00009436 * 229.4 + 0.0117$$
$$= -0.021646 + 0.0117$$
$$= -0.01016 \text{ m}^{-1}$$

Hence, the value of *dy* is given by

$$\Delta y = (dy/dz) * z$$
$$= -0.01016 * 0.01$$
$$= -0.0001016 \text{ m}^{-1}$$

This is only a value and not a converged value. The new value of tube wall temperature has to be substituted as the new assumed value and iterated until it converges, and *dy/dz* becomes positive, since there is no carbon dioxide in the feed. Then the steps have to be repeated for the next cm (after reducing the pressure as per pressure drop), and so on for each cm, until the final values are obtained. Please note that numerical values have been incorporated only to give an idea of the procedure.

4. To calculate pressure drop

The pressure drop across the 1 cm long bed can be calculated using the Erguns equation[4] or any other suitable equation for pressure drop through packed columns with catalyst particles in the form of rings.

To find the dimensionless number for packing

$$Gr_p = D_p * V_s * \rho /((1 - \epsilon) \mu)$$

where

D_p = equivalent diameter of catalyst ring
V_s = superficial velocity
ρ = density of fluid
μ = viscosity of fluid
ϵ = void fraction of the bed.

To find the friction factor (f_p) of catalyst bed

$$f_p = (150/Gr_p) + 1.75$$

Find pressure drop using the following equation.

$$\Delta p = (f_p * L * \rho * V_s^2)/(\epsilon^3/(1 - \epsilon))$$

where

L = length (1 cm)
V_s = superficial velocity
ρ = density of fluid
ϵ = void fraction of the bed

Once the pressure drop has been calculated, the new inlet pressure for the second cm is to be found.

$$p_{new} = p - \Delta p$$

The new values of various components, the new values of pressure and temperature have to be substituted and then the procedure has to be repeated for the next centimetre and so on until the bottom of the catalyst tubes are reached.

8.2 DYNAMIC MODELLING

Dynamic simulation tells an engineer how a system undergoes changes over a period of time. It is a useful tool for various purposes such as fixing the volume of accumulators, specifying control valves, and level, flow, temperature and pressure instruments, design and operation of trays, etc.

The dynamic modelling principles and practices given here are general practices of the popular simulation packages[5] for offshore and onshore oil and gas industries and refineries, etc. It is to be noted that there are only small differences between such packages and these are mainly in the use of terminologies. While going into dynamic mode, the process engineer would have already constructed a steady-state model for the process scheme and tested its validity. It is to be noted that we also have the choice to start with dynamic modelling from the beginning itself, without first building a steady-state model. Many projects, like the development of an Operator Training Simulator, follow this approach.

Steady state mode

The steady state mode uses modular operations which are combined with a non-sequential algorithm. Information is processed as soon as it is supplied. The results of any calculation are automatically propagated throughout the flowsheet, both forward and backwards.

Dynamics mode

Material, energy and composition balances in Dynamics mode are not considered at the same time. Material or pressure–flow balances are solved at every time step. Energy and composition balances are defaulted to solve them less frequently. Pressure and flow are calculated simultaneously in a pressure–flow matrix. Energy and composition balances are solved in a modular sequential fashion.

Because the pressure–flow solver exclusively considers pressure–flow (P–F) balances in the network, P–F specifications are separate from temperature and composition specifications. P–F specifications are input using the "one P–F specification per flowsheet boundary stream" rule. Temperature and composition specifications should be input on every boundary feed stream entering the flowsheet. Temperature and composition are then calculated sequentially for each downstream unit operation and material stream using the holdup model.

It is apparent that the specifications required by the unit operations in Dynamics mode are not the same as those of the Steady State mode. Unlike in Steady State mode, information is not processed immediately after being input. The integrator should be run after the addition of any unit operation to the flowsheet. Once the integrator is run, stream conditions for the exit streams of the added unit operation are calculated.

Moving from steady state to dynamics mode

We know that flow in the plant occurs because of driving forces and resistance. Before a transition from Steady State to Dynamics mode occurs, the simulation flowsheet should be set up so that a realistic pressure difference is accounted for across the plant. The following steps indicate some basic measures needed to set up a case in Steady State mode and then switch to Dynamics mode.

Adding unit operations

Identify material streams which are connected to two unit operations with no pressure–flow relation and whose flow must be specified in Dynamics mode. These unit operations include the separator operation and tray sections in a column operation.

Add unit operations, such as valves, heat exchangers, and pumps, which define a pressure–flow relation to these streams. It is also possible to specify a flow specification on this stream instead of using an operation to define the flow rate.

Equipment sizing

Size all the unit operations in the simulation using the actual plant equipment or predefined sizing techniques. Sizing of trays in columns can be accomplished using the Tray Sizing utility available from the Utilities page. Vessels should be sized to accommodate actual plant flow rates and pressures, while maintaining acceptable residence times.

General equipment sizing rules

Vessels (separators, condensers, reboilers) could be sized for 5–15 minutes of liquid holdup time. Sizing and costing calculations are also performed using the Vessel utility in the Sizing page Rating tab. Valves should be sized using typical flow rates. The valve should be sized with a 50% valve opening and a pressure drop typically between 15 and 30 kPa.

Column tray sizing rules

Tray sizing can be accomplished for separation columns using the Tray Sizing utility in the Utilities page. Any use of utilities should be done after switching to Steady State mode. The trays are sized according to the existing flow rates and the desired residence times in the tray. The important variables include:

- Tray diameter
- Weir length
- Weir height
- Tray spacing
- Space below the downcomer (vertical length)

Adjusting column pressure

In steady state, the pressure profile of the column is user specified. In dynamics, it is calculated using the dynamic hydraulic calculations. If the steady-state pressure profile is very different from the calculated pressure drop, there can be large upsets in flow in the column when the integrator (see below) is run.

A reasonable estimate of the column's pressure profile can be calculated using the Column Tray Sizing (TS) utility. This utility provides a Max ΔP/Tray value in the Results tab. The column pressure profile can be calculated using this, i.e. the Max ΔP/Tray value, and a desired pressure specification anywhere on the column.

You can change the MaxΔP/Tray value to achieve a desired pressure profile across the column.

Integrator

This is a very important tool. It allows you to dynamically watch and control some of the parameters which are used by design. Simple parameters such as the time step or the integration stop time or advanced parameters such as the execution rates of the different balances can be set from this tool.

Once a case is running in dynamics, the current simulation time and the real time factor can be viewed.

The strip charts continuously monitor on the screen the user selected variables when the integrator is running.

Event Scheduler

The event scheduler is for fixing events at allocated times. UniSim[5] Design can perform predetermined actions at given times in the simulation such as warn you by playing a sound when the temperature of a stream reaches a certain point, stop the integration once a condenser level stabilizes, or increase a feed rate after the simulation has run for a given time period.

Dynamics assistant

Using this feature is optional. The Dynamics Assistant provides a quick method for ensuring that a correct set of pressure flow specifications is used. The Assistant can be used when initially preparing your case for dynamics. The Assistant makes recommendations for specifying your model in Dynamics mode. You do not have to follow all the suggestions. It is recommended that you be aware of the effects of each change you make.

The Assistant recommends a set of specifications which is reasonable and guarantees that the case is not over specified, under specified, or singular. It has an option of doing a quick examination for potential problems that can occur while moving from steady to dynamics as well as before running the case in dynamics.

In the case of a simple separator, UniSim Design adds pressure flow specifications. However, in more complicated models such as the presence of a heater at the outlet of the separator, the Dynamics Assistant recommends the insertion of valves in some terminal streams. Although the Pressure Flow specifications added by the Assistant are adequate for starting a case in dynamics, detailed dynamic modelling can require more advanced modifications.

In cases where unit operations such as separators are directly connected through multiple streams, the flow cannot always be determined. As a temporary fix the Assistant adds a flow specification. However, it is better to add the missing unit operations (e.g. pumps, valves, etc.) to define the pressure–flow relation between the vessel and unit operations. In addition to ensuring that the correct pressure–flow specifications are used for the dynamic case, the Assistant sizes all the necessary equipment that has not yet been sized.

The parameters sized are:

- vessel volumes
- valve CVs
- k values (for equipment such as heaters, coolers, and heat exchangers)

The Assistant sizes the required equipment based on the flow conditions and specified residence times. The Assistant also checks the Tray Section pressure profile for both Steady State and Dynamics model to ensure a smooth dynamics start. It also ensures that the tray section and the attached stream have the same pressure.

Although the Dynamics Assistant ensures that the case runs in dynamics, it is not intended that the changes made are sufficient for the case. It is the designer's responsibility to ensure that an adequate control scheme is added to the case and that the model is property rated (i.e. existing vessels are adequately sized).

Logical operations

In Dynamics mode, some logical operations from the steady state are ignored (e.g. adjust operation). The Adjust operation can be replaced by PID controllers. The recycle operation is redundant in Dynamics mode.

Adding control operations

Identify the key control loops that exist within the plant. Implementing control schemes increases the realism and stability of the model. Disturbances in the plant can be modelled using the Transfer Function operation. The Events Scheduler can be used to model automated shutdowns and startups.

Adding pressure–flow specifications

Specify one pressure–flow specification for each flowsheet boundary stream. More information regarding pressure–flow specifications for individual unit operations are given in manual[5].

Controllers play a large part in stabilizing the PF Solver. For more information regarding the implementation of controllers in UniSim Design, 'Control strategy' in manual[5] may be referred.

Precautions

Pay special attention to equipment with fixed pressure drops since any fixed pressure drop specifications in equipment can yield unrealistic results, such as flow occurring in the direction of increasing pressure. Remember to check for fixed pressure drops in the reboiler and condenser of columns.

Be cautious of heaters/coolers with fixed duties. This can cause problems if the flow in the heater/cooler happens to fail the required function, or a temperature specification fails to control the temperature of a stream.

Feed and product streams entering and exiting tray sections should be at the same pressure as the tray section itself. Any large pressure differences between a feed or product stream and its corresponding tray section can result in large amounts of material moving into or out of the column.

It is necessary to isolate and sometimes converge single pieces of equipment in the plant using the ignored feature for each unit operation if there is an especially large number of unit operations in the flowsheet.

As already mentioned, run the Integrator after any unit operation is added in Dynamics mode. The Integrator should be run long enough to obtain reasonable values for the exit streams of the new operations.

Modelling a process in dynamics is a complex endeavour. From the perspective of defining the model, you must consider parameters such as vessel holdups, valve sizing, and use of pressure–flow specifications.

'View equations' tool

This provides a means of analysing cases for dynamic simulation. This tool also provides a summary of the equations and variables used by the simulation when running in dynamics. By analysing the case, it is possible to determine if there are required or redundant pressure–flow specifications. In some instances, cases which are running in dynamics fail to converge, in which case the 'View Equations' tool can be used to help determine what part of the simulation is causing problems.

Example problem—analysis of heavies removal from chilled natural gas

The natural gas obtained from wells contains, in addition to methane and ethane, costly heavier components like propane and butane. Generally, this mixture is removed by distillation at the bottom of a de-ethanizer column.

Consider the above hypothetical example shown in Figure 8.4, where the natural gas is chilled and the chilled gas removes the heat of condensation from the top of the de-ethanizer column and provides reflux to the column before being fed into the column itself for separating ethane at the top and butanes and propane at the bottom.

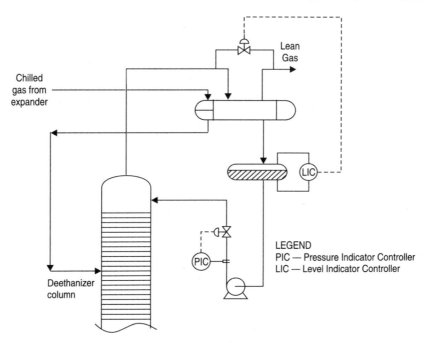

Figure 8.4 Heavies recovery from chilled natural gas.

The level of the reflux tank is maintained by adjusting the bypass to the condenser.

The problem is to fix the size and percentage opening of the control valve. We can make a chart of condensation rates for different openings of the valve based on heat transfer coefficients, which will be used to fix the control parameters.

Suppose there is a disturbance due to which the amount of chilled gas entering the system goes up temporarily. In this case, the heat removal from condenser increases and the amount of condensation also increases temporarily. The level in the reflux drum goes up and the LIC on the drum will cause the bypass valve to open. This will reduce condensation and bring back the level to normal. All this happens without upsetting the column.

The problem can be easily analysed with dynamic modelling in the following way:

(i) First, configure the system and use the procedures described. Use the chart of condensation rates for different openings of the valve.

(ii) A PID controller utility is used to fix the instrumentation parameters.

(iii) Use the event scheduler to create the disturbance.

(iv) The integrator should be set so that you can view the various changes taking place and how the column can be brought to a steady state within a time period.

The above is only a simplified example to analyse a particular disturbance. The dynamic modelling utility can be used for analysing more complicated systems which are generally encountered in practice.

REVIEW QUESTIONS

1. What are the basic equations required to model a hydrogen plant reformer?
2. How will you model an equilibrium chemical reaction?
3. How will you model a rate controlled reaction?
4. Make a flowchart for building a simulation program for a DCDA sulphuric acid plant reactor with four catalyst beds.
5. What are the different equations available for calculating the pressure drop across a catalyst bed? (for further reading)

USER-WRITTEN PROGRAM EXAMPLE

9.1 INTRODUCTION

A typical user-written example program in C++ based on the model described in Chapter 8 is given here. This is only a prototype and hence given here for illustrative purpose only. Changes based on reformer geometry, actual feed C_nH_m to be considered (only methane is considered here), the number of tubes, fuel used and flame radiating temperatures and any other necessary changes in the program have to be incorporated.

The program is suitable for terrace-walled reformers with maximum length of reactor tubes fixed at 1250 cm. In this program, we can simulate seven different configurations retaining the input and output files for each, so that they can be used for comparison to select the best configuration and optimal parameters. The program can be modified to suit both side-fired and top-fired reformers as described in Section 4.3 (Figure 4.8), the main difference being in the variation in flame and radiating gas temperature profile. This program was developed and tested on Turbo C++ Version 3.0.

9.2 TERRACE-WALLED REFORMER SIMULATION PROGRAM

```
        /* ------------Beginning of Program---------------------*/
/*reformer   simulation   program—terrace-walled    reformer    with
maximum  effective  tube  length  of  1252  cm—two  radiating  flames—
balance  decreasing  flue  gas  temperatures—inputs  are  taken  from
'.dat' files—outputs written to '.out' files*/

/*         list of variables used
```

Variable	Description	Units	Used in
f[602]	total feed	kgmol/h	All except rungex
f1[602]	methane feed	kgmol/h	"
f2[602]	co feed	kgmol/h	"
f3[602]	co2 feed	kgmol/h	"
f4[602]	h2 feed	kgmol/h	"
f5[602]	h2o feed	kgmol/h	"
f6[602]	n2 feed	kgmol/h	"

```
t[602]    bulk temperature            K
p1        pressure                    kg/cm2A
od        tube od                     m
id        tube id                     m
tublen    tube length                 m
tts       tube-tube spacing           m
ntub      no of tubes                 -
furlen    furnace length              m
furbre    furnace breadth             m
as        catalyst surface
          to volume ratio             inch-1
ac        catalyst activity
acten     activation energy
am        f.g.flow rate               kg/h
eps       catalyst void frac.
tb        bridge wall temp.           K
x[602]    methane conv.
y[602]    co conversion
t1        radiating temp              K   main,temp
q1[602]   outside heatflux            J/h·m2
dxdz      rate of change x
dydz      rate of change y
dtdz      rate of change t
dpdz      rate of change p
dy        delta y for 1 cm
z         step length                 m
l         tublen/z                    m/m     main
u         expr in value
v         expr in value
two[602]  tube skin temp.             K
n         integer-two loop            -
i         integer-z incre.            -
Variables in ratex()
```

k	reaction rate const.	
rg	univ. gas const.	kJ/kgmol·K
z1	(1000/t[i])-1	
e1	equi.const. of reaction	atm2
r	rate of reaction	lbmol/h·lbcat·atm
rowc	catalyst density	lb/ft3
ch4	partial pressure of methane	
co2	partial pressure of co2	
co	partial pressure of co	
h2	partial pressure of h2	
h2o	partial pressure of h2o	

Variables in value()

w	dxdz	
g,h, |		
p,q, |	expressions	-
r,s |		
g2 |	const in equi.	
b2 |	cnstant calc.	-

Variables in temp()

h1	heat of reaction of methane reactn	J/kgmol
h2	heat of reaction of shift reactn	J/kgmol
molwt	molwt of gas mixture	
de	eq. dia of chamber	m
cp	sp. heat flue gas	kcal/kg.C
vsc	viscosity flue gas	kg/m.h
thco	th.cond. flue gas	kcal/h.m.C
t2	assumed t.w.temp.	K
k	metal th. cond.	W/m·K
emiss	gas emissivity	
re	reynolds no.	-
pe	prandl no.	-
hc	outside h.t.coef.	kcal/h.m2.K
q	inside heat flux	J/h.m2
sigma	Stephan Boltz. const.	W/m2.K4
alph |		J/m.h
bet |		J/m.h
gam |	expression	J/h.K
del |		J/m.h
omeg |		J/h.K
temp1	deltat	K

kg	proc.gas th. con.	kcal/h.m.C (fps as used)
g1	gas mass flow rate	kg/h.m2
dp	catalyst particle dia.	m
vis	viscosity proc. gas	cP (kg/h.m as used)
cpp	sp. heat proc. gas	kcal/kg.K
hi	inside h.t. coeff.	
flux1	flux*m	W/h.m
flux2		k.h.m/W
flux3		K
vi	log(od/id)	end of list */

```cpp
#include<iostream.h>
#include<conio.h>
#include<stdio.h>
#include<math.h>
#include<string.h>
#include<fstream.h>
#include<dos.h>
   /* global variables */
   float  dxdz,x[1252],y[1252];
   double dtdz,delt,dydz,dy,dpdz;
   float  z,u,v,q1[1252],two[1252],t1;
   int    unit,n,i,l;

   /* Input variables */
   float f[1252],f1[1252],f2[1252],f3[1252],f4[1252],f5[1252],f6;
   float t[1252],p1,od,id,tublen,ntub,tts,furlen,furbre;
   double ac,as,acten,am,eps,tb;
   char s1[50],proj[50],date[50];

class top
{
public:
   void readdata();
   void getdata();
   void writedata();
   void rmcoma();
   void ratex();
   void increment();
   void value();
   void calculate();
   void temp();
   void prdrop();
```

```cpp
    void rungex();
    void runget();
    void output();
};

    void main()
    {
        clrscr();
        cout<<"    unit      code\n";
        cout<<"    ------    ----\n";
        cout<<"    abcd      1    \n";
        cout<<"    efgh      2    \n";
        cout<<"    ijkl      3    \n";
        cout<<"    mnop      4    \n";
        cout<<"    qrst      5    \n";
        cout<<"    uvw       6    \n";
        cout<<"    xyz       7    \n";
        cout<<"Note : Limit tube length to 12.5 metres\n\n";
        cout<<"\nEnter unit code :";
        cin>>unit;

        x[0]=0.0;
        y[0]=0.0;
        q1[0]=0.0;
        z=0.01;

        top medium;
        medium.readdata();
        medium.getdata();
        medium.writedata();
        f[0]=f1[0]+f2[0]+f3[0]+f4[0]+f5[0]+f6;
        two[0]=t[0];
        t1=tb+100;
l=tublen/z;
    cout<<'\n'<<"Length in cm "<<l<<" OK?\nPress enter";
        getch();
        for(i=0;i<=l;i++)
        {
            medium.rungex();
            medium.value();
            medium.temp();
            medium.increment();
        }
        medium.output();
```

```
              if(unit>0&&unit<6)
              cout<<"Job Over. Full Output in 'proj name.out' file.\n";
else
              cout<<"\nJob Over. Full Output in corresponding. out'file.\n";
              cout<<"Press any key to exit\n";
              getch();
                         }

              void top::getdata()
                         {
              clrscr();
              float fa,fb,fc,fd,fe,ff,ft;
              int m;
              flushall();
              ifstream filin;
              if(unit==1)filin.open("abcd.dat");
              else if(unit==2)filin.open("efgh.dat");
              else if(unit==3)filin.open("ijkl.dat");
              else if(unit==4)filin.open("mnop.dat");
              else if(unit==5)filin.open("qrst.dat");
              else if(unit==6)filin.open("uvw.dat");
              else filin.open("xyz.dat");
              cout<< " Welcome to the Reformer Simulation Program !!!\n";
              cout<< " ---------------------\n";
              cout<< "Enter Name of project\n";
              filin>>proj;
              cout<<proj<<'\n';
              gets (s1);
              if(s1[0]!=0) strcpy(proj,s1);
              cout<< "Enter date\n";
              filin>>date;
              cout<<date<<'\n';
              gets(s1);
              if(s1[0]!=0) strcpy(date,s1);
              cout<<"Keep pressing Enter.Changes made will be made in
              .dat files\n";
              cout<< "Enter feed flow rates in kgmol/h\n of ch4,co,";
              cout<< " co2,h2,h2o,n2\n";
              filin>>f1[0]>>f2[0]>>f3[0]>>f4[0]>>f5[0]>>f6;
              cout<<f1[0]<<" "<<f2[0]<<" "<<f3[0]<<" "<<f4[0]<<" "
                 <<f5[0]<<" "<<f6<<"\n";
              gets(s1);
              if(s1[0]!=0)
```

```
{
rmcoma();
sscanf(s1,"%f%f%f%f%f%f",&fa,&fb,&fc,&fd,&fe,&ff);
f1[0]=fa;f2[0]=fb;f3[0]=fc;f4[0]=fd;f5[0]=fe;f6=ff;
}
cout<< "Enter inlet temp,pressure to the reformer,od,id,";
cout<< "tublen,ntub,tts,furlen,furbre\n";
filin>>t[0]>>p1>>od>>id>>tublen>>ntub>>tts>>furlen>>furbre;
cout<<t[0]<<" "<<p1<<" "<<od<<" "<<id<<" "<<tublen<<" "
    <<ntub<<" "<<tts<<" "<<furlen<<" "<<furbre<<"\n";
gets(s1);
if(s1[0]!=0)
{
rmcoma();
sscanf(s1,"%f%f%f%f%f%f%f%f%f",&ft,&p1,&od,&id,
       &tublen,&ntub,&tts,&furlen,&furbre);
t[0]=ft;
   }
   cout<< "Enter catalyst ac,as,e,am,eps,tb\n";
   filin>>ac>>as>>acten>>am>>eps>>tb;
   cout<<ac<<" "<<as<<" "<<acten<<" "<<am<<" "<<eps<<" "
               <<" "<<tb<<"\n";
   gets(s1);
   if(s1[0]!=0)
   {
   rmcoma();
   sscanf(s1,"%lf%lf%lf%lf%lf%lf%lf",&ac,&as,&acten,&am,&eps,
               &tb);
   }
   filin.close();
   }

void top::writedata(void)
    {
      // all entered data is entered into file master.dat.
    ofstream filout;
    if(unit==1)filout.open("abcd.dat");
    else if(unit==2)filout.open("efgh.dat");
    else if(unit==3)filout.open("ijkl.dat");
    else if(unit==4)filout.open("mnop.dat");
    else if(unit==5)filout.open("qrst.dat");
    else if(unit==6)filout.open("uvw.dat");
    else filout.open("xyz.dat");
    filout<<proj<<'\n';
```

```cpp
filout<<date<<'\n';
filout<<f1[0]<<" "<<f2[0]<<" "<<f3[0]<<" "<<f4[0]<<" "
    <<f5[0]<<" "<<f6<<"\n";
filout<<t[0]<<" "<<p1<<" "<<od<<" "<<id<<" "<<tublen<<" "
    <<ntub<<" "<<tts<<" "<<furlen<<" "<<furbre<<"\n";
filout<<ac<<" "<<as<<" "<<acten<<" "<<am<<" "<<eps<<" "
    <<" "<<tb<<"\n";
filout.close();
  }

void top::readdata()
  {
  //data from file read to prompt user with previous run inputs.
ifstream filin;
if(unit==1)filin.open("abcd.dat");
else if(unit==2)filin.open("efgh.dat");
else if(unit==3)filin.open("ijkl.dat");
else if(unit==4)filin.open("mnop.dat");
else if(unit==5)filin.open("qrst.dat");
else if(unit==6)filin.open("uvw.dat");
else filin.open("xyz.dat");
filin>>proj;
filin>>date;
filin>>f1[0]>>f2[0]>>f3[0]>>f4[0]>>f5[0]>>f6;
filin>>t[0]>>p1>>od>>id>>tublen>>ntub>>tts>>furlen>>furbre;
filin>>ac>>as>>acten>>am>>eps>>tb;
filin.close();
  }

  void top::rmcoma(void)
  {
  // if data is enetred with commas same is removed.
  int n,i;
  n = strlen(s1);
  for(i=0;i<n;++i)
  if(s1[i] == ',') s1[i] = ' ';
  }

  void top::increment()
  {
  //update all parameters for next cm.
  x[i+1]=x[i]+dxdz*0.01;
  y[i+1]=y[i]+dydz*0.01;
```

```
        f1[i+1]=f1[0]-x[i+1]*f[i];
        f2[i+1]=f2[0]-(y[i+1]-x[i+1])*f[i];
        f3[i+1]=f3[0]+y[i+1]*f[i];
        f4[i+1]=f4[0]+(y[i+1]+3.0*x[i+1])*f[i];
        f5[i+1]=f5[0]-(x[i+1]+y[i+1])*f[i];
        f[i+1]=f1[i+1]+f2[i+1]+f3[i+1]+f4[i+1]+f5[i+1]+f6;
t[i+1]=t[i]+dtdz*0.01;
        }

        void   top::ratex()
        {
        // calculate dxdz.
          double k,rg,z1,e1,r,rowc,ch4,co,h2,h2o;
          acten=-44000.0;
          rg=8.314;
          k=3.02/p1*as*ac*pow(2.71828,(acten/(rg*t[i])));
          ch4=f1[i]/f[i]*p1;
          co=f2[i]/f[i]*p1;
          h2=f4[i]/f[i]*p1;
          h2o=f5[i]/f[i]*p1;
          z1=(1000/t[i])-1;
          e1=pow(2.71828,(z1*(z1*(z1*(0.25132*z1-0.3665)-0.58101)+
          27.1337)-3.277));
          r = k*(ch4-(co*(pow(h2,3))*e1/h2o));
          rowc=68.685;
          dxdz=(1/z)*rowc*(3.14*pow((id/.3048),2)/4)*(1-eps)*r
               /(f[i]*2.2/ntub)/30.48;
        }

    void top::value()
    {
     // calculate values of u and v.
     double w,g,h,p,q,r,s,g2,b2;
     if(t[i]<(600.0+273))
          {
          g2=0.01585;
          b2=4466.27;
          }
     if(t[i]>=(600.0+273)&&t[i]<(645.0+273))
          {
          g2=0.018238;
          b2=4312.912;
          }
     if(t[i]>=(645.0+273)&&t[i]<(695.0+273))
```

```
            {
            g2=0.019528;
            b2=4250.11;
            }
        if(t[i]>=(695.0+273)&&t[i]<(745.0+273))
            {
            g2=0.02087;
            b2=4185.75;
            }
        if(t[i]>=(745.0+273)&&t[i]<(800.0+273))
            {
            g2=0.022267;
            b2=4119.81;
            }
         if(t[i]>(800.0+273))
            {
            g2=0.022267;
            b2=4119.81;
            }
w = dxdz;
g =((-1)*g2*b2/(pow(t[i],2)))*pow(2.71828,(b2/t[i]));
h=1.0*pow((f2[i]-(y[i]-x[i])*f[i]),2)*
pow((f5[i]-(x[i]+y[i])*f[i]),2)/pow(ntub,4);
p=f[i]*(f3[i]+f[i]*y[i]+f4[i]+(y[i]+3.*x[i])*f[i])*
(f2[i]-(y[i]-x[i])*f[i])*(f5[i]-(y[i]+x[i])*f[i](pow(ntub,4));
q=(f3[i]+y[i]*f[i])*(f4[i]+y[i]*f[i]+3.0*x[i]*f[i])*
(2.0*y[i]*f[i]-f2[i]-f5[i])*f[i]/(pow(ntub,4));
u =(g/(p-q))*h;
r = 3.0*(f3[i]+y[i]*f[i])*(f2[i]-y[i]*f[i]+x[i]*f[i])*
(f5[i]-y[i]*f[i]-x[i]*f[i])*f[i]/(pow(ntub,4));
s =(f3[i]+y[i]*f[i])*(f4[i]+y[i]*f[i]+3.0*x[i]*f[i])*
(f5[i]-f2[i]-2.*x[i]*f[i])*f[i]/(pow(ntub,4));
v =w*(s-r)/(p-q);
}


        void top::temp()
{
// calculate values of dtdz,dydz and t[],two[].
    double h1,h2,molwt,de,cp,vsc,thco,t2,k,emiss,fg,re,pr,hc,q,
        sigma,alph,bet,gam,del,omeg,temp1,temp2,temp3,temp4,
        kg,g1,dp,ri,ro,vis,cpp,hi,flux1,flux2,flux3;
        double vi;
        z=0.01;
        h1=-206000;
        h2=41000;
```

User-Written Program Example

```
    molwt =(f1[i]*16+f2[i]*28+f3[i]*44+f4[i]*2+f5[i]*18+
        f6*28)/f[i];
    de=4.*furlen*furbre/(2.*(furlen+furbre)+3.14*od*ntub);
     cp=9.3079/27.5714;
     vsc=0.054185*3.6;
     thco=0.098501;
     t2=1150;
     k=21.*1.7307;
     emiss=0.1;
        if(i>90&&i<=450)emiss=.225;
        if(i>450)emiss = .145;
     emiss = emiss*pow(.25/tts,.2);    //view factor correction
     fg = am/(furlen*furbre);
     re=de*fg/vsc;
     pr=cp*vsc/thco;
     hc=(thco/de)*0.023*pow(re,0.8)*pow(pr,0.333333);
     sigma=0.0000000567;

// Setting radiating teprs.
    if(i<=100)t1 = t1+0.80;      // for terrace wall
        else if(i>100&&i<=450) t1 = t1-.5;
        else if(i>550&&i<=650) t1 = t1+.25;//for terrace wall
           else if(i>850) t1 = t1-.1;// for terrace wall
    //else if(i>450&&i<=600) t1 = t1-1.1; //.3;corrected
    //else if(i>600) t1 = t1-.12;         // 23-4-02
    // Printing rad temp to screen
  if(i%100 == 0)
                {
                    cout<<"\nRadiating Temperature\n";
                    cout<<"cm = "<<i<<" t1= "<<t1-273<<"\n";
                }
    two[i+1] = 1000; // dummy value to start loop
    n=0;
while(abs(two[i+1]-t2)>.01&&n<=10)
{
    ++n;
    t2 = two[i+1];
    q1[i]=(sigma*3600.0*emiss*(pow(t1,4)-pow(t2,4))+
        hc*4196*(t1-t2));
    q =q1[i]*(od/id);
    alph=(f[i]/ntub)*h1*dxdz+2.*3.14*q*id/2;
    bet=(f[i]/ntub)*h2*v;
    gam=(f[i]/ntub)*33899.5-(f[i]/ntub)*u*h2;
    del=alph+bet;
```

```
            omeg=2.*(f[i]/ntub)*7.5;
            temp1=(del/(gam+omeg*t[i]))*0.01;
            temp2=(del/(gam+omeg*(t[i]+temp1/2)))*0.01;
            temp3=(del/(gam+omeg*(t[i]+temp2/2)))*0.01;
            temp4=(del/(gam+omeg*(t[i]+temp3)))*0.01;
            delt=(temp1+2*temp2+2*temp3+temp4)/6;
            dtdz=delt/z;
            dydz=u*(dtdz)+v;
            if(dydz<=0.0) dydz = .01;
            dy =dydz*z;
            kg=(0.0001622*t[i]+0.0283439)*0.66; //  conv. to fps units
            g1=f[i]*molwt/(ntub*3.14*pow(id/2,2)); // in kg/m2.h
            dp=0.016;
            vis=(0.00002087*t[i]+0.013248)*0.01*3600; // conv to kg/h.m
            cpp = .0017648*t[i]+8.20477;
/*          Not Used- unclear source
            hi=(kg/(dp*.3048))*0.41*pow((dp*g1/(vis*eps)),0.5)*
            pow((cpp*vis/kg),0.3);
*/
            hi = (kg/(id*.3048))*(.813/2)*pow(2.71828,(-6*dp/id))*
            pow((dp*g1/vis),0.9);
            vi=log(od/id);
            flux1= q1[i]/3600.*od/2.;              // heat flow
            flux2=1./(hi*5.6783*id/2.)+(vi/k);     // resistance
            flux3=flux1*flux2;                     // temp. gradient
            two[i+1]=t[i]+flux3;

//          if(i%10==0){cout<<t2<<' '<<n<<'\n';getch();}
    }
        //writing on screen during calculations.
        //if(i==0)cout<<u<<' '<<v<<' '<<x[i]<<' '<<y[i]<<'\n';
        //if(i==1)cout<<u<<' '<<v<<' '<<x[i]<<' '<<y[i]<<'\n';
                        if(i%100==0)
            { cout<<u<<' '<<v<<' '<<x[i]<<' '<<y[i]<<'\n';
                cout<<"Press Enter\n";getch();
            }

    }

            void top::rungex()
            {
              //numerical integration of dxdz.
              double k0,k1,k2,k3;
```

```cpp
        ratex();
    }

void top::output()
    {
        double efft=1.02,efftw=1.1,effq=1.7;
        ofstream filout;
        if(unit==1)filout.open("abcd.out");
        else if(unit==2)filout.open("efgh.out");
        else if(unit==3)filout.open("ijkl.out");
        else if(unit==4)filout.open("mnop.out");
        else if(unit==5)filout.open("qrst.out");
        else if(unit==6)filout.open("uvw.out");
        else filout.open("xyz.out");
    //printing inputs to output file.
        filout<<"             INPUTS\n"<<proj<<"\n";
        filout<<date<<'\n';
        filout<<f1[0]<<" "<<f2[0]<<" "<<f3[0]<<" "<<f4[0]<<" "
              <<f5[0]<<" "<<f6<<"\n";
        filout<<t[0]<<" "<<p1<<" "<<od<<" "<<id<<" "<<tublen<<" "
             <<ntub<<" "<<tts<<" "<<furlen<<
             "<<furbre<<"\n";
        filout<<ac<<" "<<as<<" "<<acten<<"
"<<am<<" "<<eps<<" "
         <<" "<<tb<<"\n\n             OUTPUTS\n";
    //printing 1st set of outputs to file.
        filout<<"i"<<"    "<<"x[i]"<<"    "<<"y[i]"<<
 "   "<<"t(C)"<<"   "<<"two(C)"<<"   "<<" q1kcal"
                <<"\n";
        for(i=1;i<=2;i= i+1)
        {
        filout<<i<<"       "<<x[i]<<"        "<<y[i]<<"
              "<<((t[i]-273) *efft)*efft<<"
              "<<(two[i]-273)*efftw<<"
              "<<(q1[i]/4200)*effq<<"\n";
        }
        if(i==10)
        {
    filout<<i<<"        "<<x[i]<<"        "<<y[i]<<"
            "<<(t[i]-273)*efft<<"
            " <<(two[i]-273)*efftw<<"
            "<<(q1[i]/4200)*effq<<"\n";
        }
     for(i=20;i<=200;i= i+20)
        {
```

```cpp
            filout<<i<<"        "<<x[i]<<"        "<<y[i]<<"
"<<(t[i]-273)*efft<<"       "
<<(two[i]-273)*efftw<<"        "<<(q1[i]/4200)*effq<<"\n";
        }
    for(i=300;i<=l;i= i+100)
        {
          filout<<i<<"        "<<x[i]<<"        "<<y[i]<<"
"<<(t[i]-273)*efft<<"       "
<<(two[i]-273)*efftw<<"        "<<(q1[i]/4200)*effq<<"\n";
        }
//printing 2nd set of outputs to file.
    filout<<"\n";
    filout<<"i"<<"       "<<" ch4"<<"        "<<" co"
<<"     "<<" co2"<<"     "<<" h2"<< "     "<<" h2o"<<"\n";
    for(i=1;i<=2;i= i+1)
        {

filout<<i<<"        "<<f1[i]<<"    "  <<f2[i]<<"
"<<f3[i]<<"     "<<f4[i]<<"        "<<f5[i]<<"\n";
        }
    if(i==10)
        {
            filout<<i<<"        "<<f1[i]<<"          "
<<f2[i]<<"     "<<f3[i]<<"       "<<f4[i]<<"   "<<f5[i]<<"\n";
        }
    for(i=20;i<=200;i = i+20)
        {
            filout<<i<<"        "<<f1[i]<<"          "
<<f2[i]<<"     "<<f3[i]<<"       "<<f4[i]<<"   "<<f5[i]<<"\n";
        }
    for(i=300;i<=l;i = i+100)
        {
            filout<<i<<"        "<<f1[i]<<"          "
<<f2[i]<<"     "<<f3[i]<<"       "<<f4[i]<<"     "<<f5[i]<<"\n";
        }
    filout<<"\n u   "<<u<<"   v   "<<v<<'\n';
    i=l;
filout<<" ch4 slip % = "<<(f1[i]/(f[i]-f5[i]))*100<<'\n';
    filout<<"masvel kg/h.m2="<<f[i]
*((f1[i]*16+f2[i]*28+f3[i]*44+f4[i]*2+f5[i]*18+
      f6*28)/f[i])
    /(ntub*3.14*pow(id/2,2))<<
    "\nmol vel= "<<f[i]/(ntub*3.14*pow(id/2,2));
```

```cpp
        filout<<"\nSpace Vel= "<<f[i]*22.414/
     (3600*ntub*3.14*pow(id/2,2));
        filout.close();

       // printig to screen just before ending simulation.
       //   cout.precision(2);
            cout<<"\ni"<<"   "<<" ch4"<<"     "<<" co"
                 <<"   "<<" co2"<<"    "<<" h2"<<
                    "   "<<" h2o"<<"\n";
/*     for(i=1;i<=2;i= i+1)
         {

          cout<<i<<"  "<<f1[i]<<"        "
         <<f2[i]<<"    "<<f3[i]<<"    "<<f4[i]<<"  "<<f5[i]<<"\n";
         }
         if(i=10)
         {
              cout<<i<<"       "<<f1[i]<<"         "
         <<f2[i]<<"    "<<f3[i]<<"    "<<f4[i]<<"  "<<f5[i]<<"\n";
         }
*/              for(i=50;i<=200;i = i+50)
         {
          cout<<i<<" "<<f1[i]<<"  "  <<f2[i]<<"      "<<f3[i]<<"
   "<<f4[i]<<"     "<<f5[i]<<"\n";
         }
         for(i=300;i<=1;i = i+100)
         {
           cout<<i<<"        "<<f1[i]<<"        "  <<f2[i]<<"
    "<<f3[i]<<"      "<<f4[i]<<"      "<<f5[i]<<"\n";
         }
         if(i==0)cout<<" 0   "<<u<<' '<<v<<' '<<x[i]<<'
             '<<y[i]<<'\n';
                if(i==1)cout<<" 1   "<<u<<' '<<v<<'
             '<<x[i]<<'    '<<y[i]<<'\n';
        cout<<"\n   u    "<<u<<"   v   "<<v<<'\n';
         i=1;
          cout<<i<<"       "<<f1[i]<<"         "
              <<f2[i]<<"    "<<f3[i]<<"    "<<f4[i]<<"
             "<<f5[i]<<"\n";
            cout<<" mass vel kg/h.m2 = "<<f[i]
    *((f1[i]*16+f2[i]*28+f3[i]*44+f4[i]*2+f5[i]*18+
              f6*28)/f[i])
            /(ntub*3.14*pow(id/2,2))<<
    "\nmol vel= "<<f[i]/(ntub*3.14*pow(id/2,2));
```

```
cout<<"\nch4 slip % = "<<(f1[i]/(f[i]-f5[i]))*100<<'\n';
cout<<" Space Vel= "<<f[i]*22.414/
(3600*ntub*3.14*pow(id/2,2));
}
```

```
/* --------------End of Program----------------*/
```

One set of inputs and outputs, with explanation in brackets, is given below for illustration.

INPUTS
abcd
(configuration name)
10/05/2012
(date)
398.2 1 11.3 6.1 1397.40 7.7
(kgmol per hour of CH_4, CO, CO_2, H_2, H_2O, N_2)

883   23.5 0.121 0.1 12.5 96 0.29 9.4 9
(Temperature inlet, pressure, tube od, tube id, tube length, number of tubes, tube spacing, furnace length, furnace breadth)

15 11.64 -44000 62000 0.5 1323
(catalyst activity, catalyst surface to volume ratio, activation energy, flue gas, catalyst void ratio, bridge wall temperature)

OUTPUTS
(conversions of CH_4 and CO and temperatures of bulk gas and tube wall and heat flux)

i	x[i]	y[i]	t(C)	two(C)	q1kcal
1	0.000656	0.000231	634.950519	702.658771	28400.054286
2	0.001309	0.000363	635.257229	703.054218	28466.351048
20	0.012665	0.002085	628.257869	710.23631	29677.894095
40	0.024265	0.004085	634.43614	718.379694	31064.482095
60	0.034367	0.006085	640.750192	726.710394	32493.991048
80	0.042545	0.008085	647.22119	735.250836	33966.589333
100	0.048932	0.010085	657.310873	791.707141	75654.656381
120	0.054746	0.012085	671.312278	803.279633	72395.87619
140	0.060288	0.014085	684.482878	813.847729	69298.489143
160	0.065584	0.016085	696.886923	823.692474	66256.369905
180	0.070642	0.018085	708.5589	832.867981	63292.528381
200	0.075461	0.020085	719.539215	841.413123	60406.576
300	0.09588	0.030085	765.324976	875.894336	47117.514095

i	x[i]	y[i]	t(C)	two(C)	q1kcal
400	0.110434	0.040085	798.663611	899.268262	35625.345714
500	0.119902	0.050085	819.435776	904.807471	19935.946952
600	0.125598	0.060085	834.454504	921.37998	20378.094286
700	0.130364	0.070085	850.023699	938.563989	20787.056
800	0.134577	0.080085	864.897869	953.556995	19847.608476
900	0.138176	0.090086	878.791135	966.978821	18397.155524
1000	0.141177	0.100086	891.215039	978.346069	16494.57219
1100	0.143663	0.110086	902.199836	988.283667	14696.198857
1200	0.145761	0.120086	911.861946	996.911792	12999.494429

(component flowrates)

i	ch4	co	co2	h2	h2o
1	397.005402	1.77441	11.720211	10.104072	1395.785156
2	395.811981	2.725884	11.962159	13.926288	1394.349854
20	374.558807	20.749849	15.191341	80.914909	1369.867554
40	351.798126	39.590782	19.111111	153.116791	1343.187012
60	331.045105	56.265026	23.189884	219.45462	1318.355225
80	313.561401	69.554955	27.383669	276.099548	1296.677734
100	299.463593	79.38723	31.649202	322.658478	1278.314331
120	286.286804	88.209473	36.003735	366.543365	1260.783081
140	273.401642	96.642838	40.455524	409.650604	1243.446167
160	260.78244	104.715645	45.001938	452.054688	1226.280518
180	248.438904	112.421623	49.639484	493.722778	1209.299438
200	236.403473	119.732941	54.36359	534.553223	1192.539917
300	182.211029	149.217102	79.071892	721.838867	1113.63916
200	236.403	119.732941	54.36359	534.553223	1192.539917
300	182.211	149.217102	79.071892	721.838867	1113.63916
400	140.094	165.419373	104.985786	874.10125	1045.609131
500	110.942	168.26532	131.292618	987.866455	990.149475
600	92.698	160.351562	157.44957	1068.75293	945.749329
700	77.012	149.513229	183.974274	1142.336792	903.538269
800	62.805	136.804077	210.890854	1211.87561	862.414246
900	50.4024	122.046181	238.051422	1276.24426	822.851013
1000	39.868	105.29641	265.334656	1335.12793	785.034302
1100	31.007	86.821411	292.671112	1389.04870	748.836426
1200	23.436	67.011604	320.051727	1439.1417	713.884949

REVIEW QUESTIONS

1. Modify the program given in this chapter by adding a module to calculate the pressure drop across the catalyst bed.
2. How will you make a program similar to that in this chapter to include small quantities of ethane, propane, butane, carbon dioxide and nitrogen in the feed?
3. What equations are required to modify the program given in this chapter to receive input feed hydrocarbon as C_nH_m where n and m can be any whole number or fraction?

Chapter 10: COST ESTIMATION EXAMPLES

10.1 INTRODUCTION

The estimation of the cost of a chemical plant project is essential for the commercial evaluation of the project as well as for checking the feasibility/viability of the project. The accuracy of cost estimation plays a vital role in the commercial appraisal of the project, hence it is needless to say that utmost care and diligence should be applied in cost estimation.

There are many methods for estimating the costs of chemical plants. Some of the methods are given below.

1. Make the estimates based on the rough designs of equipment and based on the cost of material and labour along with the cost of the bought-out equipment obtained from manufacturers.
2. Make the estimates based on enquiries floated to different vendors and constructors and make the cost of plant based on the figures made available by them.
3. Make a list of items along with important specifications such as those of materials of construction and heat transfer areas for heat exchangers, volumes of vessels, horsepower of pumps and compressors, etc. Various published cost indices are available based on which the estimares can be made. Details are available in the books [1] and [2].

The first method is more applicable to natural and bioprocess chemicals and some of the inorganic and organic chemicals which are specific to particular regions due to availability of local raw materials and infrastructure.

For large petrochemical complexes and refineries the last method is more suitable.

For a typical plant and machinery project (excluding offsites) the cost estimation procedure is illustrated in this chapter by way of the following two examples.

1. Cardanol Distillation Unit
2. Industrial Alcohol Distillation Plant

The cost estimations worked out in this chapter are based on the prices prevalent in the year 2010.

10.2 CARDANOL DISTILLATION UNIT

A very small scale facility for cardanol production is considered for cost estimation as a case study. The process of cardanol distillation has been explained in Section 2.1. Here, a 500-litre per batch 'cashew nutshell liquid' distillation section along with its design and cost estimation is given.

10.2.1 Reactor (Distillation Vessel)

The following data is given. See also Figure 10.1

$$\text{External pressure} = 760 \text{ mm} - 5 \text{ mm internal pressure}$$
$$= 755 \text{ mm Hg}$$
$$= \text{Approximately 1 kg/cm}^2$$

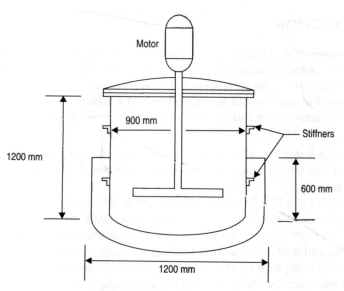

Figure 10.1 Reactor.

The vessel is designed for external pressure as per 'ASME, Section VIII' procedure and terminologies.

Assume cylindrical wall thickness, $t = 4$ mm

$L/D_o = 1200/900 = 1.33$

$D_o/t = 900/4 = 225$

Factor B = 3700 (ASME, Section VIII)

Allowable pressure = $3700/14.22 * 225$ Pa
$= 1.156 \text{ kg/cm}^2$

Hence, 4 mm wall thickness is enough since the allowable pressure is more than the actual external pressure ($\approx 1 \text{ kg/cm}^2$).

Dished ends under external pressure

    Assume dished-end thickness, $t = 4$ mm
    Radius = 900 mm
    $R_i/100t = 900/100 * 4 = 2.25$
    Factor B = 5500

Allowable pressure = $5500/14.22 * 225$ Pa
                        = 1.72 kg/cm²

    Hence, 4 mm dished-end thickness is sufficient since the allowable pressure is more than the actual external pressure.

Weight of SS316

Shell of 900 mm diameter and 4 mm thickness.
(The weight of 4 mm SS316 sheet is 33.15 kg/m² — Refer to the Appendix)

1. Shell
   $1.2 * 3.0 * 33.15$ (length * width * weight/m²)   = 119.34, say, 120 kg
2. Top and bottom dishes
   $1.1 * 1.1 * 2 * 33.15$ (length * width * weight/m²) = 80.223, say, 80 kg
3. Paddle   = 5 kg
4. Branches hand holes   = 10 kg
5. Shaft   = 5 kg
6. Total weight   = 220 kg
7. Take 15% extra   ≈ 30 kg
8. Final weight   = 250 kg

Weight of mild steel

    Jacket shell 1200 mm diameter
    (The weight of 6 SWG MS sheet is 40.2 kg/m²)

Jacket shell
    $4 * 1.0 * 40.2$ (length * width * weight/m²) = 165 kg

Top and bottom
    $1.2 * 1.2 * 40.2 * 2$ (length * width * weight/m²) = 115.8, say, 116 kg
    Support angles (approx.)   = 40 kg
    Take 10% extra   = 30 kg
    Final weight   = 351, say, 350 kg

Cost of distillation vessel

Price of SS316	= ₹ 336/kg
Cost of SS316 (250 * 336)	= ₹ 84,000
Price of carbon steel	= ₹ 42/kg
Cost of MS (350 * 42)	= ₹ 14,700
Price of motor, 2 HP	= ₹ 20,000
Price of reduction gear	= ₹ 23,000

Price of oil immersion heaters = ₹ 12,000
Fabrication charges (20%)   = ₹ 32,000
Total   = ₹ 185,700
20% profit   = ₹ 37,140
Final cost   = ₹ 222,840 say, ₹ 223,000

Notice an important point to be considered. The cost of SS316 is ₹ 336/kg whereas that of SS304 is about ₹ 200/kg. There will be cost variations as well as variations in plant life while using different materials of construction, based on which judicious decisions will have to be taken.

10.2.2 Condenser

The following data is given. See also Figure 10.2.

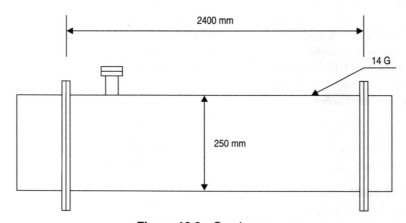

Figure 10.2 Condenser.

Calculation of number of tubes
 Heat transfer area = 4 m^2
 Area of one tube = $\pi * 0.025 * 2.4 = 0.1885$ m^2
 (1 inch dia 8 ft long)
 Number of tubes = 4/0.1885 (area/tube area) = 22

SS sheets
 Shell 250 ϕ * 8ft long (length * breadth * kg/m^2)   = 2.5 * 0.785 * 17.1
 (The weight of 14G, SS316 sheet is 17.1 kg/m^2)   = 33.56 kg, say, 40 kg
 Tube plates 350 ϕ * 16G thick = 0.35 * 0.35 * 136 * 2 nos. = 33.32, say, 40 kg
 Baffles, liners, etc.   = 10 kg
 Extra   = 10 kg
 Total   = 100 kg

Cost of condenser
Price of SS316 = 100 * (₹ 336/kg) ≈ ₹ 33,600 say, ₹ 34,200
176 ft of tubes 1" o/d * 16G = 176 * (₹ 100/ft) = ₹ 17,600
Mild steel = 100 * (₹ 42/kg)   = ₹ 4200
Fabrication, 20%   = ₹ 11,160, say, ₹ 11,200
Total   = ₹ 67,000
20% profit   = ₹ 13,400
Final cost   = ₹ 80,400, say, ₹ 81,000

10.2.3 Cooler

The following data is given. See also Figure 10.3.

Figure 10.3 Cooler.

Heat transfer area   = 1.3 m^2
Number of tubes (area/area of 1 tube)   = 7 nos
(calculation not shown)
SS316 sheets   = 100 kg
Cost of sheets   = ₹ 34,000
Cost of tubes 70 ft   = ₹ 7000
Total cost   = ₹ 41,000
Fabrication charges   = ₹ 8,000
20% profit   = ₹ 9,800
Final cost   = ₹ 58,800, say, ₹ 59,000

10.2.4 Receiving Vessels

Receiver 1

Volume of tank required   = 300 litres
Take a tank of 600 mm diameter and 600 mm length

Weight of SS316
Shell of 600 mm diameter and 2.5 mm thickess.
(The weight of 2.5 mm SS316 sheet is 21 kg/m^2)
3.14 * 0.6 * 0.6 * 21 (length * breadth * kg/m^2)   = 23.7, say, 30 kg

Top and bottom dishes

$0.8 * 0.8 * 2 * 21$ (length * breadth * kg/m^2)   = 26.88, say, 30 kg
Branches, handholes, etc.   = 10 kg
Total   = 70 kg

Cost of Receiver 1

Price of SS316   = ₹ 340/kg
Cost of SS316 (70 * 340) = ₹ 23,800
Fabrication charges   = ₹ 4760
Profit 20%   = ₹ 5710
Total cost   = ₹ 34,270, say, ₹ 35,000

Receiver 2

Volume of tank required   = ₹ 30 litres
Take a tank of 250 mm diameter and 250 mm length

Weight of SS316

Shell of 600 mm diameter and 2.5 mm thickness
(The weight of 2.5 mm SS316 sheet is 21 kg/m^2)
$3.14 * 0.25^2 * 21$ (length * breadth * kg/m^2)   = 4.12, say, 5 kg

Top and bottom dishes

$0.25 * 0.25 * 2 * 21$ (length * breadth * kg/m^2)   = 2.63, say, 3 kg
Branches, handholes, etc.   = 2 kg
Total weight   = 10 kg

Cost of Receiver 2

Price of SS316   = ₹ 340/kg
Cost of SS316 (10 * 340) = ₹ 3400
Fabrication charges   = ₹ 2000
Profit 20%   = ₹ 2000
Total cost   = ₹ 7400

A typical plant and machinery cost summary (general offsite facilities, taxes, etc. not included) is given in Table 10.1.

Table 10.1 Cost summary of a typical cardanol distillation unit plant and machinery

Item	Weight (₹)	Cost + Fabrication (₹)	Purchased items (₹)	As %	Total cost (₹)
Reactor	600	223,000			223,000
Condenser	200	23,000			81,000
Cooler	100	14,000			59,000
Receiver 1	70	7600			35,000
Receiver 2	10	3000			7400
Pump 1			100,000		100,000
Pump 2			100,000		100,000
Hot oil heater			550,000		550,000
Subtotal (cost)					**1155,400**

Item	Weight (₹)	Cost + Fabrication (₹)	Purchased items (₹)	As %	Total cost (₹)
Piping items				20	231,080
Insulation				10	115,540
Instruments				20	231,080
Electrical				20	231,080
Civil				20	231,080
					21,95,260
Total cost					(say 22 lakhs)

The cost given is approximate, based on the already suggested conditions, and for illustration only. Note that, in this case, plant cost is only a small portion of the project cost.

10.3 INDUSTRIAL ALCOHOL DISTILLATION UNIT

A 35,000 litres per day industrial alcohol (rectified spirit) distillation section is covered here. Refer to Chapter 2, Section 2.3 for process description.

10.3.1 Wash Column (Main material of construction—deoxidized copper)

See Figure 10.4. The material of construction can be taken as deoxidized copper, since even though costly it has an excellent resale value. In the above calculation, a two-metre square plate is taken, and multiplied by its weight in kg/m² obtained from the weight charts for 14G plate. The weight chart is included in the Appendix. This holds well for all the calculations given below. Also note that fabrication wastages are considered. The symbol ϕ is used to indicate the diameter.

1. Base liner 1800ϕ, 14G (length ∗ breadth ∗ kg/m²)
   = 2.0 ∗ 2.0 ∗ 19 = 76 kg
2. Bottom shell 8G (πD ∗ height ∗ kg/m² ∗ no. off)
   = 5.66 ∗ 1.2 ∗ 37.4 ∗ 2 = 508 kg
3. Normal shells 9 nos 10G (πD ∗ height ∗ kg/m²)
   = 5.66 ∗ 1.2 ∗ 30 ∗ 9 = 1834 kg
4. Top dish 10G (length ∗ breadth ∗ kg/m²)
   = 2.3 ∗ 2.3 ∗ 30 = 160 kg

Total weight of shell plates
   = 2578 kg

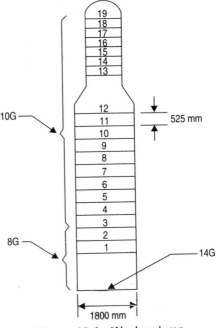

Figure 10.4 Wash column.

Hood type plate (Figure 10.5) 1800ϕ —19 numbers

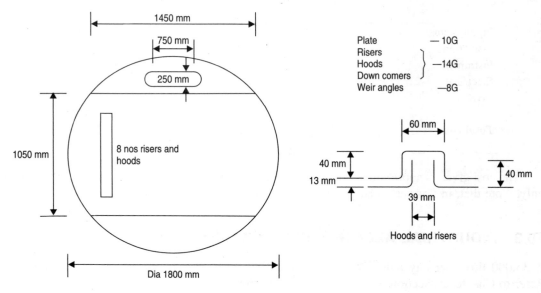

Figure 10.5 Wash column plate.

Thicknesses

Plate = 10G
Risers, hoods, down comers, etc. = 14G
Weir angles = 8G

1. Plates(1.9 m × 1.9 m)—19 nos. (length * breadth * kg/m^2) = 1.9 * 1.9 * 30 * 19 = 2060 kg
2. Risers and hoods 8 * 19 nos. (length * breadth * kg/m^2) = 0.3 * 1.05 * 19 * 8 * 19 = 910 kg
3. Down comers—19 nos. (length * breadth * kg/m^2) = 0.5 * 2.3 * 19 * 19 = 416 kg
4. Weir angles—2 * 19 nos. (length * breadth * kg/m^2) = 3.2 * 1.5 * 37.5 = 180 kg

Total weight of tray plates = 3566 kg
Total weight of copper = 2578 + 3566
= 6144 kg, say, 6200 kg

10.3.2 Rectifying Column 1500 ϕ

The following data is given. See also Figure 10.6.

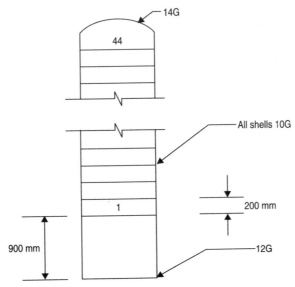

Figure 10.6 Rectifying column.

1. Base dish 12G = 1.9 ∗ 1.9 ∗ 24.5 (length ∗ breadth ∗ kg/m²) = 89 kg
2. Base shell 10G (πD ∗ height ∗ kg/m²)   = 4.75 ∗ 1.0 ∗ 30 = 143 kg
3. Seven normal shells 10G (πD ∗ height ∗ kg/m²)   = 4.75 ∗ 1.3 ∗ 30 ∗ 7 = 1297 kg
4. Top shell (πD ∗ height ∗ kg/m²)   = 4.75 ∗ 0.7 ∗ 30 = 100 kg
5. Top dish (length ∗ breadth ∗ kg/m²)   = 1.9 ∗ 1.9 ∗ 19 = 69 kg
    Total weight of shell plates   = 1698 kg

Hood type plates 1500ϕ (Figure 10.7) — 44 nos.

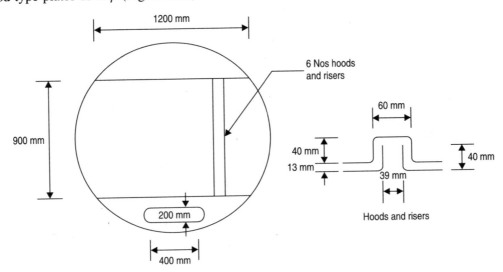

Figure 10.7 Rectifying column plate.

Thicknesses

Plate = 12G
Risers = 12G
Perforated plates, down comers, weir angles = 14G

1. Plates—44 nos. (length * breadth * kg/m²) = 1.6 * 1.6 * 24.5 * 44 = 2760 kg
2. Risers—6 * 44 nos. (length * breadth * kg/m²) = 0.04 * 0.9 * 6 * 44 * 24.5 = 233 kg
3. Perforated plates—44 nos. (length * breadth * kg/m²) = 0.9 * 2.0 * 19 * 44 = 1505 kg
4. Down comers—44 nos. (length * breadth * kg/m²) = 0.2 * 1.5 * 19 * 44 = 251 kg
5. Weir angles—2 * 44 nos. (length * breadth * kg/m²) = 5.3 * 1.2 * 19 = 121 kg
   Total weight of tray plates = 4870 kg
   Total weight of copper = 1698 + 4870
   = 6568 kg, say, 6600 kg

10.3.3 Wash Preheater (Figure 10.8)—84 m²

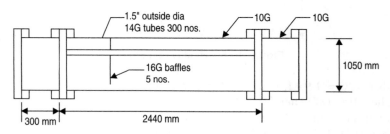

Figure 10.8 Beer preheater.

1. Shell 1050ϕ * 10G (πD * height * kg/m²) = 3.3 * 2.6 * 30 = 260 kg
2. Headers (length * breadth * kg/m²) = 3.3 * 0.8 * 30 = 80 kg
3. Tube plates 1150ϕ * 16 mm (length * breadth * kg/m²) = 1.2 * 1.2 * 151 = 220 kg
4. Baffles—5 nos. (length * breadth * kg/m²) = 1.1 * 0.8 * 5 * 15 = 66 kg
5. Pass partition plates (length * breadth * kg/m²) = 0.3 * 12 * 30 = 108 kg
   Total weight of sheets = 734 kg, say, 750 kg
   Weight of one tube = 4.95 kg
   (1 ½" o/d * 14G)
   Weight of 293 tubes = 293 * 4.95 = 1450 kg

10.3.4 Final Condenser (Figure 10.9)—64 m²

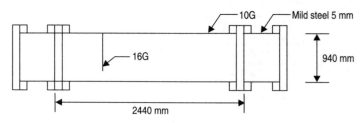

Figure 10.9 Final condenser.

1. Shell plate 940ϕ ($\pi D *$ length $*$ kg/m^2) $= 3.0 * 2.6 * 30 = 234$ kg
2. Tube plates (length $*$ breadth $*$ kg/m^2) $= 1.2 * 2.4 * 151 = 435$ kg
3. Baffles (length $*$ breadth $*$ kg/m^2) $= 3.5 * 1.0 * 15 = 53$ kg
   Total weight of plates $= 722$ kg, say, 730 kg

Tubes

   Weight of tubes $= 270 * 3.32 = 900$ kg
   (No. of tubes $= 270$ and weight of 1 tube 3.32 kg)

      Total weight $= 730 + 900 = 1630$ kg

10.3.5 Vent Condensers (Figure 10.10)—10 m^2

1. Shell 10G ($\pi D *$ length $*$ kg/m^2) $= 0.95 * 2.5 * 30 = 75$ kg
2. Tube plates (16 mm thick) (length $*$ breadth $*$ kg/m^2) $= 0.4 * 0.8 * 151 = 50$ kg
3. Baffles (length $*$ breadth $*$ kg/m^2) $= 0.3 * 1.3 * 15 = 6$ kg
   Total weight of plates $= 131$ kg, say, 150 kg
   Weight of one copper tube $= 3.32$ kg
   (1¼" o/d $* 16$G $* 8'$ long)
   Weight of 42 nos. tubes $(42 * 3.32)$ $= 140$ kg
   Total weight $= 290$ kg
   Total weight of final and vent condensers $(1630 + 290) = 1920$ kg

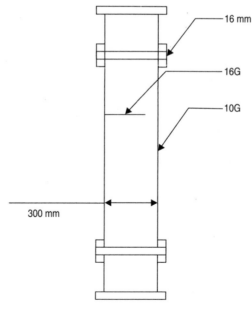

Figure 10.10 Vent condenser.

10.3.6 Product Cooler—23 m^2

The product cooler shell is 1.5 ft diameter and 8 ft in length.

1. Shell ($\pi D *$ length $*$ kg/m^2)   = 1.45 $*$ 2.6 $*$ 19 = 75 kg
2. Tube plates (16 mm) (length $*$ breadth $*$ kg/m^2) = 0.5 $*$ 1.0 $*$ 151 = 76 kg
3. Baffles (length $*$ breadth $*$ kg/m^2)   = 0.5 $*$ 2.0 $*$ 15 = 15 kg

Total weight   = 166 kg, say, 170 kg

Weight of copper tube   = 3.32
(1¼" o/d $*$ 16G $*$ 8' long)

Weight of 96 nos. copper tubes   = 320 kg

Total weight of product cooler   = 320 + 170 = 490 kg

10.3.7 Reflux Tank (Figure 10.11)

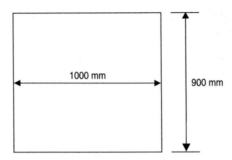

Figure 10.11 Reflux tank.

Side shell ($\pi D *$ length $*$ kg/m^2)   = 3.2 $*$ 0.9 $*$ 19 = 55 kg
Bottom and top (length $*$ breadth $*$ kg/m^2) = 1.0 $*$ 1.0 $*$ 19 $*$ 2 = 38 kg
Total weight   = 55 + 38
                                            = 93 kg, say, 100 kg

10.3.8 Rectified Spirit Plant (Plant and Machinery)—Cost Summary

A typical plant and machinery cost summary (general offsite facilitie, taxes and duties etc. not included) is given in Table 10.2.

The cost of copper is taken as ₹ 600/kg as per the year 2010 cost. As mentioned, copper is a costlier material but a good investment since it has an excellent resale value. The plant cost given is approximate and for illustration only. Note that the plant and machinery cost is only a portion of the project cost.

What we have done here is a detailed estimation. For a much quicker estimation of modern plants the book by Peters and Timmerhaus[1] can be referred to by the readers.

Table 10.2 Cost summary of a typical rectified spirit plant (plant and machinery)

Item	Weight (kg)	Cost (₹)	Fabrication and erection (%)	Items as %	Total cost (₹)
Wash column	6200	3,720,000	20		4,464,000
Rectifying column	6600	3,960,000	20		4,752,000
Wash preheater	2200	1,320,000	20		1,584,000
Condensers	1920	1,152,000	20		132,400
Cooler	490	294,000	20		352,800
Reflux tank	100	60,000	20		72,000
Pumps etc.					900,000
Subtotal					**13,507,200**
Piping items				20	2,701,440
Insulation etc.				15	2,026,080
Instruments				20	2,701,440
Electrical				20	2,701,440
Civil				20	2,701,440
					26,339,040
Total cost					(say, 2.7 crores)

WEIGHT OF PLATES

Weight of Plates in kg/m²

For various materials and thicknesses based on SWG and thickness in mm

SWG	mm	Al	Cu	C. steel	S. steel	Lead	Cast iron
	20	56.1	189	164.85	170.5	232.7	145
	10	28.3	94.2	82.42	85.3	117.42	72.1
	8	22.4	75.4	65.94	68.2	93.94	59.1
1	7.62	21.8	70.6	62.9	64.9	89.51	
2	7.01	19.6	64.7	57.9	59.6	82.3	
3	6.401	17.83	58.8	52.8	54.45	75.17	
	6	16.78	54.3	49.7	51.1	70.47	
4	5.893	16.49	53.25	48.7	50.2	69.21	
5	5.385	15.2	50	44.5	45.8	63.24	
	5	14.1	46.4	41.4	42.1	58.73	
6	4.877	13.6	45.1	40.2	41.45	57.28	
7	4.37	12.5	41.1	36.9	38	52.5	
8	4.064	11.38	37.4	33.5	33.7	47.14	
	4	11.2	36.8	33	33.15	46.03	
10	3.25	9.07	29.9	26.8	26.9	38.02	
	3	8.36	27.6				
12	2.64	7.4	24.5	21.3	22.2		
	2.5	7	23.2	20.6	21		
14	2.03	5.68	18.8	16.72	17.1		
	2	5.6	18.5	16.5	16.85		
16	1.626	4.53	15	13.4	13.68		
	1.5	4.17	13.8				
	1.25	3.45	11.5				
18	1.22	3.38	11.25	10.1	10.45		
	1	2.77	9.25				
20	0.914	2.55	8.48	7.54	7.64		

REFERENCES

Chapter 1

[1] Narasimhan, M.S. (Tata Institute of Fundamental Research), Engine of Growth, *Science Today*, Times of India Publication, Mumbai, 1974.

[2] VDI Heat Atlas, VDI Verlog GmBH, Dusseldorf, 1993.

[3] Agarwal, Rajeev, et al., Uncovering the Realities of Simulation, *Chemical Engineering Progress*, May 2001.

[4] Srikumar, K., Process Simulation Science with a Bit of Art, *Chemical Industry Digest*, January-February 2003.

[5] Carlson, Eric C., Don't Gamble with Physical Properties for Simulations, *Chemical Engineering Progress*, October 1996.

Chapter 2

[1] Peters and Timmerhaus, *Plant Design and Economics for Chemical Engineers*, McGraw-Hill, 2003.

[2] Lichts, F.O., World Fuel Ethanol Production, *Industrial Statistics*, Renewable Fuels Association, 2010.

[3] Robinson and Gilliland, *Elements of Fractional Distillation*, McGraw-Hill Publication, 1950.

[4] Mathew Van Winkle, *Distillation*, McGraw-Hill Publication, 1967.

[5] Brownell, Lloyd E. and Edwin H. Young, *Process Equipment Design—Vessel Design*, Wiley Eastern.

[6] Production of Seaweeds, Food and Agriculture Organisation Publication.

[7] Nandakumar, C.K., Years of experience in production of Agar, *Agar and Agarose*.

[8] Warren Lee McCabe, Julian Cleveland Smith, and Peter Harriott, *Unit Operations of Chemical Engineering*, 5th ed., McGraw-Hill Publication, 1993.

[9] Connemann, J. and J. Fischer, *Biodiesel Processing Technologies, Biodiesel in Europe*, 1998.

[10] Jessica Ebert, quoting Robert Babcock, chemical engineer, Supercritical Methanol for Biodiesel Production, *University of Arkansas*, 2008.

Chapter 3

[1] Browning, F.M., Vapour Phase Processes Serve Industry Well, *Chem. Engg*, 59, No. 10, 158–61.

[2] Ray, A.B., *Chem. and Met. Engg*, Volume 47, 329–32.

[3] Courouleau, P.H. and R.E. Bensen, *Chemical Engineering*, 55, No. 3, 112–115.

[4] Mantell, C.L., *Adsorption*, McGraw-Hill Publication, 1952.

[5] Bhaskara Rao, B.K., *Modern Petroleum Processes*, 5th ed., Oxford & IBH, 2008.

[6] Brockmeier, N.F., Gas Phase Polymerization, *Encyclopaedia of Polymer Science and Engineering*, Wiley, New York, 1987.

[7] Website 'mcdonough.com', Depolymerisation of nylon 6.

[8] Nauman, E., Polymerization Reactor Design, *Encyclopaedia of Polymer Science and Engineering*, Wiley, New York, 1994.

[9] Well, G.M., *Handbook of Petrochemicals and Processes*, Gower Publishing Company, 1991.

[10] Macgreavy, C., *Polymer Reactor Engineering*, Blackie Publishers, 1994.

Chapter 4

[1] Chen, C., et al., Solid–Liquid Equilibrium of Aqueous Electrolyte Systems with Electrolyte NRTL model, 1986; Local Composition Model for Excess Gibbs Energy of Electrolyte Systems, 1982.

[2] Lirohoff, B., Pinch Analysis—A State of the Art Overview, *Trans Ichem E*, Vol. 71, Part A, 503–523, 1993.

[3] David, P. Dewitt, *Fundamentals of Heat and Mass Transfer*, 4th ed., 2000.

[4] Robinson and Briggs, Heat Transfer, *Chem. Engg. Prog. Symposium*—Series No. 64, Los Angeles, 1965.

[5] Greer, Lindsay, A. and N. Mathur, Material Science—Changing Face of the Chameleon, *Nature*, 437(7063), 2005.

[6] Deuker and West, Manufacture of Sulphuric acid, *ACS Monograph 144*, Reinhold, New York, 1959.

[7] George T. Austin, *Shreve's Chemical Process Industries*, 5th ed., 1984.

[8] Ghosh, S.N., *Advances in Cement Technology*, Tech Books International, 2002.

[9] Taylor, H.F.W., *Cement Chemistry*, Thomas Telford, 1997.

Chapter 5

[1] Bhaskar Rao, B.K., *Modern Petroleum Refinery Processes*, 5th ed., Oxford & IBH, 2008.

[2] Packie, J.W., Transactions AIChE, Vol. 37, 1041, p. 51.

[3] Watkins, R.N., *Petroleum Refinery Distillation*, 2nd ed., Gulf Publishing Co., 1979.

[4] Ahmed, A. Faegh and John Collins, Lummus Technology, *Hydrocarbon Engineering*, USA, February 2010.

[5] Lummus, ABB, Delayed Coking Technology, Houston, Texas. (At present ABB and CB&I Lummus have separate technologies).

[6] Nelson, W.L., *Petroleum Refinery Engineering*, 4th ed., McGraw-Hill Kogakusha, 1958.

[7] Reza Sadeghbeigi, *Fluid Catalytic Cracking Handbook*, 2nd ed., Gulf Publishing Co., 2000.

[8] Wilson, J.W., *Fluid Catalytic Cracking, Technology and Operation*, PennWell Publishing, 1997.

Chapter 6

[1] American Petroleum Institute, *API Specification* 12J, 14C, 14E, 520, 521, 2200, 2350, 21610.

[2] Campbell, John M., Robert A. Hubbard, and Robert A. Maddox, Gas Conditioning and Processing, *Campbell Petroleum Series*.

[3] Srikumar, K., Oil and Gas Developments and Optimisation, *Chemical Industry Digest*, December 2007.

[4] Arnold K.E., Water Droplet Size Determination, *Society of Petroleum Engineers Technical Conference*, New Orleans, USA.

[5] Srikumar, K. and V. Sajish, Separators in Oil and Gas Production, *Chemical Industry Digest*, July–August 2001.

[6] Arnold, K. and M. Stewart, Designing Oil and Gas Production Systems, *World Oil*, Gulf Publishing Co., November 1984.

[7] Ken Arnold and Maurice Stewart, *Surface Production Operations—Volumes 1 and 2*, Gulf publishing Co., 1998.

[8] Perry, John H., *Chemical Engineering Handbook*.

[9] Srikumar, K., Oil and Gas Pipeline Developments and Optimisation, *Chemical Industry Digest*, December 2007.

[10] Muravyov, I., R. Andriasov, and V. Poloskov, *Development and Exploitation of Oil and Gas Fields*. Peace Publishers, Moscow.

[11] Mohitpur, M., et al., Pipeline Design and Construction as well as Operation and Maintenance, A Practical Approach, 2005.

[12] Jandjel, D. Gregory, Select the Right Compressor, *Chemical Engineering Progress*, July 2000.

- [13] Peters and Timmerhaus, *Plant Design and Economics for Chemical Engineers*.
- [14] Coatings for Corrosion Prevention, *Colarado School of Mines*, April 14–16, 2004.
- [15] Parrish and Praunitz, Dissociation Pressures of Gas Hydrates Formed by Gas Mixtures, *Industrial Engineering Chemistry*, ACS Publication, Washington, 1972.
- [16] van der Waals and Platteuw, Thermodynamic Properties of Gas Hydrates, *Molecular Physics Journal*, Taylor and Francis, U.K., 1958.
- [17] Ng, Robinson, et al., Hydrate Formation in Mixtures Containing Methane, Ethane, Propane, Carbon Dioxide in Presence of Methanol, University of Alberta, Canada.
- [18] Katz, Prediction of Hydrate Formation in Natural Gas, *Society of Petroleum Engineers*, Dallas, USA, 1945.
- [19] Khalid Ahmed Abdel Fattah, Evaluation for Natural Gas Hydrate Prediction, *King Saud University*, Saudi Arabia, 2004.

Chapter 7

- [1] Yuepeng Wan, Specially Shaped Filaments for CVD Reactor. US Patent.
- [2] William, C. O'Mara, Robert B. Herring, and Lee P. Hunt, *Handbook of Semiconductor Silicon Technology*, 1990.
- [3] Solarelectronics Inc., 'Hydrochlorination Process', DOE /JPL-1012-71, 1982.
- [4] Eicke R. Weber, Fraunhofer ISE Germany, *Solarcon India Conference*, 2009.
- [5] Jain, M.P., et al., Studies on Hydrochlorination of Silicon in a Fluidized Bed Reactor, October-November 2009.

Chapter 8

- [1] Shreyashi Kanunjna, Paper presented at AspenTech South Asia Engineering & Innovation User Group Meeting, Mumbai, May 30–31, 2006.
- [2] Grover, Sham and M.G. Sadasivan, Basis for Development of Model, *Fact Engineering and Design Organisation*, 1981.
- [3] Srikumar, K. and K.S. Preethakumari, An Overview of Reformer Design in Hydrogen, Ammonia and Methanol Plants, *Chemical Industry Digest,* September 1998.
- [4] Ergun, S., *Chemical Process Engineering*, 1952.
- [5] Honeywell Corporation, *Unisim Design—Dynamic Modelling Reference Guide*, 2009.

Chapter 10

- [1] Peters and Timmerhaus, *Plant Design and Economics for Chemical Engineers*, McGraw-Hill, 2003.
- [2] Navrrate and Cole, *Planning Estimation and Conrol of Chemical Construction Projects*, Marcel Dekker, 2001.

INDEX

Absolute alcohol, 33, 36
    azeotropic distillation, 34, 36
    mass balance, 38
    molecular sieve method, 34
Acephate, 126
Acetaldehyde, 19, 113
Acetic acid, 19
Acetone, 89, 92
    adsorber design, 95
    cost estimation, 97
    distillation column design, 95
    production of, 91
Acetone recovery, 93, 97
    flowsheet, 94
Acetone water system, 96
Acid leaching, 198
Acrylonitrile, 112
    production of, 113
Acrylonitrile butadiene styrene (ABS), 112
Acylonitrile butadiene rubber (ABR), 112
Adiabatic pre-reforming, 85
Agar, 39
    chemical structure of, 39
    gelling mechanism of, 41
    production of, 42
    properties of, 40
    uses of, 42
Agarose, 39, 45
    chemical structure of, 40
    CPC method, 46
    manufacture of, 46

Agarose gel, 46
Albert Einstein, 3
Alcohol, 19
Aldehyde tanning, 69
Alkali metals, 163
Alkaline earth metals, 163
Alkaline leaching, 198
Alkyl esters, 72
Alloxan, 30
Alumina, 157, 164, 185
Aluminium, 164
    alumina from bauxite, 164
    electrolytic process, 164
    production from alumina, 165
    reduction pot, 165
Alum tanning, 69
Ammonia, 150, 177, 182, 183, 228
    plant, 150, 153
Ammonia synthesis, 150, 153
    inert gas purging, 151
    refrigeration circuit, 151
    synthetis loop, 151
Ammonium sulphate, 20, 183
AMOCO, process, 111
Amorphous, 157
Ampicillin, 27, 28, 29
    enzymatic production of, 29
Ampicillin anhydrous, 29
Ampicillin sodium, 29
Ampicillin trihydrate, 29
Animal oils, 70

338 Index

API 12J standard, 233, 235
API RP 14E, 250
API SL X65, 248
Aspen Plus, 141, 179
ASTM gap, 207
Atmospheric distillation, 203
Aviation Turbine Fuel (ATF), 203
Azeotropic distillation, 36

Bakers' yeast, 22
Bayer's process, 164
Beet sugar, manufacture of, 53
Beneficiated iron ore, 167
Benzene, 36, 47, 49
Bessemer converter, 168
Biodegradable plastics, 25
Biodiesel, 72
    batch process, 73
    continuous process, 73
Biofuel, 19
Bitumen, 203
Black oil reservoir, 230
Blast furnaces, 167, 171
Boroaluminosilicate glass, 159
Borosilicate glass, 159
Brain tanning, 69
Bronze, 171
Butadiene, 63
    production of, 64
Butanol, 19
Butyl rubber, 62, 66

Calcium, 166, 184
Calcium carbonate, 130
Cane sugar, 19
Capex, 258
Caprolactam, 107
    economics of, 108
    manufacture of, 108
Carbon dioxide, 25
Carbon fibre, 112
Carbon neutral, 72
Carbon steel, 168
Cardanol, 15
    cost estimation of, 317
    distillation unit, 16
    latent heat of, 17
    mechanical design, 18

Cashew nut shell liquid, 15
Catalytic cracking, 223
Catalytic naphtha, 227
Catalytic reforming, 227
    catalyst-regeneration process, 227, 228
    raw materials for, 227
Caustic soda, 159, 161
    electrolysis of brine, 159
    production of, 161
Cellulose polymer, 106
Cellulose xanthate, 106
Cement, 184
    components of, 185
    compressive strengths, 186
    production of, 187
Centrifugation, 239
Chalcogenide glass, 158
Chamois tanning, 69
Chemical vapour deposition, 265
Chlorine, 159
    production of, 161
Chloroprene rubber, 62, 66
Chrome tanning, 67, 69
Citric acid, 23
    by submerged fermentation, 24
Claus process, 50
Claus sulphur recovery system, 51
Clay adsorbent beds, 70
Clean technologies, 172
Cleavage reactors, 91
Coal, 47
    anthracite coal, 47
    bituminous coal, 47
    lignite, 47
Coal gasification, 48
    chemical reactions, 48
    Koppers Totzek gasifier, 50
    Lurgi moving bed gasifier, 48
    Winkler gasifier, 49
Coal tar products, 47
CO conversion, 143
Coke, 47, 171, 218, 221
    material balance, 220
Coker fractionator, 213, 214
Coker gas, 221
Coking, 211, 212
    decarbonizing efficiency, 211
Column Tray Sizing (TS), 294

Index

Combined cycle power generation, 50
Complex fertilizer, 153, 182
    NPK plant, 183
Compostable plastics, 126
Compressed yeast, 22
Condenser, 320
Contact coking, 222
Continuous catalyst regeneration, 227, 228
Continuous deglycerolization (CD) technology, 73
Controlled crystallization, 158
Cooler, 321
Copolymerization, 121
Copolymers, 125
Copper, 168, 171, 267
    electrowinning, 168
Coriolis meters, 248
Corundum, 274
Cracked LPG, 101
Crude oil furnace, 206
Crude phenol, 91
Crystal diameter, 275
Crystal manufacture, 272
Crystal pullers, 272
Crystal-shaping, 274
Cumene hydrogen peroxide, 90, 92
Cyclic amide (lactam), 107
Cyclohexane, 36, 107
Cyclohexanone, 107, 109
Cypermethrin, 126

D-alpha-aminophenylacetic, 29
DDT, 2, 126, 127
DDVP, 126
Decoking, 212, 218
Dehydrogenation, 227
Delayed coker yields, 218
    effect of API gravity, 218
    effect of recycle ratio, 218
Delayed coking, 211
    technologies, 212
Dense gas, 74
Design 2 for Windows, 84, 99, 102, 141, 179
Detergents, 60
    linear alkyl benzene sulphonate (LABS), 60
    sulphonation of LAB, 60
Diammonium phosphate, 20, 183
Diamond, 274
Diaphragm cell, 159, 160

Dicofol, 2, 127
Dihydrate process, 137, 138
Dimethyl ketone, 92
Distillation curves, 202
Distillation vessel, 318
    cost of, 319
Doping, 273
Double barrel separator, 237
Down's cell, 163
D-rebose, 30
D-sorbitol, 29
Dynamic modelling, 292
Dynamic simulation, 292
Dynamics assistant, 295

Economy of scale, 6
Electrolysis cells, 159, 161
Electrolyte NRTL, 141, 179
Electrolytic cell, 163, 166
Electrowinning, 168, 172
Emulsion polymerization, 66, 124
Enamel paints, 193
Enamel varnishes, 193
Endosulphan, 126
Environmental Impact Assessment (EIA), 6
Environmental planning, 6
Enzymes, 26
Epitaxy, 275
Equilibrium flash vaporization curve, 205, 209
Essential oils, 70
    benzaldehyde, 70
Esterification reaction, 72
Ethanol, 83
Ethyl alcohol, 19
Ethylene dichloride, 117
    conversion to vinyl chloride monomer (VCR), 118
Ethylene Vinyl Acetate (EVA), 118
Event Scheduler, 294, 297
Extractive distillation, 179

Fatty oil, 59
Fermentation, 19, 20
    aerobic submerged, 27
    heterolactic, 25
    homolactic, 25
Fermenter, 22

Flexi coking, 222
Float glass, 158
Fluid catalytic cracker unit, 223
Fluid coking, 221
Foaming crude, 237
Foster Wheeler design of steam hydrocarbon reformers, 278
Fractionator overhead system, 215
Frasch process, 172
Free radical polymerization, 124
Free water knockouts, 238
Fructose, 26
Fuel alcohol, 33
Fuel oil, 203
Fullers Earth, 70
Fusel oil, 32, 79, 83
Fusion bonded epoxy, 252

Gas condensate reservoir, 230
Gas/oil ratio, 230, 231, 236
Glass(es), 156
    cooling, 158
    fused silica, 158
    photochromatic, 158
    types of, 158
Glass blowing, 157
    glass ceramics, 158
Glass-making furnaces, 157
    pot furnace, 157
    tank furnace, 157
Glass rolling, 158
Glucose isomerase, 26
Glycerine, 20, 59, 60, 73, 193
Glycerol, 73
Gracilaria, 43
Green coke, 212
Gur, manufacture of, 53

Haldor Topsoe design of steam hydrocarbon reformers, 278
Hall Heroult, 166
HDPE, 121
    autoclave method, 121, 122
    manufacture of, 122
Heat exchange reforming, 150
Heavy coker gas oil, 212, 214
    production of, 215

Heavy crude, 239
Heavy gas oil, 223
Heavy water, 198, 199
    production of, 200
Hemidihydrate process, 137
Hemihydrate process, 137
Hemihydrate recrystallisation (HRC) process, 137
Henkel technology, 73
Hevea Brasiliensis, 62
High fructose corn syrup, 26
    production of, 26
High Integrity Pressure Protection (HIPPs), 251
High Speed Diesel (HSD), 203
Horizontal separator, 242
Houdry process, 64
Hydrate formation, 253
Hydrated line, 189, 190
Hydraulic jet decoking, 218
Hydrazine, 181, 182
    Ketazine process, 182
    Olin Raschig process, 182
    Pechiney Ugine-Kuhlman process, 182
Hydrazine hydrate, 181
Hydrochloric acid, 162
Hydrocracking, 226
    two-stage process, 226
Hydrocyclones, 239
Hydrodesulphurization, 77
Hydrodesulphurizer, 78
Hydrogen, 141, 168
    production plant, 142
    purification, 143
Hydrogenated oil, 71
Hydrogenation reaction, 71
Hydrogen chloride, 162, 269
Hypo components, 206
Hypothetical components, 219
HYSYS, 8, 219

IGCC (Integrated Gasification Combined Cycle), 50
Immobilized enzymes, 26
Industrial Alcohol (Rectified Spirit), 30
    cost estimation of, 317
    cost summary, 328
    distillation system, 31, 323
    final condenser, 326
    mechanical design, 32
    product cooler, 328

Index

rectifying column, 324
reflux tank, 328
vent condensors, 327
wash column, 323
wash preheater, 326
Invert sugar, 52
Iron, 168
direct reduction, 168
production of, 167
Isobutyl alcohol, 83

Jatropha, 72
Jatropha oil, 74
John D. Rockefeller, 113
Jordan engine, 57

Kellogg design of steam hydrocarbon reformers, 278
Kraft process, 55

Lactic acid, 25
production of, 25
Lactonization, 29
L-ascorbic acid, 29
production of, 30
LDPE, 121
manufacture of, 122
pipe reactor method, 123
Lead, 170, 228
production of, 171
Leather, 67
hair-on, 69
manufacture of, 67
tanning, 67, 69
Leva correlation, 280
Light coker gas oil, 212, 214
production of, 215
Light crude, 239
Light metals, 163
Lime, 188
Limestone, 130, 131, 184
beneficiation of, 186, 187
Line pack, 249
Liquefied Petroleum Gas (LPG), 97, 202, 221
case study, 98
production by absorption technique, 100
production by chilling, 98

Liquid leakages, 14
LLDPE, 121, 123
production of, 123
L-sorbose, 29

Magnesia, 157
Magnesium, 166, 184
Malathion, 126, 127
production of, 128
Mancozeb, 126
Mathematical modelling, 7, 278
Medium crude, 239
Membrane cell, 159, 160
Membrane separator, 22
Mercury cell, 159, 160
Metalloid, 163, 171
Metals, 162
groups of, 163
Methane, 248
Methanol, 76
a case study, 84
distillation, 82
process plant, 77
Methanol reactor, 81
Methanol synthesis gas, 76
Mineralization, 126
Mini plant, 2, 271
Molasses, 19, 22, 27, 53
fermentation of, 21
Molten iron, 168
Mono-ammonium phosphate, 183
Monocrophos, 126
Mono-olefins, 65
Motor spirit, 203
Mutations, 22

Naphtha, 203, 221, 226
Natural gas, 248
Natural gas liquid, 98
Natural rubber, 62
Nitric acid, 177
dual pressure process, 177, 179
single pressure process, 177
Nitrile rubber, 62, 66
Nylon, 6, 107, 109
manufacture of, 108, 110
Nylon salt, 109

342 Index

Octane number, 227
Oil contaminents, 225
Oleo resinous, 192
Opex, 258
Orthophosphoric acid, 136
Oxidative dehydrogenation, 64
Oxo-alcohols, 141
Oxo-biodegradable polyethylene, 125
    standards for, 126
Oxylene, 30
Ozone, 6

Paints, 190
Paper, 57
    production with Fourdrinier machine, 57
Papermaker's alum, 57
Parathion, 127, 128
Penicillin, 27
    production of, 27
Pesticides, 126
Petrochemicals, 114
    production by steam cracking of naphtha, 114
    naphtha cracker unit, 115
    pyrolysis of the naphtha, 115
Petroleum, 202
    atmospheric distillation, 203, 204
    refinery products, 203
    vacuum distillation, 209
Petroleum coke, 203
Phase separation, 230
Phenol, 89
    production of, 91
Phenol formaldehyde, 90
Phenolic resins, 90
Phorate, 126
Phosphate rock, 135
Phosphatic fertilizers, 136
Phosphoric acid, 136, 183
    production of, 138
    simulation of plant, 139
    wet process, 136, 137
Phosphorus, 135, 182
    electrochemical process of manufacture, 136
Photolithography, 275
Photovoltaics, 276
Phthalic anhydride, 119
    manufacture of, 120
Pinch technology, 148
Pipeline configuration, 258

Pipeline, looping, 254
Pipeline network, 256
    a case study, 256
    pipelines, optimization of, 250
Pipenet, 251
Pipesim, 251
Planning,
    approaches to, 3
        bottom-up approach, 3
        MS project, 5
        object-oriented approach, 4
        PERT chart, 4
        Primavera, 5
        software, 5
        top-down approach, 3
Plate-to-plate calculation, 37
Polarimeter, 52
Polybutadiene rubber, 62, 66
Polyester, 111
    production by PTA route, 111
Polyethylene, 126
    biodegradability of, 126
Polyhydroxy butyrate, 126
Polylactic acid, 25, 126
Polymers, 120
    branched chain, 120
    cross linked, 120
Polymerization, ring opening type, 109
Polysilicon, 262, 263
    from magnesium silicide, 267
    production from MG-Si, 263, 264
    reactor, 266
Polysilicon plant, 269
    capital cost of, 269
    operating cost, 269
Polysulphide rubber, 66
Polyurethane rubber, 66
Poly vinyl acetate, 124
    polymerization of, 125
Poly vinyl chloride, 124
    emulsion polymerization, 124
    suspension polymerization, 124
Portland cement, 185
    types of, 185
Potable spirits, 19
Potable water, 195
    electrodialysis process, 197
    reverse osmosis process, 196
    vapour recompression process, 195
Potassium, 164, 182

Index

Precipitated calcium carbonate, 130
    a case study, 133
    plant, 132
Pre-fermenters, 20
Pressure swing adsorption, 34, 36, 143
Process design documents, 5
Process plants, design of, 7
Process simulation software, 8
    ASPEN PLUS, 8
    DESIGN2 FOR WINDOWS, 8
    FLARENET, 9
    HYSYS, 8
    OLGA, 9
    PIPELINE STUDIO, 9
    TGNET, 8
    TLNET, 8
    UNISIM, 8
Propylene, 101, 113, 116
Prosim, 179
Pseudo components, 206, 219
Pulp, 54
    black liquor recovery, 57
    sulphate process, 55
Pyrolysis, 115

Quadruple effect vacuum evaporator, 53
Quicklime, 189

Receiving vessels, 321
Rectified spirit, 19, 30
Recuperative heat exchangers, 167
Red phosphorus, 135
Refinery fuel system, 217
Refining column simulation, 206
Reformed gas economizer, 80
Reformers,
    design considerations for, 144
    heat balance, 279
    modelling, 278
    mole balance, 281
    pressure drop through bed, 282
    reaction kinetics, 281
    simulation program, 283
Riboflavin, 30
Rubber, 61
    natural, 62
    synthetic, 62
    vulcanization, 62, 66

Saltpeter (potassium nitrate), 177
Scandium, 166
Seaweeds, 43
Secondary nutrients, 184
Semiconductors, 271
Side-fired reformer, 145, 299
Silent spirit, 31
Silicon, 1, 269
Silicone rubber, 62, 66
Simulation, 13, 278
    multidisciplinary software, 13
Slaked lime, 57
Slaking, 130
Slenderness ratios, 241
Slurry oil, 223
Soaps, 58
    production by continuous saponification, 59
Soda ash, 154
    manufacture of, 155
    dual process, 156
    Solvay process, 155
Sodalime glass, 159
Sodium, 163
    production of, 163
Sodium carbonate, 154, 193
Sodium phosphate, 135
Sohio process, 113
Solar energy, 1
Solar panels, 271, 276
Solvay process, 155
Spherical catalysts, 228
Spherical separator, 237
Sponge iron, 168
Spreadsheets, 8
    a case study, 242
Starch, 53
    production of, 54
Steel, 166
    production of, 168
Streptococcus lactase, 25
Streptomyces griseus, 28
Streptomycin, 28
    production of, 28
Styrene, 63
    production of, 64
Styrene butadiene rubber (SBR), 62
    production of, 62
    polymerization coagulation, and drying, 65
Succinic acid, 20

Sugar, 51
    manufacture of, 52
Sugarcane, 51
Sugar syrup, 52
Sulphur, 67, 114, 172, 184
    by modified Claus process, 173
    production by Frasch process, 172
    production by modified Claus process, 173
Sulphuric acid, 174, 177, 179, 183
    process design, 176
    production by OCDA process, 174, 175
Supercritical fluid, 74
Supercritical technology, 73
Superior Kerosene Oil (SKO), 203
Supervisory Control and Data Acquisition (SCADA), 248, 251
Surfactant, 61
Suspension polymerization, 124
Syntan tanning, 69
Synthetic rubbers, 62

Terrace-Walled reformers, 145, 299
    simulation program for, 299–315
Tgnet, 251
Thermodynamic model, 9
Thermodynamic module, 254
Three-phase oil-gas separator, 236
Three-phase separation, 232, 239
    coalescers, 234
Tin, 171
Titanium dioxide, manufacture of, 190
TLNET, 251
Top-fired reformers, 145, 299
    temperature distribution, 146
Trans-esterification reaction, 73
Transition metals, 163
Tray sizing, 293, 294
Triazaphos, 126
Trichlorosilane (TCS), 269
    experimental study, 269
Triglycerides, 69
True boiling point curve, 205
True gas reservoir, 230
Two-phase flow, 251
Two-phase oil–gas separator, 235
Two-phase separators, 231

Ultrasonic flow meters, 248
UNISIM, 8, 219
Unsaturated hydrocarbons, 114
Uranium, 197, 198
    production from ores, 199
Uranium carbides, 198
Uranium dioxide, 198
Uranium tetrafluoride, 198
Urea, 180
    Montecatini total recycle process, 181
Urea hydrolyzer, 181

Vacuum distillation column, 209
    fuels type, 210
Vacuum filtration, 24
Vacuum residue, 220
Valve trays, 98, 103
    specifications for, 104
van der Waals, 253
Varnishes, 190, 192
VDI heat atlas, 9
Vegetable oils, 69, 70, 141
    extraction of, 70
    hydrogenation of, 71
Vegetable oils, 69, 70
Vegetable tanning, 67, 69
Vessel utility, 294
Vinyl acetate, 118
Vinyl chloride, 116
    production of, 117
Vinyl chloride monomer, 118
Viscose rayon, 105
    production of, 106
Vitamin B_2, 30
Vitamin C, 29
Volatile oil reservoir, 230

Wafer cutting, 274
Wafering, 274
Wafer production, 274
Walking the pipeline, 249
Water, 193
    ion exchange treatment, 194
    municipal treatment, 194
Waxy crude, 237
Wet gas compression, 215

Wet gas compressor, 216
Winkler generator, 49
Winsim Inc, 84, 99

Xanthates, 168
Xanthation, 106
Xylenol, 49

Yeast, 19
    production of, 23
Yellow phosphorus, 135

Zeolites, 194
Zero effluent, 6
Zinc, 169, 171, 253
    production of, 170